The Underworld of Rural China

"I have read this book 3 times. The author has worked on some of the impetuosity and puffery in the first draft and has brought to us a good result: a great book with a fresh and important topic, hard-to-come-by and abundant information and a prudent and multi-faceted analysis. This is a ground-breaking academic monograph."

—Philip C. C. Huang, *Famous International Scholar, Professor of UCLA, Visiting Changjiang Scholar at Renmin University of China*

"This is an excellent doctoral dissertation. It centers on a central research topic and offers us a unique perspective on some major social and theoretical issues. With sufficient academic learning, the author knows well the relevant theories and literature but did not limit himself to explicating the theories. The author applies multi-disciplinary knowledge in an analysis of social life: the analysis was meticulous, in-depth and down-to-earth; his thoughts were articulated fluently and coherently. This dissertation offers valuable theoretical and practical contributions."

—Su Li, *Famous Scholar, Changjiang Scholar and Distinguished Professor at Peking University Law School*

"This is one of the best doctoral dissertations I've evaluated in recent years. With his strong sense of social responsibility, the author sets himself an aim to understand rural China during a transitional period, rather than simply to be 'on par with' or engage in a 'dialogue' with the West. The author based his research on an important yet easily neglected issue that offers a fresh perspective on understanding social changes in rural China, and thus brings us novelty in his dissertation. With detailed information, coherent language and a reasonable structure, it is indeed an excellent dissertation. It is properly-informed, well-structured and remarkably coherent; it is indeed an excellent dissertation."

—Cao Jinqing, *Famous Expert on the Three Rural Issues, Professor at Social and Public Administration School of East China University of Science and Technology*

"This book offers a unique perspective on rural issues. With in-depth investigation and keen perception, the author brought our attention to 'rural gangs', a very important social group that has been overlooked by the academia. His analysis is based on his thorough understanding of the rural society. 'Rural gangs'—the central topic of this dissertation—deserves more social attention."

—Li Changping, *Famous Expert on the Three Rural Issues, Author to Being Truthful with the Premier*

Baifeng Chen

The Underworld of Rural China

Research on Rural Gangs on China's Jianghan Plain and Dongting Lake Plain

華中科技大學出版社
Huazhong University of Science and Technology Press

Springer

Baifeng Chen
School of Law
Zhongnan University of Economics
and Law
Wuhan, Hubei, China

Translated by
Guo Jun
School of Foreign Languages
Zhongnan University of Economics
and Law
Wuhan, Hubei, China

https://doi.org/10.1007/978-981-19-8710-6

Jointly published with Huazhong University of Science and Technology Press
The print edition is not for sale in China (Mainland). Customers from China (Mainland) please order the print book from: Huazhong University of Science and Technology Press.

Translation from the Chinese Simplified language edition: "乡村江湖: 两湖平原"混混"研究" by Baifeng Chen,

This Springer imprint is published by the registered company Springer Nature Singapore Pte Ltd.
The registered company address is: 152 Beach Road, #21-01/04 Gateway East, Singapore 189721, Singapore
(www.springernature.com/mycopy)

Translator's Foreword

This book is a translation of the second published edition of Prof. Chen's doctoral dissertation *Xiangcun Jianghu*. Professor *Chen* finished his dissertation in 2008 and revised it in 2010 before the first edition was published in 2011. He then revised his dissertation and republished the revised edition in 2019. The first published edition of his dissertation includes 9 chapters. In the second edition, he added Chap. 10 as a supplementary chapter to overcome the limitations he believed to be there with the first edition and to discuss the latest developments in the rural gangs.

This will be the first published translation by the translator, and the translator has personally typed in and checked every single word in this English version of the book. Because of the pandemic, the author and the translator have not met personally, but have engaged in numerous online talks to discuss what the author intends to express when things become not so clear. She has checked and rechecked the technical terms and has pondered on the collocation, choice of word, sentence pattern, etc. After completing the first draft, she has reviewed her own work many times to correct spelling, grammatical mistakes and to ensure accuracy and improve language fluency.

The translator has spent a great deal of time and effort on the translation of citations. There are mainly two kinds, direct citations quoted from a Chinese source and indirect citations from the Chinese version of a non-Chinese source. For the first kind, the translator will have to translate by herself, but for the second kind, the translator must find the English version of the non-Chinese source. Sometimes the non-Chinese source is originally also written in English, but for those sources that are not written in English, the translator will firstly try to find an English translation of the source. There are also occasions when the translator fails to find accessible sources and must translate the source by herself. But even so, the translator would find the key words used by the cited source first.

The translator has always kept the foreign readers in her mind. For instance, she added notes for geographical location. When comparing gangster influence in different regions of China, the author would mention villages of different regions, but simply transliterating the village names is not enough, because foreign readers do not know where the village lies. The translator has also clearly marked such notes as "translator's note" to differentiate them from the author's notes. Since this book has been translated from Chinese, many expressions do not have correspondents in English, or some readers may be interested in knowing more about the source text; the translator has added a glossary at the end of this translated book, showing the *Hanyu Pinyin* version and the corresponding Chinese expressions. This glossary is added for interested readers to refer to and also to avoid a mixture of Chinese and English in the main body of the translation.

Before beginning to translate, the translator has revised the logical reasoning structure of the source text. The difference between Chinese and English in terms of writing styles is already well known by all. The Chinese paragraphs are often written in a loosely structured manner. So to clarify the logical reasoning in the source text, the translator often has to sort out the main points for the author. For instance, based on social stratification of the gangsters, the translator divided up Sect. 4.4 of Chap. 4 "Social Stratification and Unlawful Interests" into three sub-sections, each about the methods used by a certain class to make illegal gain.

This dissertation is a real masterpiece. The author pointed out that the academia has not paid enough attention to the rural gangs issue. The same could be said about the translator. The translator had not paid any attention to this issue before she begun to translate, but later she came to experience the influence of gangsters in rural China. In a casual visit to *Suizhou*, she has a personal encounter with them, when she had to pay one of them a hundred yuan for entering a scenic area that was supposed to be free simply because the place was guarded by them. All tourists must deal with the gangsters before entering the scenic area.

The translation task is commissioned by the Law School of our university. The Law School has found me as the translator through the School of Foreign Languages, so I would like to thank Prof. *Cai Shengqin*, the dean of our School, for entrusting me with this important task. I would also like to thank the author of the book for his full trust with me. He has patiently answered my questions and has completely believed in me. I hope they would be satisfactory with my work. The translator has used a simple, concise and straightforward style in her language. That has benefited from her former Prof. *Li Changshuan*'s teaching. So I must also thank my professor for his helpful guidance during my postgraduate program.

Wuhan, China
November 2022

Guo Jun

Guo Jun is a lecturer of the School of Foreign Languages, *Zhongnan* University of Economics and Law at Wuhan, China. She graduated from Graduate School of Translators and Interpreters, Beijing Foreign Studies University. She has almost ten years of experience in oral and written translation. In those tasks, she has worked with all sorts of Chinese and overseas organizations, such as Chinese government agencies, the Transportation Security Administration of the United States and famous companies of the private sector.

Foreword by Prof. He Xuefeng

When he was about to republish the revised version of *Xiangcun Jianghu*, *Baifeng* asked me to write a foreword. The book has been revised and published based on his doctoral dissertation and was included in the "Rural Governance in Central China" series with me as the editor in chief. Since the book series already has a general foreword, I did not write a separate foreword for *Xiangcun Jianghu*. But for its republished edition, the general foreword does not apply, so I decided to write a separate foreword for it.

Baifeng was the first doctoral candidate under my supervision. When guiding my students, I lay special emphasis on the building of a solid foundation for their studies, especially in terms of reading classics and field investigation. I refer to this as "twofold training" (training through reading and experience). Without a good mastery of the basic theories and strict theoretical training, a student could never make it as a good scholar. In the same way, without long-term field investigation and a good understanding of one's experience, one could never become a good scholar either. *Baifeng* has done a good deal of classics reading during his postgraduate program and a good deal of field investigation in his doctoral program and thus may be regarded as a student who has undergone proper "twofold training". Based on extensive field investigation, *Baifeng* selected "rural gangs" as his research topic for doctoral dissertation and quickly finished it. His dissertation received favorable comments in the dissertation defense and has had some impact after its publication. *Xiangcun Jianghu* has a special perspective for revealing the nature of the Chinese rural society after the beginning of the twenty-first century. Many ideas in the book may be regarded as groundbreaking in that they have facilitated the study of rural sociology and broadened the scope of the study on "Three Rural Issues". When revising the book for republication, *Baifeng* has added his views for the latest developments in rural areas of suburbs and the broader rural China after the lift of the agricultural tax, offering a more comprehensive view of rural China.

Baifeng studied law in both his bachelor's and master's degree programs and pursued a degree in sociology in his doctoral program. The advantage of sociology is its emphasis on experience and speaking the truth. *Xiangcun Jianghu* is based on in-depth investigations and thus also speaks the truth, but in addition, it also attempts to reason things out and has offered us many inspirational abstractions. The concepts that he has derived from "the acquaintance society" and "rural gangs" have clearly demonstrated his remarkable academic skills.

I believe the Chinese researchers of social sciences must "be inspired in the fields, think boldly and write straightforwardly". Researchers must strive for ingenuity in social sciences studies. For a relatively long time, the Chinese researchers of social sciences must all rush into field research before social science researches in China may experience a stage of rapid growth. *Xiangcun Jianghu* may be regarded as an example for the kind of field research that I've always recommended. Though it has its own limitations, it is definitely a book worth reading.

I hope *Baifeng* will continue to be inspired by the fields in his subsequent researches and present us with more groundbreaking works.

At home
July 2018

He Xuefeng
Wuhan University
Wuhan, China

Preface

Since the 1990s, great changes have taken place to rural China. Such changes permeate major aspects of the society, including the social structure of country subdivisions, mechanisms of the maintenance of public order, life ethics and farmer values. These unprecedentedly profound and radical changes are still yet to be thoroughly understood by the academia. This book attempts to deepen such understanding.

When radical changes are occurring, the marginal groups in a society are often the first to feel and adapt to the changes. If they ally with the perceptive and aspiring intellectuals, they may even bring about social changes. That's why the marginal groups deserve attention from social sciences researchers. Through research on the rural gangs on the *Jianghan* Plain and *Dongting* Lake Plain, this book attempts to deepen the academia's understanding on the nature of China's rural society during a transitional period.

The author's investigations into rural gangs, farmer ethics and gang influence on rural governance reveal that the nature of the rural society on the *Jianghan* Plain and *Dongting* Lake Plain has fundamentally changed. This book refers to such change as "moral ambiguity in the rural society." At the same time, due to gang interference and involvement, the trend of "rural governance involution" becomes increasingly noticeable.

This book attempts to approach the topic from the following 3 perspectives.

1. The "rural gangs" perspective describes the history, organizational structure, social strata and social mobility of rural gangs since reform and opening up.
   Rural gangs have gone through 2 phases of development: the 1980s and the period since the 1990s. During the 1980s, farmers are freed from the people's communes; gangs are formed by the young people who are freed from labor. Prevalent among them are the misconceptions of heroism, mainly characterized by bellicosity and vanity. When they threaten village order, the country has to "crack down upon" them and that leads to their decline for some time. In the 1990s, gangs are reformed, but instead of seeking vanity or fame, they would rather seek real gain. Great opportunities have been created by the ongoing reform and the thriving economy. Seizing such opportunities, rural gangs expand through

networking into stable organizational structures and form an alliance spanning both urban and rural areas. By then, the rural underworld is characterized by social stratification and social mobility.

2. The "public order" perspective discusses the grassroots government's coping tactics and its predicament.
   In the 1980s, in its effort to maintain rural order, the government inherits the same "mass line" as in the collectivization period: responding directly to the masses' demands, relying on the masses to maintain order and giving the masses a sense of security by being identified with them. The government ensures public security by following main stream values of an "acquaintance society" and relying on "local resources." The young dare not to act willfully, with themselves and their families being constrained by the "acquaintance society." When an alliance has been formed in the underworld in the 1990s, public security is greatly undermined, for the upper class in the alliance could effectively evade government punishment, while the lower class has an endless supply and thus may not be eliminated. As a result, the "mass line" is simplified to "being identified with the masses," but the other requirements—to "respond directly to the masses' demands" or "rely on the masses to maintain order"—become greatly restricted by law enforcement protocols. As the "mass line" changes, progress has been made in "special campaigns" such as police-people public security alliance and the informant system, meaning that how state power works at the grassroots level has changed tremendously. What used to go hand-in-hand with the mass line, "governance through hands-on operations" and "supererogation (going beyond the call of duty)" has gradually become ineffective, whereas governance through technology that should have worked with special campaigns has proved to be not as effective either. Therefore, the government finds itself in an awful predicament.
3. The "village" perspective describes the interactions between rural gangs and village life. When rural gangs develop into organized networks, they become the "super power" and have fundamentally changed the state of affairs in the village. Compared with the "super power" of the gang alliance, villagers seem to have meager power; village officials and community are generally impotent; sometimes even the grassroots government is rather weak. Now and then, rural gangs become the dominant role in public order, and under certain circumstances, they are even by stealth the really useful role in village governance and life. They become the most "honorable group." That changes interpersonal relationships in the acquaintance society and the basic relationship principles, for "competent yet unethical" officials and "honorable" gangsters are oddly both active players in rural governance. When both rural life and governance are controlled by rural gangs, violence gradually becomes commonplace.

When rural gangs start to interfere with village life, familiarity and intimacy gradually diminish, and patterns of behavior start to change. Reciprocity is no longer an element of social conduct; people do not care as much about their relationships and are no longer tolerant toward each other. Instead, to protect their interests, people would often seek help from rural gangsters and resort to violence. With such change in people's patterns of behavior, villages could no longer be described as "the acquaintance society" or "the intimate community." Villagers may still know each other, but the basic principle of reciprocity, their behavior pattern and order maintenance mechanism have all changed fundamentally. Such change is summarized as "moral ambiguity" (*hui se hua* literally means "turning gray or ambiguous").

At the same time, due to interference and intervention from rural gangs, after abolition of the agricultural tax, with resources being allocated to rural areas, rural governance starts to grow in an involutionary mode. Due to its dependency on rural gangs to govern the country, the grassroots government has to adopt a policy of "strategic tolerance," allowing rural gangsters to use resources to their good advantage. Their unlawful acts of lining their pockets with national resources that could have been shared by the rural people have led to complaints against injustice and diminishing acceptance of the grassroots government by the rural people. As a result, the allocation of resources to rural areas backfired, lowering the legitimacy of grassroots government. And that is what is referred to as the involution of rural governance.

The sources of information for this book include: interviews and archives that the author has consulted for dissertation-writing purposes, interviews for non-investigation purposes and fellow researchers' fieldwork reports. This information mainly focuses on rural areas on the "*Lianghu*" (literally double-lake) Plains (*Jianghan* Plain and *Dongting* Lake Plain) in the middle and lower reaches of Yangtze River of central China. My fieldwork reveals that rural societies there have similar social structures and regional cultures, similar experiences in implementing the same policies, laws or systems, and would react similarly to external pressure. Therefore, this book uses the research method usually used by researchers of social history and offers a holistic view on villages of the same region by combining information for different villages. Based on previous research on "village governance models," this attempt to go beyond the village level, to investigate regions and do a regional comparative study may turn out to be the right method for research on rural topics.

This book may be regarded as representative of one particular stage of my research. Since 2005 when I first started rural study, I have stuck with empirical research by "being inspired in the fields, thinking boldly and writing straightforwardly." The rural issues I've investigated include justice, land ownership systems, suicide and the nature of the villages. When I was about to select a topic for doctoral dissertation, I shifted my focus to rural gangs and finished the first draft. The focus on the rural gangland does not constitute an end in itself, but rather, through this particular lens, I intended to delve into rural society and politics at the grassroots level.

When I look back and ahead, I found the empirical research approach very effective, and it may be used to further rural studies. That's also the kind of approach adopted by our academic community. We hope our continuous efforts will deepen people's understanding on rural China and the wider Chinese society and politics.

Through empirical research, we hope to eventually put forward social science theories and other conceptual systems with strong explanatory power for a better understanding of China and make our bit of contribution to the development of social sciences in China.

Wuhan, China Baifeng Chen

Acknowledgments

When I'm about to finish the final draft of my dissertation, I have to make a long list of people who I should thank. First, I would like to thank my supervisor, Prof. *He Xuefeng*, for providing guidance throughout the whole project. We have discussed the details in my dissertation. Many are thoughts jotted down from our talk. Some have been directly proposed by my supervisor, and I'm just an awestruck writer of his views. He has deep feelings for his nation and keen interest in the country's future and destiny. He makes a clear distinction between what he loves and hates; he detests evilness as if it were his enemy. He is strict with himself and his students. Due to his continuous efforts and guidance, China Rural Governance Research Center of Huazhong University of Science and Technololy (HUST) has developed into a grouping of academics who work closely together as committed members of our country, our nation, and social sciences. As one of the members to this academic group, Prof. *He* hopes to conduct a study into rural China in an effort to explore pathways of developing social sciences that suits China. As one of his disciples and also a member to the research center, I often feel that great responsibilities rest with me; as I may not shy away from fulfilling them, I must redouble my efforts.

Professor *Dong Leiming* is eight years older than me. He is both a teacher and a friend to me. We would talk freely and openly for a long time. His quick academic thinking is a great inspiration to me. Professor *Luo Xingzuo* is a kind and honest person. He is a skilled investigator and interviewer. I learned from him the rules of conduct for both research and life. All of the three teachers have carefully and specifically instructed me on methods of rural investigation in the same way a veteran carpenter trains an apprentice. Words fail me when I want to express my gratitude to them. Professor *Wu Yi* also showed great interest in my project and gave me useful advice. When we disagreed on some academic issues, he would gently and patiently share with me his opinions. Professor *Wang Ximing* is a humble and hard-working person and an example for all of us. He is also a great inspiration and help for me in both study and investigation. My investigation for the dissertation has greatly benefited from Farmer *Yang Songgang*'s assistance. He is also a great inspiration to me in writing. As he knows well what I need as a researcher and has provided me with extraordinary teachings, he is nothing short of an instructor to me, so I call him the "Farmer Master." Mr. *Li Changping* has worked at the town and township government for many years. He is a keen observer of current events and has provided me with many useful inspirations and instructions. He inspired me to approach major issues of social sciences by investigating some inconspicuous problems.

The members to HUST Research Center have brought me plenty of academic inspirations and happiness in life. Many scholars have become close friends. I enjoyed having intense debate with them at the center's "Master and Doctor's Forum." I should thank the outspoken ones among them for having frequent interaction and communication with me. They've greatly inspired and motivated me. They are *Liu Qin,Yan Qingfeng, Shen Duanfeng, Lv Dewen, Li Derui, Yang Hua, Guo Liang, Zhang Shiyong, Tian Xianhong, Ouyang Jing, Gong Chunxia, Song Lina, Di Jinhua, Zhao Xiaofeng, Liu Yanwu, Yuan Song, and Wang Hui.* Special thanks go to *Yang Hua*, a younger scholar of the center, for having plenty of discussions with me over the writing of my dissertation. He also read my first draft and gave me advice on how to improve it. I would like to also thank other scholars of the center, including *Liu Yue, Mao Gangqiang, and Ding Wei*, for exchanging ideas with me despite their busy work. I've benefited a lot from such exchange of ideas. I'd like to thank Postdoctoral Researcher/Associate Professor *Wang Qiliang*. We are thousands of miles apart, but we talk like old friends over phone and letter. I've benefited a lot from exchanging ideas with him.

My thanks also go to the teachers of Law School of *Zhongnan* University of Economics and Law (ZUEL), including *Chen Jinliang, Fan Zhongxin, Zhang Jicheng, Zhang Demiao, Liu Zhuo, and Zhang Zhengping*, for their continuous support to me in both life and research. Compared with the cohorts in graduate school, I was lucky to have stayed at our university after graduation and may keep learning from the wise teachers. Mr. *Chen Jinliang* is the first teacher for me as a scholar. He has been a loving guidance to me in both life and research. For months, this dissertation has been the focus of discussion between Mr. *Zhang Jicheng* and me during our walk. He gave me many specific suggestions. I should also thank *Liu Renshan*, the dean of law school, and *Cai Minghui*, the party secretary to Law School, and other school leaders for creating a good teaching and research environment.

I'd like to thank Prof. *Li Yuanxing* from School of Sociology of *Anhui* University, Professor *Xu Diyu* from Law School of *Hunan* University, and *Qin Dongxiong* from the College of Continuing Education of *Tsinghua* University for their continuous attention, assistance, and suggestion. They have helped me with establishing connections with my investigation destinations. My thanks also go to Prof. *Zhu Bingxiang, Jiang Lihua, Qi Haibin, Yi Jiming, Lei Hong, Sun Qiuyun, and Zhang Tingguo* for providing useful advice for my dissertation proposal and defense. Special thanks go to Prof. *Deng Zhenglai* who funded my project in its final stage with his "Academic Workshop" Program funded by *Fudan* Institute for Advanced Study in Social Sciences. I'd like to also thank my assignment editor *Fu Lipeng* and my copy editor for contributing wisdom and effort in publishing my dissertation.

I'd like to thank Prof. *Cao Jinqing* and *Zhu Suli* as external examiners of my dissertation. I respect them and feel great pride in gaining their approval and praise. I'd like to thank Prof. Phillip Huang, a renowned international scholar, for making valuable suggestions on revision after reading my first draft. His suggestions were very helpful for improving my dissertation. I'd like to thank Prof. *Lin Duan* from National *Taiwan* University for sharing with me *Taiwan*'s experience with "Anti-Hooligan Act" and for providing me with relevant information. I'd like to thank Assistant Professor *Lv Jie* from American University for copying needed foreign language materials and studying and discussing the materials with me, which has directly benefited the first draft writing. I'd like to thank my alumnus *Hou Meng*. We both graduated as masters from ZUEL's Department of Jurisprudence, and he graduated 4 years earlier. We did not meet at school, but as a graduate candidate to the same department, he showed continuous support for me. His image as an honest and generous person is a fine model for alumni. I'd also like to thank Doctor *You Chenjun*—a dear friend of mine. He is a great encouragement and competitor to me and vice versa.

I'd like to thank my beloved wife. Her constant nagging has kept me company, and I may not bear a lonely day without her doing so. She is the very first reader of my draft and gave me plenty of specific advice on writing and spending a lot of time on draft polishing. I'd like to thank my parents and siblings. For my prolonged stay in school, I may have appeared more naive compared with my peers. I failed to fulfill some of my obligations to them and may have transferred some of urban-living pressure onto my family.

Perhaps at this moment those who I should thank the most are the rural officials and residents whose names remain unknown and impossible to be enumerated. Being fully aware that my investigation would not bring them any real benefit, many of them still had warm-hearted talks with me until our lips were dry. Many of them were very candid and outspoken to me, because they have faith in our ruling Party. They even regarded having talks with me as a way of contributing to the Party's cause. I often felt dwarfed by their generous contribution.

Written on April 2008
Revised on May 2010

Baifeng Chen

Contents

About the Author

Baifeng Chen is a professor, a doctoral advisor, and the vice dean of *Zhongnan* University of Economics and Law's Law School. He has published 5 academic monographs, i.e., *Legal Management of the Watchdog Role of the Media, The Underworld of Rural China, Rural Justice, Violence and Order, Farmer's Life and Values*, and over 100 treatises on journals such as *Chinese Journal of Law, China Legal Science, Sociological Study, CASS Journal of Political Science*, and *Social Sciences in China*. He has conducted several key projects and annual projects funded by the National Social Science Fund of China and annual projects and final-stage projects funded by Ministry of Education (MOE). He gained basic research funding from the *Fok Ying Tung* Education Foundation. He was selected into the China's "Ten Thousand Talents Program" MOE *Changjiang* Scholars Program the New Century Excellent Talents Program and the Hubei Young Talents Program. He was nominated for China's Top 10 Young and Middle-aged Jurists at the 8th Form of Chinese Young Jurists; he was awarded Young and Middle-aged Expert with outstanding contributions to *Hubei* Province and the 4th *Hubei* Top 10 Young and Middle-aged Jurists. He received Higher Education Outstanding Scientific Research Output Award, Hubei Social Sciences Research Outstanding Achievement Award, Outstanding Achievement Award in Law, *Dong Biwu* Youth Law Achievement Award, *Wuhan* Outstanding Achievement Award in Social Sciences, and the award for China's "Top Ten Outstanding Young Jurists" at the 9th Forum of Chinese Young Jurists.

Chapter 1
Research Topic and Approach

1.1 Initiation of Research and Selection of the Central Topic

In July 2005, I went on an investigation trip led by Professor *Luo Xingzuo* to our fieldwork base at *Jingmen* to investigate drought-relief efforts and water supply there. We got information on a "gangster" case: embroiled in a dispute when scrambling for more water in their drought-relief efforts, two farmers sent for bandits from towns or cities, and the two sides reached a deadlock while the village officials and police officers were present (Luo 2006, p. 282). Such incidents are nothing new in urban areas, but the fact that it happened in a pure agricultural village has shocked both of us, because we did not see it coming and its occurrence does not match our understanding on pure agricultural villages. Later when we talked with Professor *He Xuefeng* about this incident, he also described it as unexpected and called for "a thorough investigation into the rural society" (He 2005a). To anyone with living experience in the countryside after the launch of reform and opening-up, gangsters are so familiar that they are consigned into oblivion. However, our researches have often overlooked their existence, or have never treated them as a serious academic topic.

In the following years, I had several trips to the "Double-lake Plains" (the *Jianghan* Plain and *Dongting* Lake Plain) for intensive fieldwork and also went to other rural areas for comparison purpose. I talked to my research partners and exchanged fieldwork experience. On trips to the vast land of the Chinese country, we constantly heard about "rural gangs" and their stories. Their pervasive existence in rural China is directly pertinent to a variety of issues, such as interpersonal relationship, the nature of the rural society, regional governance, governance framework, ethics and values. Professor *He Xuefeng* asserted that rural gangs constituted a different "base" for rural governance and policy execution; hence the bases for rural governance are twofold: an official and public base and a concealed gray obscure one. Researchers may take rural gangs as the perspective to further research on rural governance (He 2006a).

B. Chen, *The Underworld of Rural China*,
https://doi.org/10.1007/978-981-19-8710-6_1

The period since reform and opening-up differs from any previous historical periods in at least the following 4 aspects. First, based on the economic achievements after the founding of new China, economic growth sustained after reform and opening-up is unprecedented and unrivaled by any other time period. Second, as the economy grows, the widening economic disparity between regions and between urban and rural areas, coupled with accelerated urbanization, have resulted in more frequent temporary and permanent migration of rural residents into urban areas. The number and frequency of migrants into urban areas is unrivaled by any other time period. Third, compared with the collective commune system, administrative apparatus of the state has constantly backed off from rural governance and weakened in control. The mismatch between the weakening administrative apparatus and the state's demand for rural resources has once led to grave governance crisis at the end of the 1990s (Li 2002). The crisis was not alleviated until after the agricultural tax was lifted. Forth, while communism and socialism remain the dominant political doctrines, with the dismantling of people's communes and radical transformations in the economic systems, communism and socialism gradually withdraw from people's daily lives. At the same time, the failure to restore Confucianism and the rapid dissemination of market-economy consumerism in rural areas resulted in farmer's confusion of ethical values about marriage, wealth, labor and interpersonal relationships. Such level of chaos may not be found in other historic period except Confucius' time, or the Spring and Autumn (*chunqiu*) and the Warring States periods (*zhanguo*). Some scholars even argued that rural China was undergoing an "ethical crisis" (Shen 2007a).

The above-mentioned 4 aspects sum up the research background, with the third and the fourth being the most pertinent to the gang issue. My dissertation aims to deepen the understanding on the nature of rural society during a transitional period by looking into the influence of rural gangs on rural life in the Double-lake Plains. My fieldwork on rural gangs, farmer ethics and gang influence on rural governance reveals that the nature of the rural society in the Double-lake Plains can no longer be described by the classic model of "the acquaintance society". Rural gangs have fundamentally altered the intimate relationships, patterns of behavior and the moral order in the traditional acquaintance society; it has become everyday business for rural gangs to wield their dictatorial power, and as a result, villages in the Double-lake Plains enter into a state of what is referred to as "moral ambiguity".

1.2 Theoretical Model and Concept Analysis

1.2.1 Public Order Maintenance Mechanisms in Rural China

Mechanisms of the maintenance of public order in rural China have been discussed a lot. Prasenjit (1988) states that "the trend that seeks to articulate the complex relationships between state and society… begins with Karl Marx and Max Weber", who have been overly "impressed by the immensity of the imperial Chinese state, and viewed the forms of local society and local elites largely as outgrowths of this overwhelming state structure. The authority structures of local communities seemed to be entirely controlled by the imperial state through the examination system, the bureaucracy, and official ideology".

This old paradigm for state-society relations was superseded in the 1960s by the "gentry-society" paradigm, which saw the gentry as mediators between state and society. It holds that in the traditional period, maintenance of rural order below the county level mainly relied upon the gentry, for the state did not penetrate through to this level. In this system, the imperial examination system provided guarantee to the sustainability of the gentry. John King Fairbank (1983) argues that, in the past millennium, the gentry have played a dominant role, assuming leadership and important social management responsibilities, such as expropriation and taxation, spreading Confucianism teachings, planning civil defense, building dams and roads, or handling disputes. Therefore, direct involvement of the state in rural affairs was not necessary. It is generally believed that state power did not penetrate to the county level, which allowed the gentry to step in and take charge of local affairs. They assumed leadership and enjoyed privileges. When faced with government officials, they stood for local interests and assumed management responsibilities (Chang 1955). Qu (2003: 282) argues for a gentry-government symbiosis with inevitable clashes: the gentry had an unofficial authority to manage local affairs, while local governments were the authorized managers of social affairs; the gentry were the only legitimate representatives of the local social groups that engages in negotiation of local affairs and political consultation; they acted as the mediators between local officials and the populace, making policy suggestions to the officials and engaging in activities like civil engineering, social welfare, education, *baojia* management and civil group activities. Hsiao (1960) and Ho (1962) also adopted the gentry paradigm in their discussion on social structure and public order maintenance mechanisms in China.

Since the 1970s, many scholars have come to terms with the weakness of this paradigm in explaining China's public order. The inflexible and across-the-board gentry paradigm may not cover all the local elites, especially certain groups whose influence on the social and historical courses of modern China should not be neglected, such as the merchant groups, military groups and powerful landlords, etc. Therefore, some of these scholars replaced the "gentry" concept with "local elites": for example, Kuhn (1970); others regarded the gentry as a highly segmented and all-encompassing entity: for example, Zhou (2000). The "local elites" paradigm is nothing more than some fine-tuning since it has inherited some important propositions from the previous one. Kuhn (1990) pointed out that the decline of *Qing* dynasty was not the same to the decline of the traditional Chinese civilization. He argued, in the second half of the nineteenth century, the administrative apparatus of the *Qing* dynasty may have fallen apart, while local elites, as the base for public order, have remained unharmed. When faced with internal enemies, the central authority could survive the crisis by harnessing the power of local elites, though the result may be a weakened central power and stronger local elites.

In the 1980s, in his research on state power and rural order in the northeast villages in the first half of the nineteenth century, Duara (1988) broke out of the gentry paradigm and coined the new term "cultural nexus of power" to explain public order in rural China. The "nexus" encompasses a complicated series of interactive and mutually influential hierarchical organizations (such as markets, clans, religious organizations, and flood control organizations) and unofficial networks (patron/client, kinship ties, and friendship bonds). The nexus constitutes the base for wielding power and authority. The term "culture" in "cultural nexus" refers to the symbols and norms embedded in networks and organizations and deeply held by the participating people. Because of these shared symbols and norms, the nexus is respected as an authority; it also encourages people to have social responsibility, status, and prestige and thus to seek leadership within the nexus. Prasenjit Duara argues that, at the end of the nineteenth century, both local and central governments relied heavily on the cultural nexus of power for establishing their authority in the northeast villages, and that the state's attempt in the twentieth century to penetrate into rural China while discarding or even destroying the nexus was bound to fail. Duara used "the brokerage paradigm" instead of the gentry paradigm to depict state control on rural society and rural order. This paradigm avoided regarding the gentry stratum as an indivisible whole or making an absolute distinction between the gentry and the other elite groups.

An important reason for Duara to succeed in his theoretical breakthrough may be that the issues he had investigated were highly pertinent to rural order, and that he had collected information from everyday life. Previous analysis based on the gentry paradigm mainly consulted official historical records, among which the most unorthodox ones are county records (*xianzhi*) and historical notes by the gentry class (*shishen shiliao biji*). These records mainly include important events, while the everyday business was often left out. In other words, the important events would be treated as the whole picture, and the gentry as the sole role in maintenance of rural order. On the contrary, Duara based his empirical research on information from *Reports of Regular Survey on China* (*Zhongguo Guanxing Diaocha Baogao*) compiled by Japanese South Manchurian Railway Company (Mantetsu). Those reports are mostly based on farmer interviews, inscriptions, deeds and authentic records of indictments (*bingzhuang*) issued by villagers to the county *Yamen*, the law enforcement department. Duara also consulted ethnographic research done by some anthropologists on the villages of north China. What's recorded by these sources is the everyday rural life. As a matter of fact, because the gentry or the elites accounted for a relatively small share of the total population, they may be directly involved in major events but were often not involved in everyday business, which inevitably will lead to the limitation of the gentry paradigm in providing an explanation for everyday business.

Once we shift our focus on day-to-day business in rural life, we would find Duara's "nexus" paradigm with a strong explanatory power, but it also has some weaknesses. His explanation on the rural order in China requires further investigation in the following 2 aspects. First, with all the success in defining and explaining the "nexus", Duara did not elaborate on the definition of "power". His research centers on the rural elites, their power relations with ordinary villagers and their battle over resources that came by with power. However, as a matter of fact, varied power relations also exist among ordinary villagers, some regular and others occasional, which should not be overlooked in any attempt to form a whole picture of the rural order. Second, due to Duara's limited focus on elite-villager power relations, his investigation into the nexus has been restricted to looking into how power relations work between the elites and villagers, while what's behind other types of power relations has been left out.

This dissertation attempts to deepen the understanding of the nature of the rural society during a transitional period through investigating rural gangs and gang-related rural life. Because the acquaintance society is viewed as an ideal model for the traditional rural society, I must first investigate the model in terms of its behavior pattern, public order maintenance mechanisms and nexus of micro-level power relations. In the third Section of this Chapter, I will justify viewing "the acquaintance society" as the basic unit in looking into the social nature; in the fifth Section of the second Chapter, I will attempt to furnish an answer to this question: for the acquaintance society, what has been brought about by the chaos in some rural areas during changing rule of dynasties and by collectivization of the countryside after the founding of PRC?

1.2.2 Acquaintance Society and Its Behavior Pattern

"Acquaintance society" is a classic model given by Fei (1992) to interpret the nature of rural society in China. Based on this model, Su (1996, 2000) discussed the processes and results of implementing modern laws and systems in rural societies. Some other researchers, such as Qiang (1997, 1998) and Zhao (1997), have also used the same strategy in their discussions. They were very perceptive and have had wonderful discussions on law enforcement in the acquaintance society. However, we have good reasons to inquire into the validity of still using the acquaintance society as the backdrop for law enforcement today. Can we still use *Fei*'s model to epitomize the present rural society?

In the way He (2003a, 2003b) sees it, after the economic reform, administrative villages in China have changed into "half acquaintance society" in nature. His further research reveals that with increased social mobility, diversified employment, and division of labor, the rural population is marked by increased heterogeneity and is experiencing radical changes in their personal and public lives. These facts suggest the nature of society has changed (He 2008). Considering the influence of social mobility on rural governance, Wu (2005) coined the term "baseless society of acquaintance" to describe the less developed rural areas. Nevertheless, is the change from the acquaintance society to a "half acquaintance society" or to "a baseless society of acquaintance" a fundamental one? What on earth are the key elements of the acquaintance society?

In my view, the acquaintance society is characterized not only by "familiarity", but also "intimacy". Intimacy is developed on the basis of "*renqing*" (reciprocity or responsiblity).[1] "*Renqing*" has connotations in 4 aspects—emotion, connection, convention and mechanism. *Renqing* shapes the acquaintance society and develops it into a nexus of micro-level power relations, or an intimate community characterized by few internal disputes and unity against the external world. The formation of intimate community serves as the precondition for Confucianism to regulate relations of reciprocity or responsibility with ritual and to establish social order. This can be referred to as the process of ritualization (incorporating *renqing* into the Confucian ritual system). In this way, "*renqing*" no longer just refers to natural human feelings, but has blended with the ritual system and finally been incorporated into the system. It also becomes the basic thinking pattern of people in the acquaintance society. The successful incorporation of reciprocity in the ritual system may be attributed to the fact that the Confucianism theory caters to practical needs of the acquaintance society. In the acquaintance society honoring ritualized *renqing*, people's behaviors are influenced by *renqing* (reciprocity or responsibility); they embrace reciprocity or responsibility as their code of conduct, which may be summarized as "*xiangtu luoji* (literally the traditional behavior pattern accepted on the rural land)". People's behavior pattern may be summarized into 4 principles, with the first one serving as the basis for the other three. The first principle is the "face" (do no harm) principle: in daily interpersonal relationships, village fellows should care to bond and save face in the spirit of give and take; the second is the "not going too far" principle: one should refrain from an uncompromising mindset and always leave room for maneuver; the third is the "discriminatory" principle: one may discriminate against strangers and disregard their interests; the forth is the "attachment" principle: one must be sentimentally attached to one's home town, and one's personal honor and grace are closely linked to it. Behavior pattern of rural land (*xiangtu luoji*) may provide applicable standards for measuring the current changes in rural order.

1.2.3 Rural Gangs and Moral Ambiguity

The Chinese character "*hun*" has 4 connotations: to mix, to deceive, to muddle along, and carelessly (the Dictionary Editing Office of the Institute of Linguistics at Chinese Academy of Sciences, 1983: 508; 2005:616). These meanings have remained largely the same since the launch of reform and opening-up; in "*xiangcun hunhun*"—the Chinese expression for "rural gangs", the Chinese character "*hun*" should be interpreted in its third meaning: to muddle along. Contemporary Chinese Dictionary (2nd edition) published in 1983 does not contain entries for "*hunhun*", whereas the 5th edition published in 2005 includes 2 entries for "*hunhun*": (1) "*hunhuner*", defined as "(dialect) hooligan, rogue", and (2) "*hunshi mowang*", defined as "a figure of

---

[1] Translator's note: The Chinese expression "*renqing*" literally means "human feelings". It may also be used to refer to reciprocity in personal relationships or exchanges of favors and gifts. People often feel obligated or have the responsibility to honor past favors. Therefore, the concept of "*renqing*" may be explained by reciprocity or responsibility.

speech for villains who disturbs the world order and causes serious harm to people". The reason why the 2nd edition does not include any entries related to "*hunhun*" may be that, by the time it was published, the expression "*hunhun*" had almost entirely vanished. Based on the entries provided in the 5th edition, "*hunhun*" is a strongly negative phrase. In my research, I came to understand that people started to use the expression "*hunhun*" towards the end of the 1980s, before which time the name given to village transgressors was "*liumang*" (hooligan). Besides "*hunhun*", many other names are also used, such as *"hun hun zi", "hun hun er", "liu zi", "er liu zi", "liu da gui", and "lan gan zi", "lan tong zi".*[2] All of them are strongly negative names, with no exceptions.

As an ordinary farmer sees it, rural gangsters have no legitimate profession, gain benefits by extortion or deception, intimidate farmers by coercion, harm their personal and property safety, and disturb the rural life order. This is also the view of rural gangs I have adopted in this book. I refer to their world as the rural *Jianghu* (the underworld of mobs or gangs). I investigated rural gangs in terms of their characteristics, strategies, organizational structure, and expansion patterns and processes. In short, rural gangs have indeed greatly harmed the rural society and caused many serious consequences. They have generally gone from "bearing the name" to "living up to the hype": at first, they are known for showing a belligerent tendency, but later they start to gain real benefits through violence or threats of violence.

In rural areas of the "Double-lake Plains", the pervasive gang influence extends to both in and out of villages and to both daily and political life. They are the "strangers at people's doorsteps", leading to estrangement in the acquaintance society. The gang problem is endogenous; its relation with the rural society is one of interactivity and mutual causality. Rural gangs have formed in some loosely-structured villages with a low degree of interpersonal closeness, and in turn have loosened up the social structure and weakened social ties; in some other villages, gangs do not form from within, but the entry of external rural gangsters has also led to changes in social structure and nature. My research reveals that the gang influence would fundamentally change rural life and gradually bring about the collapse of the ethical order in the acquaintance society.

Through network expansion, rural gangs in rural areas of the "Double-lake" Plains have now built up a super power structure, with every gang member as the center forming a loose alliance with the others. This alliance, also featured by social stratification, shields them against risks of being in the gangs. Compared with the "super power" of the gang alliance, villagers seem to have meager power; village officials and community are generally impotent; sometimes even the grassroots government is rather weak. Currently, village communities and grassroots state power both find it difficult to handle the gang threat on public order.

---

[2] Translator's note: Literal meanings of these expressions are: hooligan *for "liumang"*; an idle person with no official employment *for "hun hun zi", "hun hun er", "liu zi", "er liu zi", "liu da gui"*; a rough stick *for "lan gan zi"*; a broken bucket *for "lan tong zi"*. The correspondent Chinese expressions may be found in the "Glossary for Chinese terms".

As gang activity mainly relies on force or threat, social order becomes increasingly chaotic and interpersonal relations no longer bound by the traditional behavior pattern accepted on the rural land. The locally born and bred rural gangsters treat local villagers as strangers and disregard the face and discriminatory principles followed by members of the acquaintance society. When involved in disputes, villagers also tend to seek help from rural gangs and resort to violence, rather than sticking to the former principles for handling relations. To protect their interests, people have become more inclined to go to extremes and abuse privileges. No longer attached to the land, villagers spare no efforts to severe their links with the village and try to leave the villages. This is even truer with rural gangsters. They would not go back to the villages unless expecting to make some gain. Because people no longer feel attached to the land, the principles that have grown out of this attachment do not apply any more. Drastic changes in people's patterns of behavior have come as a shattering blow to moral order, as well as people's perception of labor and wealth. On one hand, the gang network becomes more attractive and thus recruits more members including many young people. To work as a gangster even becomes an ideal job. On the other hand, people also grow dependent on the rural gangs for resolving daily conflicts or financial disputes. As gang influence on rural life grows stronger, moral order in the country is gradually falling apart.

When the above changes have taken place to patterns of behavior and moral order, the rural society may no longer be regarded as "the acquaintance society". The villagers may still know each other, but fundamental changes have taken place to their basic principles for interpersonal relationships, behavior pattern, power relations and public order. Such change may be summarized as "moral ambiguity in the rural society" (the Chinese term "*hui se hua*" literally means "turning gray" or "becoming ambiguous"), as I found the term "half acquaintance society" inappropriate. Public order under the influence or control of rural gangs may be referred to as "the morally ambiguous order".

Here a distinction must be made between rural gangs and other outlaws in historical and contemporary periods, such as *kuang kun (thugs)*, "*e ba*" (tyrants), "*daofei*" (bandits), "*shehuixing daofei*" (social bandits), "*xiashi*" (martial arts heroes who defends the weak), "*Mimi huishe*" (secret society), and "*Heishehui zuzhi*" (organized crime groups), etc.

Thugs are overbearing and reckless singles, who will stop at nothing for they have no spouse or children. Tyrants are bullies and oppressors supported by some political or social forces. The key difference between them and rural gangs is that both thugs and tyrants lack organizational structures. However, just like rural gangs, they are also involved in illegitimate businesses. Therefore, thugs and tyrants are occasionally also classified into the category of rural gangsters.

Bandits are professional thieves and robbers operating in organized groups in rural areas and oppress villagers by establishing predominance. Unlike the rural gangs with an organizational structure, bandits often organize into independent groups. Generally speaking, the bandits are even more immoral. They often hide in the dark and engage in secret operations (for example during night). During historical periods of environmental degradation and harvest failure, bandit activities may become pervasive, public, and justified (Perry 1980). Currently in the rural areas on the North China Plain, bandit activities show the tendency to go public, and that's maybe why some villagers put them in the same class as rural gangs.

Social bandits may be characterized as illegal by law and regarded as criminals by the government and the upper class, but "are not regarded by public opinion as simple criminals, but rather as champions of social justice, as avengers or as primitive resistance fighters" (Hobsbawm 2010). The same can be said about martial arts heroes. The key difference between them and rural gangs is that in the villagers' eyes, social bandits and martial arts heroes have a righteous and moral image.

Organized crime groups are illegal organizations that take a stand against the mainstream society. They are well structured with detailed assignment of responsibilities, strict discipline, and harsh punishment. They would resort to serious violence and accumulate wealth through illegal means. They often collude with government officials and assume hegemony in their territory. Both rural gangs and organized crime groups seek financial gain, but rural gangs do not disrupt social order in a well-organized manner. Their behaviors are less violent. Rural gangs would often rely on psychological intimidation to achieve their goals.

1.3 Methodology

1.3.1 Research Traditions for Rural Governance in Central China

This study belongs to the broader research on rural governance in central China, whose origin may be dated back to a study on villager autonomy in the 1980s. Having reflected upon the social sciences research of their day, the elder generation of political scholars, such as *Zhang Hou'an*, *Xin Qiushui* argued against sheer theoretical research, proposed to shift focus from state governance to rural governance, and further to "engage in agriculture with theories (theorists should serve agriculture, rural areas and farmers)" (Zhang 2001). That was when *Zhang* initiated his rural research; he was the first to investigate villager's autonomy—a rarely discussed topic then. Based on his research results, he compiled *Rural Grassroots Governing System in China* (Zhang 1992). What really sent theorists down to the grassroots level was, as Wu (2005) put it, the 1989 political movement. It was like an awakening call for politics to see the real world, while previously it had been working under the banner of rationalism and enlightenment. As *Lv Dewen* put it (2006a), *Xu Yong* added footnotes to this academic history with his efforts and achievements by intentionally shifting the academic focus from state governance to local governance (Xu 1991) and by including rural politics in the research horizons of politics, claiming its due place in this science (Xu 1992).

In the 1990s, research on rural governance in central China gained unprecedented progress, because the government attempted to advance villager's self-governance. In the wake of the Commune era, through the 1982 amendment of the constitution, the state designated villager's self-governance as the political system in place of the commune system in the rural areas. In particular, in the mid-1990s, villager's self-governance obviously took over the responsibility to establish "democracy", which has later led to a politically and socially influential "public academic movement" in the field of politics (Shen 1998; Wu & Li 2007). Villager's self-governance practice opened up great opportunities for research on rural governance in central China and paved the way for the dominant position of this research topic in the broader rural studies. One of the representative works was *Villager's Self-Government System* by Xu (1997). In the process of establishing villager's self-governance, *Xu* and *Xiang Jiquan* inherited *Zhang*'s work and broadened the reach of self-governance research to villager-level governance research (Zhang et al. 2000).

At the turn of the century, researchers proposed two closely connected prospective areas of interest, restructuring the domain of discourse for research on rural governance in central China. One was to understand the daily rural life, as proposed by *Wu Yi* et al. after the *Huangmei* pilot program[3] (a villager's self-governance pilot program) has failed; the other was to analyze the nature of the rural society through an observation of village committee elections, as proposed by *He Xuefeng* and *Tong Zhihui* (Wu et al. 2005).

The *Huangmei* pilot program failure prompted *Wu Yi* et al. to make a dramatic shift in their areas of interest. Despite the failure to put into practice the theories on villager's self-governance, the program was enlightening to all the people that have witnessed it. It was just after this failure that *Wu*'s group found villager's self-governance ill-suited to the rural society as a top-down policy. They have later come to terms with the fact that the rural researchers lacked a good understanding of the rural areas and therefore made up their mind to bid farewell to what was later referred to as "research on rural politics that associates everything with ideology" (Wu, 2005). They shifted their focus to providing an interpretation of village politics and governance. The main research results were included in *Wu*'s book published in 2002.

In 1998, with the promulgation of *The Organic Law of Village Committees of the People's Republic of China*, committee elections took place in nationwide villages, providing researchers with opportunities for field observation and participation in rural politics. They went to dozens of villages around the country. The elections broke the daily routine for village politics, reinforced connections and escalated conflicts, bringing out all that drama for researchers to see. At the same time, researchers have also examined how the election system has evolved in different social fields. The real political system has deviated from the established rules that were supposed to

[3] Translator's note: Funded by the *Hubei* provincial government, "the *Huangmei* Pilot Program" was a villager's self-governance reform conducted at *Shuiyuean* Village, *Xiaochi* Town, *Huangmei* County, *Hubei* province. The program aimed at putting into practice the theories on villager's self-governance.

govern rural politics and was also different from the textbook election system. This prompted the researchers to shift their areas of interest to "providing an interpretation and understanding of rural China during a transitional period", or to understand the reality in the peaceful everyday country life (Feng, 2002). Main research results were included in the works of He (2003a, 2003b) and Tong (2004).

That was when "self-governance (*cunmin zizhi*)", abbreviated as "*Cunzhi* (*cun* means village; *zhi* governance)", was redefined as "rural governance (*xiangcun zhili*)"; and "rural governance research" as "research on the nature of rural China during a transitional period". Later, as the research became more inclusive, more topics were incorporated. *Wu Yi* et al. attempted to provide a deep understanding of the changes in villages (Wu 2002, Student Tan 2007), whereas *He Xuefeng* et al. to further shift their focus from "basic research of rural policies" to a research on "the nature of the rural society during a transitional period". Having done extensive fieldwork, he gradually came to terms with the vagueness and lack of feasibility for research on "the nature of the rural society during a transitional period", and instead, proposed "a research on the social foundation for village governance" as a new approach for "basic research of rural policies". This new approach attempts to conduct an in-depth investigation into rural areas across the nation to understand regional differences in terms of the processes, mechanisms and consequences of bringing policies, laws and systems to the rural areas. Only when they have a better understanding of the rural society, when they have a good grasp of the consequences of the laws, policies and systems, would they know better what laws, policies and systems should be in place. The key question was "what and how different policies have been implemented in different types of Chinese villages?".

Since 2003, researchers have gradually moved onto special focuses and the stage of fieldwork for regional comparison. Special focuses were water resources (Luo 2006), welfare benefits for older people (Wang 2007), dispute settling (Dong 2006a, 2006b, 2006c, 2008; Chen 2005, 2008a, 2009c; Dong et al. 2008), rural medical care and cooperation between farms (He Xuefeng, 2009b), how village governance works (He 2009a), and rural political systems, etc. Researchers mainly focused on central and western agrarian regions for an understanding of "80% of rural China and 80% of the things that happen there". To this point, research on rural governance in central China has taken on an interdisciplinary approach with an aim to understand the maintenance of rural order and the development of the rural society. The research will be conducted from the following 3 perspectives: macro-level, meso-level and micro-level. The macro-level perspective intends to understand the social environment, the historical context and the reality for village governance; the meso-level perspective: the processes, mechanisms and results of top-down policies, systems and laws; the micro-level perspective: how the society functions internally and farmer outlook on rural life (He et al. 2007).

1.3.2 The Basic Unit of Rural China

My study offers a narrower perspective into the broader research on rural governance in central China and takes "the village" as the subject. Therefore, I will demonstrate herein "the village" reasonably as the basic unit of the rural society. Classic paradigms include "village community" proposed by Japanese scholars and grassroots "market town" proposed by the American scholar G. William Skinner. The former holds that the basic unit of the rural society is the village characterized by insularity and solidarity; the latter argues that the basic unit is not the village, for "the village" could not qualify as a complete structure and lacks certain functions, but the marketing system with standard market towns at the center of the system. Later scholars have conducted extensive discussions on the two opposing views, and offered different perspectives for what factors to consider when determining the basic unit of rural China.

In 1941, based on *Reports of Regular Survey on China* by Japanese Mantetsu, Japanese scholar Hirano Yoshitaro identified the collective nature of the Chinese villages: in the rural areas of China and other parts of Asia, rice agriculture carried out through mutual assistance among family members required cooperation and cordiality, which in turn, consolidated the foundation of village communities. The collective nature of villages may be reflected by such collective activities as agricultural production, security and defense, worship and sacrifice, entertainment, weddings and funerals, and by the commonness of moral conscience and norms. Hirano pointed out that the rural Chinese societies are organizations of collective life where temples are not only a place for worship but have become a center of social life. Before him Shimizu Morimitsu has also noted the collectivity in Chinese villages. After reviewing Hirano's theory, Hatada Takashi not only insisted on the distinctive collective nature of Chinese villages, and at the same time, regarded it as the social foundation for despotism. However, the "village community" theory has been refuted by Kaino Michitaka and Fukutake Tadashi. Besides, Hatada Takashi argued that Hirano's assertions on collectivity derived from his intention to form the basis for a Great East Asian Co-prosperity Sphere, and that Kaino's intent to argue against village community was to advocate the path of development that he had idealized for the West (Li, 2005; Zheng, 2006; Masao, 2001). It may be said that the Japanese scholarship has been influenced by their ideological preferences, but has generally regarded the village as the basic unit of rural China.

In the 60s, American scholar G. William Skinner put forward the standard market system theory. He contended that the basic unit of the traditional Chinese country was the standard market towns, rather than open villages that lacked a complete structure or functions. "The effective social field of the peasant", he argues, "is delimited not by the narrow horizons of his village but rather by the boundaries of his standard marketing area." "Chinese marketing systems have important social as well as economic dimensions… also as a cultural-bearing unit (Skinner 1964)".

Philip C.C. Huang summarized the contrasting views as an opposition between Western formalist and Japanese substantivist tendencies, and attributed the contrast to regional differences among the North China plain, *Chuanxi* Plain and the Yangtze River Delta. He argued that American scholars have studied mainly the most highly commercialized areas of China (e.g. the *Chuanxi* Plain and the Yangtze River Delta), whereas the Japanese scholarship has focused on the North China plain during the war. Differences in degrees of commercialization, levels of social stratification, and the power of clans or the lacking of it and kinship structures have led to great disparity in peasant lives and social structures (Philip Huang, 1988). Having considered the merits of the two opposing analyses, Prasenjit Duara shifted the perspective to the building of state power and argued that the penetration of state power has resulted in changing power structures, and proposed a more comprehensive paradigm: "the cultural nexus of power" (Duara 1988).

The controversy between the village community and the market town theories centers on what constitutes the basic unit of the rural Chinese society. In the 1930s, *Fei Xiaotong* applied ethnographic methods that were commonly adopted in studies on primitive societies to research on Chinese villages, and developed the famous "community research model". This model was recognized by Bronislaw Malinowskia, a pioneer in social anthropology then, as "a landmark in the development of anthropological fieldwork and theory" (Fei 1939, Malinowskia's preface 13). "The Modern Chinese School of Sociology" *Fei* has established had a long-lasting influence on Chinese studies and had ambitious academic goals, i.e., to give a comprehensive picture of China at large through the scope of village community. Therefore, the community research model has, from its very beginning, has raised doubt from academia about its representativeness. Two questions have been proposed: (1) as Edmund Leach put it, can individual community model the whole country? And, (2) was the standard marketing community proposed by Skinner the basic unit of the Chinese society? Before Skinner, Maurice Freedman has expressed similar doubts by arguing that even a holistic analysis of a small place may be inadequate to present a whole picture for such a complex civilization as China, and that one must draw on long-term sinology researches on civilization history, look beyond communities and study China over longer time and in broader space (Wang, 2005:32). Maurice Freedman, G. William Skinner and their disciples formed a school of "anthropological research into the Chinese society", exerting a long-lasting influence on subsequent research.

I will leave the first question for the next section and deal with the second one first. The lack of a comprehensive view of rural China on the part of the researchers may lead to a fuzzy picture of the basic unit of the Chinese country. As Philip Huang puts it, the controversy may be attributed to the fact that researchers have based their propositions on empirical data with stark regional differences. This is a very reasonable argument, and has been proven by extensive rural research. Empirical research also reveals that the rural areas where the *Han* population is concentrated may be segmented into five regions: south China, north China, central China, western *Sichuan* and the *Yangtse* River Delta (He 2009a: 3, 2007b). This segmentation has taken into consideration all the economic, social and cultural imbalances, as well as historical and geographical differences, which may be summarized as differences in: (1) the distance from the central government, (2) the beginning of development,

(3) physical environment and natural resources, (4) the influence of crop failure caused by natural disasters and the influence of consequent migrations, (5) internal structure of villages, (6) land ownership and use, and (7) residential structures. In the classic works on the subject, Japanese scholars Itaro Hirano, Takashi Shimizu, Hatada Takashi and American scholars Philip Huang, Prasenjit Duara and Ramon H. Myers have focused on north China villages; *Fei Xiaotong* and Philip Huang on the *Yangtse* Delta; Maurice Freedman and many subsequent American scholars on south China; G. William Skinner on *Chuanxi* Plain; Fukutake Tadashi on central China; and my research focuses on typical central China villages in the rural areas of *Hubei* province along the *Yangtse* River, and thus use *Hubei* Province as an archetype for understanding other rural areas.

On *Chuanxi* (western *Sichuan*) Plain, abundant rainfall and sufficient water supply from its river system are great for rice production, but in traditional agricultural periods, due to high soil stickiness, farmers may not use wheel barrows for transporting manure or grain like the people in the north and have to rely on manual labor instead. Therefore, cottages are built as close as possible to the farmland, which may lead to dispersed settlements. The problem in this is vulnerability to theft or robbery, and that has led to "absent landlords": landowners and rich farmers move to towns and start to do some business there. Two things have taken place consequently: firstly, the standard market towns have come into being; secondly, while being away, the elite groups are less committed to their clans, which causes identity and influence of those organizations to gradually diminish. Meanwhile, the renting and trading of land are more likely to take place as long as a higher price is on the table, generating a land market. As land is traded more frequently, tenant farmers constantly move in pursuit of cheaper land. Such a lack of physical proximity for the people has in turn led to looser kinship ties (for example brotherhood); other ties and a sense of identity that could have developed out of geographic closeness are also no longer there (He 2007c). In historical periods, almost all the villages form a well-developed standard market system, making it hard for villages to stay as independent communities. However, a noteworthy fact is that the standard marketing system is unique to *Chuanxi* plain. Therefore, Skinner's hasty conclusion that this paradigm may apply to the whole Chinese country with unbalanced regional development, has obviously biased researchers' judgment.

Markets exist in different areas of China; thus, we may not regard markets as an enclosed space. Villager activity is not restricted to market towns and takes place on the level of standard market systems. Other regions do not have well-developed market towns as *Chuanxi* Plain; their villagers are also much less influenced by market towns. Take the *Yangtze* Delta. What the *Yangtze* Delta and *Chuanxi* Plain have in common is: flatland, high land productivity, highly sticky soil, inconvenient land transport as well as premature and well-developed market economy. However, in the case of the *Yangtze* Delta, what came with the network of tributaries and canals is convenient water transportation and a higher flood vulnerability. Transport convenience allows villagers to live together; while frequent floods force villagers to live on highlands. Therefore, concentrated housing on highlands there are much larger than courtyards at *Chuanxi* Plain. The Yangtze Delta has a higher level of economic growth with well-developed handicraft industry and commerce. Cities quickly develop into "clusters". A unique "dual ownership" is formed: land is owned by city dwelling landlords, and tilled by farmers on a long-lasting tenancy. Landlords keep contact with tenants through intermediaries; tenants may live together for generations and thus develop geographical consciousness. However, village tenants and city dwelling landlords are two opposing social classes divided by serious class conflicts (He 2007b; Lv 2007). Therefore, instead of forming a market system that covers several villages as in *Chuanxi* Plain, villages in the *Yantze* Delta remain independent living communities. Besides the *Yangtze* Delta, the same could be said about rural areas in south, north and central China. The degree of openness, the level of geographical consciousness and kinship closeness may be different, but "village" may be regarded as the basic unit for rural living in all those regions.

Whether villages constitute a collectivity is not our concern; it obviously depends on the definition of a collectivity, and our research area for gathering empirical data. Undoubtedly, "village" has been the basic unit for daily life and for the rural society in most regions of China, with *Chuanxi* Plain as the only exception. As a matter of fact, once *Chuanxi* Plain has gone through the era of collective farming, the "village" also becomes the fundamental unit of the rural society there. Just like the other regions, during era of collectivization, rural areas in western *Sichuan* are reshaped into independent communities, and the standard market towns decline. The village, therefore, becomes the basic unit and develop into the acquaintance society. After reform and opening-up, despite of rejuvenation in market towns, the collectivized village still plays a major role in the supply of public goods, such as road building and water infrastructure development. Tea houses that used to be found only in grassroots markets are also opened in the collectivized villages; production brigades and production teams are still by nature the acquaintance society. In other words, after the era of collectivization, state power has penetrated to the local village level (Yang 2007a). Therefore, once collectivization is over, the "village" may be regarded as the basic unit of the overall China country.

The time frame of this research starts with the 1980s, when there is nothing logically wrong with taking "the village" as the basic unit. However, due to regional differences, "the village" mainly refers to a natural village or occasionally also an administrative village. Because village forms what *Fei Xiaotong* called "the acquaintance society" or what *He Xuefeng* called "the half acquaintance society", it can be safely concluded that the rural China is made up of the acquaintance society.

1.3.3 From the "Village" to Regional Comparison

AS researchers of rural governance in central China entered a research stage of special focuses and regional comparison, I chose rural governance as my special focus. Therefore, my research was characterized by the same methodological features in this special research stage. My research adopts two perspectives, one being the village life perspective, the other being the regional comparison perspective. The village life perspective intends to find out how rural gangs fit in with rural life and provide an explanation for "the moral ambiguity of the rural society", or form a structural understanding of all sorts of phenomena in the village, and to understand the connection between moral ambiguity and other political and social occurrences. Standards of morality has become ambiguous at a particular time and in a particular place. This process has been governed by some particular pattern of behavior and is a reflection of a particular aspect of village life, inseparable from other aspects. Therefore, we could not simply sever its connection with everyday village life. We should look at it as embedded in the rural social environment.

In real practice, the village life perspective works in a way similar to the way "indexity" works in ethnomethodology. In ethnomethodological terms, social behaviors are always part of a setting, and the setting is also part of the behavior; both are then practical accomplishments on the part of members of a society. Behaviors are indexical, that is, all behaviors "depend on unstated assumptions and share knowledge for the mutual achievement of sense". Any seemingly independent "expression" and "action" is, after all, a listing in an endless index, and could never be anchored to an ultimate foundation that is not bothered by the indexity issue. That is what Harold Garfinkel called "a boat without a bottom". Behaviors are accountable and reflective: accountable means observable, reportable and understandable, and reflective means a behavior and the description of the behavior cannot be separated one from the other. Behavior, settings and descriptions of such constitute a complex practical whole (Yang 1999: 56). Based on these views, ethnomethodology attempts to adopt appropriate research methods according to the features of the social environment and local settings.

When we try to understand the rural underworld in the context of village life, what we see as specific actions and phenomena must have been done on certain occasions or in certain circumstances; the actions and phenomena are bounded by certain human relations, and constitutes a part of the broader village life. Therefore, we need to uncover the actions and meanings behind those specific actions and phenomena. That requires a careful enquiry into the people involved, asking them to give a reasonable explanation. We also need to be immersed in their situation by "seeing things through their eyes". There is no end to "indexical actions"; we may need to simply uncover the superficial reasons behind the actions and phenomena, or possibly also the underlying reasons, and even what's behind the underlying reasons. The village life perspective requires us to uncover reasons behind village governance actions and phenomena "from within" the village life and providing locally situated interpretations of those actions and phenomena. When seeking theoretical interpretations, we must keep our reasoning practical and reach justifiable conclusions on the actions and phenomena from within the village life context. The village life perspective does not encourage quickly turning to theories to search for theoretical resources to back up practical reasoning, for that involves a major risk of not just having practical reasoning dissected by the theories, but even supremacy of theories over local practices. Of course, the village life perspective does not go against looking beyond the villages, and it could not possibly do that, because in terms of time, the experience we gain from village governance is a continuity of history, and in terms of space, the village can never be isolated from the outside world. The subsumption of present villages beneath modern policies, systems and cultures renders it impossible to provide an explanation for village practices without considering the modern factors behind those practices.

Based on a good deal of field investigation, we found that villages in the same region demonstrated a lot of similarities in village governance. Therefore, regional comparison may offer us an open-ended framework for understanding the Chinese country. This dissertation focuses on the villages of Double-lake Plains and, through regional comparison, highlights the trend of moral ambiguity at the Plains. To start with, my research examines the intrinsic factors that have contributed to moral ambiguity and consider these factors in a regional comparison framework. This will not only deepen our understanding of individual villages, but also identify common intrinsic factors for moral ambiguity. Apparently, the village life perspective was an inheritance of the long-established traditions of research on rural governance in central China, while the regional comparison perspective was necessitated by the need to test villager autonomy policies.

The most controversial part in regional comparison is its feasibility, or in other words, can case study move up to the regional level, and how? This is the above-mentioned clichéd "representation" problem, or as Edmund Leach has put it, can an individual community model the whole country? In response to Leach's question, *Fei Xiaotong* acknowledged that a part could not represent the whole; therefore, the same method could not be used in all regions, and he proposed "gradual approximation" for moving from a part to a whole (Fei, 1990). The ideas behind his analysis was to start from individuality and identify village types by comparison to gradually come to an understanding of approximately all the villages. That's why he investigated three villages in *Yunnan* after studying country life in the *Yangtze* River. After the launch of reform and opening-up, he shifted his attention from the village level to townships, still adopting the same comparison method. However, *Fei* failed to elaborate on how to move from "points" to "areas" in his analysis. Therefore, "gradual approximation" lacks a clear feasible road map or clear theoretical guidance on the categorization of villages and townships. The comparison research plan does not work out for other researchers. Subsequent researchers also failed to take the research plan a step further.

Since the 1960s, influenced by Freedman and Skinner, scholars no longer looked at individual villages as China in miniature, but combined local studies with sinology to examine the individuality of regional societies, social systems, or organizations. After the launch of reform and opening-up, scholars seemed to have abandoned the ambition of representing the whole Chinese society using the community research model (Chen 2007). Therefore, no one has directly answered the question proposed by Freedman and Skinner. Nor have they responded clearly with their practical research. Case studies may still attempt to "afford wide views in individual cases" (Lu 2005: 29), but more attention was given to examining political changes in rural China against the historical background, or to look through individual cases and examine political changes in the wider China. Most of the scholars avoided the representation question, claiming that they did not intend to see the general patterns and only focus on local knowledge or issues specific to a certain locality. Some of them even denied any effort to seek an answer to the representation question. Some commented by writing that "community investigations have become a laboratory for testing hypotheses and propositions" (Lu 2005: 29). It is worth mentioning that at the end of the 1980s, when conducting fieldwork in north *Zhejiang* Province, *Cao Jinqing* et al. proposed a culture-based and ecology-based segmentation of the rural China and were ready to conduct a research on rural China by segments (Cao et al. 1995). Later Cao even did fieldwork all by himself in the reaches of the Yellow River (Cao 2000). Unfortunately, they have failed to fulfill this great ambition in practical research.

In research traditions for rural governance in central China, special policy guidance ruled out pure research; research must be done in a practice-oriented manner. We all knew China was characterized by unbalanced development, but we didn't know how, and that's the problem we had to deal with in executing any policy. In this way, there was no getting around the representation problem. In the current research framework, we move from case study to regional features through summarizing "rural governance models", and then to regional comparison (He 2007a).

Case study would always encounter the representation problem: moving from particular to general and from the micro-level to the macro-level. The description or analysis of an individual case often fails to represent the whole society, and that becomes even more so as the society grows more complex. Quantitative analysis exacerbates the problem, further undermining the representativeness of individual cases. To deal with this problem, *Zhu Xiaoyang* and *Lu Huilin* et al. favored the extended case method. Lu and Li (2007) states that the extended case method aims to establish a macro perspective for micro-sociology, examines complex macro questions through their everyday micro manifestations, apply theories thoroughly in practice, and through rebuilding the theory, establish general rules. The representation problem would be properly dealt with. Zhu (2007) argues that the traditional social sciences have been dominated by the positivism methodology of "regarding the objective world as separate from the subjective mind", which made it impossible for Chinese scholars to comprehend the "principle", "spirituality" and "human nature" in the Chinese people's everyday life and worldview, and for a long time perplexed intellectuals such as *Fei Xiaotong* and his followers. In view of this situation, *Zhu* proposed the extended case method as an important method of holism philosophy to observe and interpret social norms and public order in China, and carried it out in his research (Zhu 2008).

While acknowledging the effectiveness of regional comparison in our research methodology, Zhu (2004: Note 46) and *Lu Huilin* et al. doubted the possibility of generalizing regional features based on individual cases. My take on the question is that their research approach does not necessarily contradict with ours. In fact, the representation problem is relevant to us all, but we had different angles dealing with it. Instead of examining empirical evidence, *Zhu Xiaoyang* and *Lu Huilin* et al. focused on concepts and macrostructures in social sciences, while in their view, the processes, mechanisms, and results of policy implementation that we focused on were still on the empirical level. In our research we are trying to identify boundaries of an area to which our empirical evidence applies, whereas they have given up such efforts because they doubted the feasibility of such search. However, their research stands at the theoretical level and keeps us alert to the fact that empirical evidence is deeply rooted in the concepts and macrostructures. In fact, we've given a simple solution to this problem by proposing to "understand village governance from within the villages, and to understand villages by investigating village governance", but their research keeps us highly aware of the methodology we have been adopting in our investigations and study.

Based on our research experience, in our scope of research, it is completely possible for us to move from individual cases to regions and then to regional comparison. To do that, we must first do an in-depth investigation on individual cases, and construct case-based rural governance models. Key aspects to be included are the processes, mechanisms, and results of policy implementation. Quantitative indices are used to measure political and social phenomena. Then we compare the individual village against the regional benchmark to identify differences, discover unusual phenomena against the regional norms, and then summarize the region's general features. By comparing different rural governance models, we will spot the differences of the models and determine the right key words for depicting those models. This will make it easier for us to construct more precise models. Based on regional comparison, we will be able to build a theoretical framework that may be applied to all regions (He 2007a). When moving from individual cases to regions, we must use statistical data or sample data based on existing theoretical factors, to construct regions and to put forward propositions to facilitate regional comparison once regional segmentation is finished.

1.4 Fieldwork and Dissertation Structure

Once finalizing "rural gangs and moral ambiguity" as the research topic, I selected some localities of *Chujiang* city, *Linhu* City and *Linjiang* County on the Double-lake Plains for special field trips, and designated localities of *Shanxiang* County in south China and *Pingyu* county in north China as destinations of fieldwork for comparison.[4] All 5 fieldwork destinations are located in agricultural towns and townships far from counties or cities. I also went to some other places on the Double-lake Plains for gathering information on the topic. *Huazhong* University of Science and Technology (HUST)'s Research Center of Rural Governance in Central China has done tremendous work on Double-lake Plains: it has organized around 100 field trips to villages and has drafted plenty of papers, research reports, academic monographs, master's theses and doctoral dissertations, some not yet published. Their research has provided abundant empirical evidence that I could not have obtained and paved the way for my research. Sources of information are shown in the figure.

---

[4] Based on the anonymity rule in sociology and anthropology, the names for places and people in this dissertation have been modified.

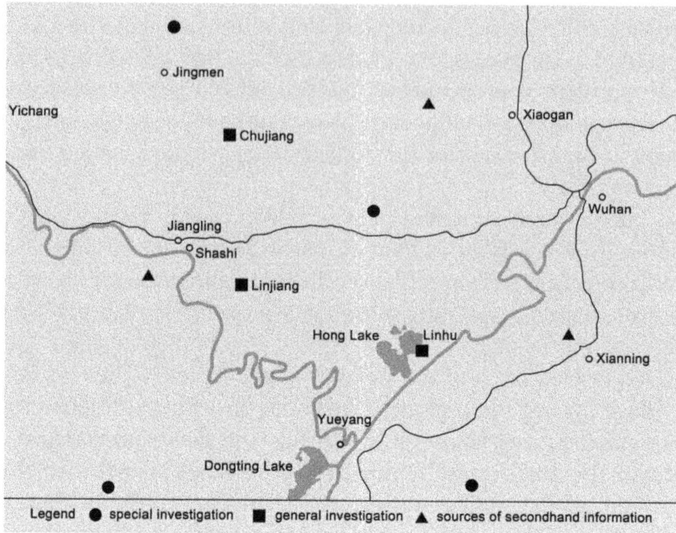

Research information mainly include 3 types: interviews and archives I have collected on special field trips, interviews for general investigation purposes, and fieldwork reports from fellow researchers. This dissertation mainly focuses on villages of the Double-lake Plains and use villages of other areas for comparison. Destinations for special field trips on the Double-lake Plains were *Shaqiao* Village, *Linsha* Village, *Liji* Village, *Wangcun* Village and *Xinwang* Village in *Chujiang* City, *Qiaotou* Village and *Fucun* Village in *Linhu* City, *Huchang* Village in *Linjiang* Vounty; destinations for general investigation purposes were *Puxing* Village in *Jingzhou* City, *Chencun* Village in *Xianning* City, *Dengwan* Village in *Jingshan* County; sources of secondhand information were villages in *Jingmen, Tianmen, Changde, Yueyang*, etc.

Those villages at *Chujiang* City and *Linhu* City were also fieldwork bases for HUST's Research Center of Rural Governance in Central China. From 2005 to 2008, I visited those villages many times and stayed for altogether 80 days. During my stay, I got into contact with farmers, rural gangsters, officials and police officers and through in-depth interviews I got firsthand information, but I did not obtain their archives. When I was in *Linjiang* City, I stayed for 10 days in *Huchang* Village and 15 days in township police stations, consulting and obtaining archives from there. I did similar investigations on *Pingyu* Village of Henan Province and *Shanxiang* County of *Hunan* Province. On general investigation trips I "mingled" with the local folks: I lived and dined with them and interviewed officials, farmers and rural gangs. My investigation did not stop with them, but rather concerns every aspect of the village. On trips to townships, I mingled with the officials and police officers.

Research methods mainly included doing interviews and consulting archives. I interviewed township officials and police officers on the rural gang issue, public security and stability and their day-to-day work. I also interviewed some available rural gangsters. The archives I consulted were mainly criminal and administrative cases records, documents released by higher public security authorities, daily briefs of county and city public security bureaus, to name just a few. Since officials and police officers would rotate between jobs within the county (city), the information I could get from police stations was indicative of the whole county (city).

The *Dongting* Lake Plain (*Hunan*) and the *Jianghan* Plain (*Hubei*) are collectively referred to as the Double-lake Plains (plain area in two neighboring provinces of *Hunan* and *Hubei*). The name derives from "**Hu**nan and **Hu**bei" (*Liang hu*: "*liang*" means two, "*hu*" means lake). Double-lake Plains, to the east of the Three Gorges Dam and the south of the *Dabie* Mountains, covers an area of 50,000 square km with an altitude of less than 50 m. It is high in the north and descends to the south. It was created by deposition in rivers such as the *Yangtze* River and the *Han* River. The area has a complex system of lakes and rivers. With fertile land, the place abounds with crops like cotton, rice, and rapeseed. Since the *Ming* Dynasty, especially after the building of a dam on the *Yangtze* River towards the end of the *Ming* Dynasty, reclaiming land from low-lying areas surrounded with dikes was very popular and this practice did not end until China's liberation. It may be said that throughout its history, people have been immigrating into the area to reclaim swamps or land from the lake. Because villages do not take shape until late in history, at the time of liberation, the villages there still lack a distinct boundary. In hilly areas, farmers would live in scattered houses along the slopes. This is true to fieldwork destinations in *Chujiang* City, *Xianning* City, *Jingshan* County, *Jingmen* City, *Yueyang* City and sources of secondhand information in *Jingmen* City, *Yueyang* City and *Changde* City. On flat lands, to guard against floods, farmers would live along the embankments or on high ground; some even set up their own abutments for supporting houses. This is true to fieldwork destinations in *Linjiang* County, *Jiangzhou* City and sources of secondhand information in *Tianmen* City. Because of a complex system of rivers and lakes, some farmers make a living on fishery or "half fishery half agriculture", and hence live in lake-side sheds or on boats. They do not live in houses on land until after liberation and therefore villages do not take shape until then. This is true to fieldwork destinations in *Linhu* City.

Generally speaking, villages on the Double-lake Plains are small and scattered, lacking a distinct boundary. There are possibly two reasons for this. First, with a complex system of waterways and hence high soil stickiness, road networks have not been established in traditional time periods.[5] For the convenience of transporting manure and harvesting crops, farmers would live not far away from the farmland. Second, the society is formed by immigrants, who often fail to quickly fit in with the local society and would choose to cultivate new farmland and live next to it. Therefore, independent and flexible families are operating as "separate atoms",[6] failing to form developed clans or formulate internal village conventions. As a result, in the recent

[5] Translator's note: "Traditional time periods" in the book may refer to times of feudalism or also contemporary time periods before the 1980s. The author's research begins with the 1980s. In the book, the author generally compares "traditional time periods" with the 1980s and since.

[6] Translator's note: Atomism or social atomism is a sociological theory arising from the scientific notion atomic theory, referring to "the tendency for society to be made up of a collection of self-interested and largely self-sufficient individuals, operating as separate atoms".

decade or so, the village communities and their facilities, especially water facilities, have been severely damaged.

After large-scale construction is completed during the people's commune period, an irrigation system has been built up, with large reservoirs as backup and electric pumps at small and medium reservoirs as the main means of irrigation. With all sizes of reservoirs, this irrigation system serves multiple functions including water storage, diversion and pumping, However, since the end of the 1990s, water facilities have been severely damaged: reservoirs, ponds and water channels could no longer be used for irrigation because they are out of repair or vandalized. After market reforms have been introduced to managing the main means of irrigation, electric pumps at small and medium reservoirs are no longer fully utilized. In addition, due to lax management by the village organ, the main irrigation facilities have failed to fulfil the needs of thousands of small farmers. Under such circumstances, the government granted subsidy for building irrigation facilities on farmland. State subsidies have effectively motivated farmers in their efforts to build small irrigation systems. The best example is subsidy for drilling pumping wells. Despite a high cost, wells are very popular for the convenience, saving farmers the trouble of having to work with others. However, since the Double-lake Plains have abundant water resources, it is incredible for farmers to pump water from underground for irrigation. The downside to the wells is a vicious cycle: farmers keep drilling to deeper ground, which entails a higher cost, and at the same time, shallower wells are quickly abandoned. Without proper planning, wells are drilled at the wrong places and are not as effective as they should have been. To make things worse, the wells disrupt the natural irrigation system and disturb the ecological balance. It could be said that vandalism to facilities and lack of cooperation in facility-building are well indicative of the "separate atoms" state of mind on the farmers and a lack of social norms at the villages.

This first edition of the dissertation consists of 9 chapters. Apart from this chapter on "the research topic and approach" and the concluding Chap. 9, the main body deals with the main topic by taking 3 steps and adopting 3 perspectives: first, the "rural gangs" perspective describes the growth and development of rural gangs (Chaps. 3 and 4); second, the "public order" perspective discusses the counteractions of the grassroots government and its predicament (Chap. 5); third, the "village" perspective describes the interaction between rural gangs and village life (Chaps. 6, 7, 8), including rural gangs and public order (Chap. 6), rural gangs and interpersonal relations (Chap. 7), and rural gangs and village governance (Chap. 8). Here is a detailed introduction to the dissertation structure and chapters:

Chap. 2, "The Acquaintance Society and Reciprocity-Based Behavior Pattern", discusses public order and pattern of behavior in the acquaintance society as the ideal model of rural China.

Chap. 3, "The Rise of Rural Gangs and the Acquaintance Society", describes the rural gangsters' transgressions, state of affairs in the rural underworld and social control that governs the rural gangs during the early days of the reform and opening up.

Chap. 4, “The Renaissance of Rural Gangs and the Acquaintance Society”, describes the transgressions and state of affairs of the rural underworld since the 1990s.

Chap. 5, “Changes and Predicament in the Public Security System”, discusses the changes in public security efforts and the local government’s difficulties in governance mechanisms and technologies.

Chap. 6, “Rural Gangs and Public Order”, discusses the gang influence on public order from 3 aspects: influence from local gangsters, from external gangsters, and local gangster’s counter-invasion efforts.

Chap. 7, “Rural Gangs and Interpersonal Relationships”, discusses the gang influence on interpersonal relationships from 3 aspects: rural gangsters becoming the most “good looking” people, relationships between local rural gangsters and their acquaintances, and gang influence on acquaintance relationships.

Chap. 8, “Rural Gangs and Village Governance”, discusses the gang issue against the backdrop of village governance predicament, and mainly focuses on the two special topics of “competent but unmoral” Officials and “moral gangsters”.

Chap. 9, “The Changing Behavior Pattern and Moral Ambiguity”, summarizes the interactions between rural gangs and village life and analyzes changes in the pattern of behavior in the acquaintance society and the change in the nature of the rural societies. Such change was referred to as “moral ambiguity”.

However, the first edition of my dissertation does have the following limitations.

Firstly, I was able to focus on only a limited number of village types. During my doctoral program, I mainly chose the traditional agricultural villages in central or western China for destinations of field research; the same could be said about my research for dissertation-writing purpose and for my discussion on gang influence on village governance. Therefore, I did not pay enough attention to the rural gangs’ activities, roles and influence in other types of villages. Rural gangs also play a very active role in villages with mineral (e.g. mining farms) or other resources and villages at suburbs where land expropriation causes demolition. My dissertation fails to devote enough attention to those villages.

Secondly, rural governance has undergone new developments. Since the abolition of the agricultural tax in 2006, many changes have taken place to rural governance in central and western areas. In particular, the state’s relationship with farmers and rural areas has changed drastically. Instead of collecting taxes from farmers, the state turns to provide resources for them in the form of direct subsidies to crop producers and for purchasing agricultural means of production. Other than those subsidies directly credited into the farmers’ accounts, state resources are often allocated through government projects, giving rural gangs plenty of opportunities to intervene and influence rural governance. I did research for the book and wrote the first draft in 2007; though in 2010 I revised the draft and added some new information, the overall framework remained the same. In 2007, when the agricultural tax was lifted just recently, the resource-allocation projects have not been widely implemented yet, and we failed to obtain an in-depth observation of the pertinent issues.

Being aware of these limitations, I revised my dissertation and in its second edition published in 2019, I added Chap. 10 as a supplementary chapter to overcome the above-mentioned limitations. This chapter discusses the roles and influences of rural gangs on current rural governance, and gives a clearer picture of grassroots governance.

Chapter 2
The Acquaintance Society and Reciprocity-Based Behavior Pattern

The main purpose of this dissertation is to understand the change in social nature of rural China through an investigation into rural gangs. Therefore, a study of social nature in traditional time periods is needed for comparison. While focusing on the current transitional period, this dissertation examines the traditional rural society as a relatively mature model. As I have argued in Chap. 1, the basic unit of rural China in traditional time periods is the village, which in nature is the acquaintance society, so the acquaintance society will be used as the ideal model in comparative analysis. To provide an understanding of the nature and public order maintenance mechanisms of the acquaintance society (or intimate society), this chapter will investigate its social order and reciprocity-based behavior pattern. I summarize how order mechanisms work in a traditional rural society as "the traditional behavior pattern of rural land" *(xiangtu Luoji)*. This concept has multiple underlying meanings, and inevitably I must touch upon the internal structure of the acquaintance society. Previous research fails to look at how order mechanisms work from within the acquaintance society. Besides, this chapter also briefly discusses the impact of collectivization on the acquaintance society.

2.1 The Acquaintance Society: "Familiarity" and "Intimacy"

In his work *Xiangtu Zhongguo* (translated as *From the Soil, the Foundations of Chinese Society*), *Fei Xiaotong* argues that the Chinese society is parochial; people are tied to the land; under the constraints of parochialism, the rural society is characterized by "familiarity". "The acquaintance society", therefore, becomes the classic model for rural society and for describing the nature of such a society. Fei wrote (1992), "… life in rural society is very parochial. Villagers restrict the scope of their daily activities; they do not travel far; they seldom make contact with the outside world; they live solitary lives; they maintain their own isolated social circle. All of

B. Chen, *The Underworld of Rural China*,
https://doi.org/10.1007/978-981-19-8710-6_2

these characteristics contribute to the parochialism of rural China." "People in rural China know no other life than that dictated by their own parochialism. It is a society where people live from birth to death in the same place, and where people think that this is the normal way of life. Because everyone in a village lives like that, a distinctive pattern of human relationships forms. Every child grows up in everyone else's eyes, and in the child's eyes everyone and everything seem ordinary and habitual. This is a society without strangers, a society based totally on the familiar", because "the land itself constrains the lives of rural people. The people they see every day are the ones they have known since childhood, just as they know the people in their own families. They do not have to select the kind of society they would live in; they are born into it; choice is not a factor".

Based on *Fei*'s theory on the "acquaintance society", *Su Li* expounded on the processes and results of implementing modern laws and systems in a rural society. In his analysis on the films "*The Story of Qiu Ju*" and "*The Accused Uncle Shangang*", Su (1996: 23) first acknowledges that the acquaintance society is characterized by familiarity and tacit agreement, and on the basis of this acknowledgement, he argues that the movie protagonists are in an extremely awkward position because familiarity and people's tacit agreement and expectations have been disturbed by the intrusion of modern legal system. It further undermines people's long-term relationships and benefits. In his analysis on a private settlement case, considering that rural China is by nature the acquaintance society, Su (1996: 41) argues that avoiding legal actions has contextual rationality, and is a special manner for laws to take effect. This is particularly true to sexual assault cases, in which victims often choose private settlement to avoid consequences incurred by "familiarity". In his systematic analysis on China's grassroots judicial system, Su (2000) incorporates the social nature and order mechanisms of the acquaintance society in his examination on specific topics concerning grassroots justice, including the manner of enacting laws while knowledge of law is being delivered to rural areas, the system of judicial committees at grassroots-level courts, judicial expertise of grassroots judges, and the approaches to grassroots justice. Nothing was discussed outside the acquaintance society context. Local leaders, who are "familiar" with the village, become "the carrier of local knowledge" (Su 2000: 44); judges need someone "familiar" with the village to be "the deliverer of legal documents" (Su 2000: 313). Apparently, *Su* also believes that the acquaintance society of rural China is characterized by "familiarity".

No doubt the acquaintance society is characterized by familiarity, but people could easily regard familiarity as its most essential feature, and thus simplify everything to an information transparency issue, and then take a step further to believe that order mechanisms could be explained by the basics of institutional economics. That would be a major misunderstanding about the acquaintance society. Part of Sang (2005: 322) criticism on *Su Li*'s arguments is based on the same opinion. *Su* mistakes "local resources" for "specific locally created laws and systems", and argues that the problems in establishing the rule of law in China should not be attributed to insufficient use of "local resources", but to a crisis of confidence that has resulted from a short supply of "social resources"; and thus the confidence crisis has been caused by—in institutionalist terms—problems in public opinion supervision and private punishment. In my take, in the broad sense of "locally created systems", the acquaintance society itself is a kind of "local resource" that the legal system deploys. It could be argued that establishing the rule of law in the acquaintance society could not be simply encapsulated in such familiarity-incurred problems as public opinion supervision and private punishment, and "local resources" could not be simply replaced by "social resources". Apparently, we should reinterpret the "familiarity" in the acquaintance society.

In He (2003a: 1, 2003b: 42) eyes, after the founding of the P.R.C. and the political reforms in rural China, the present administrative villages have evolved into half acquaintance society. Production brigades during the people's commune period become administrative villages; production teams of the lower hierarchical level become village groups (*cunmin xiaozu*). In most places, administrative villages are identical with natural villages, with 30 to 50 households or 200 to 300 people. During the commune era, production teams are the basic unit for collective production and labor, and for that, they may be regarded as the largest and most efficient acquaintance society. Frequent collaborative labor greatly enhances familiarity; egalitarian income distribution builds up more connections; more contact between men and women in collaborative labor creates romantic relationships and increases connections through marriages; mutual assistance in production and life makes exchanges of favor more frequent—production teams are, in nature, the acquaintance society. Production brigades, where villagers only look familiar to each other but do not know each other well, are too big for intimate contact and acquaintanceship. Since the reform and opening-up, production teams have been dismantled and superseded by village groups. Within village groups, people still work and live together, exchange favors and undertake cultural and recreational activities. Production teams are managed by village committees instead. Villagers have increased contact with village committee officials, but not with villagers from other village groups. Therefore, village groups are still the acquaintance society in nature, whereas administrative villages (made up with several village groups) are just half acquaintance society.

Recent studies of He (2008) further elaborates on the concept of "half acquaintance society". As *He* has pointed out, as reform and opening-up deepens, with increased mobility, diverse employments and social differentiation, farmers become increasingly heterogeneous. Great changes have occurred to their personal and public lives, for example, they have increased privacy in the family, and tend to have fewer visits and chats. As they grow less accustomed to such interactions with clear purposes but no obvious withdrawal strategy, they crave for another way of public life that allows them to participate and quit freely. All that suggests a shift from the acquaintance society to the half acquaintance society. Nevertheless, *He Xuefeng* still regards "familiarity" as the essential feature of rural society. Take collaborative irrigation projects in the rural areas of *Jingmen* City. People there seem to believe in a self-fulfilling prophecy, or in their words, "those who do not fear starving to death will not starve whereas those who do will". This mentality is described by *He* also as the "traditional behavior pattern of the acquaintance society". As acquaintances, people know each other's temperaments very well. People would count on the public-spirited and hard bargainers to behave in a certain manner. They are expected to invariably benefit very little from such public service projects. As they are bound to give more than they take in every project, they become completely marginalized and economically impoverished. Eventually they have no say in the village and could not accomplish anything. On the contrary, those who are often seeking free rides and successfully do so will benefit the most, and further gain a central position in the village (He 2004).

Apparently, this self-fulfilling prophecy cannot be explained by the traditional behavior pattern of the acquaintance society. To make a distinction between different patterns of behavior, therefore, He (2007a) attempted to distinguish between "rural society" and "acquaintance society". I believe it is very important to tell the two different behavior patterns apart, but attempts to distinguish between "rural society" and "acquaintance society" will probably lead to conceptual confusion. *Fei Xiaotong*'s classic model for "acquaintance society" still holds room for further theoretical development, providing a possible effective explanation for the two different behavior patterns. That also requires us to reinterpret "familiarity" in the acquaintance society.

Fei (1992) wrote, "In a society characterized by this level of familiarity, we achieve a level of freedom whereby we can do whatever we please without fear of violating the norms of the society. This type of freedom is unlike those freedoms defined and protected by laws. The social norms in a familiar society rest not upon laws but, rather, upon rituals and customs that are defined through practice; hence, to follow these norms is to follow one's own heart and mind (*xin*)." People often say, "We all know each other very well (*shuren*). If you need my help, you have it. You don't have to ask twice." "There is no way for laws like written contracts to develop in a rural society." "Isn't that what outsiders do?" "In rural society, trust derives from familiarity. This kind of trust has very solid foundations, for it is rooted in customary norms". "Trust in rural society is based not on the importance of contracts but, rather, on the dependability of people, people who are so enmeshed in customary norms that they cannot behave in any other way." In the acquaintance society, people trust each other based on their familiarity. However, is familiarity simply the result of "information transparency"? In urban areas, colleagues of the same workplace, or people in the same professional community, share information, know each other well (*shuren*), and thus may form the acquaintance society or community. If familiarity can be simplified as the information issue, behavior patterns between acquaintances in urban and rural societies should be the same. However, the way acquaintances in urban areas deal with their relationships is often affected by established institutions, very different from patterned behaviors between acquaintances in rural society. So what is really behind such "familiarity" and familiarity-based "trust" and "norms" in the acquaintance society?

It's intimacy! Fei (1992) wrote, "Familiarity is an intimate feeling that develops from frequent and repeated interaction occurring over a long period of time". People who are familiar with each other often omit names when addressing one another. Footsteps, voices, even smells are enough for identification (Fei 1992). In his opinion, during long periods of common and intimate living, people develop long-term social connections, which at some point may become spontaneous connections. Intimate community may develop only among people living from birth to death in the same place, because only they will know each other well. For them, neither space nor age will become barriers for mutual understanding.

Fei (1992) wrote, "In a society characterized by this level of familiarity, we achieve a level of freedom whereby we can do whatever we please without fear of violating the norms of the society. This type of freedom is unlike those freedoms defined and protected by laws. The social norms in a familiar society rest not upon laws but, rather, upon rituals and customs that are defined through practice; hence, to follow these norms is to follow one's own heart and mind (*xin*)". He argues that "knowledge acquired from familiarity is specific and is not deduced from abstract general principles. People who grow up in a familiar environment do not need such principles. They only need to know the specific relationship between means and ends within the scope of their activities. They do not seek universal truths (Fei 1992)". However, he did not elaborate on the patterns of interactions between familiar or even intimate people, or the specific principles that they have become familiar with during long periods of common living. In other words, he did not reveal how "intimate community" has formed.

2.2 Public Order Maintenance in Intimate Community

When depicting the acquaintance society, Fei (1992) described several pairs of contrasting models or concepts, among which the important pairs were: gesellschaft [a society based on ritual and customs (*lisu*)] and gemeinschaft [one based on law and reason (*fali*)]; the rule of ritual and the rule of law; *chaxu geju* (the differential mode of association) and *tuanti geju* (the organizational mode of association). Gesellschaft and gemeinschaft describe two different sets of social norms that people rely on; ritual and law describe the social order mechanisms generated by different social norms; and, *chaxu geju* and *tuanti geju* describe people's different relationships governed by different social norms. But he did not investigate the behavior pattern of the acquaintance society or compare it with that of other social forms.

In a gesellschaft, the law acts as the basic social norms. The Latin word for "law" is "jus", French "droit", and German "recht". Since all three words for law signify rights and justice, in western languages, words for "rights" and those for "law" share the same word root. In a Western gesellschaft, rights are deeply embedded in people's thinking patterns, and the "law" also indicates "rights". People tend to behave while bearing their rights in mind and act by the law. In contrast, in the Chinese acquaintance society, people follow principles of "reciprocity or responsibility" (*renqing*). With "*renqing*" deeply embedded in their thinking patterns and incorporated in the Chinese ritual and customs (*lisu*), the acquaintance society is essentially the intimate community in nature.

In the acquaintance society, people often use the expression "*renqing*" to refer to 3 things: first, it indicates people's spontaneous emotions and temperaments, which is also the basic meaning of "*renqing*"; second, it indicates human relationships, or the "affections (*qingfen*)", "feelings (*qingyi*)" and the give and take in human relationships, similar to the concepts of "rights" and "responsibilities" in a gesellschaf[1]; third, it indicates social norms, similar to the concept of "law" in a gesellschaf. The third meaning of "***renqing***" (social norms), together with "celestial reason (***tianli***)" and "***guofa*** (state law)", may be collectively referred to as "***qing li fa***" (emotion, reason and law). "*Guofa*" (state law) refers to the positive law; "*Tianli*" (celestial reason) describes how the universe works (reason describes how the tangible things work in the acquaintance society). The state law and celestial reason are both objective and universal. In contrast, as a meaning that has grown out of the second meaning "human relationship", the third meaning of "*renqing*" (social norms) is subjective and variable. Apart from the above-mentioned meanings, "*renqing*" is also used to refer to an established custom of the acquaintance society.

As previously stated, "*renqing*" may be used to refer to 3 things: emotions, relationships and norms. But first and foremost it refers to human relationships, not natural "emotions". The most classic explanation on interpersonal relationships was the one offered by Fei "chaxu geju" (1992). "It (the pattern of human relationships) is like the circles that appear on the surface of a lake when a rock is thrown into it. Everyone stands at the center of the circles produced by his or her own social influence. Everyone's circles are interrelated." "Like the ripples formed from a stone thrown into a lake, each circle spreading out from the center becomes more distant and at the same time more insignificant." In the acquaintance society, two ripples are particularly important—the ones that make distinctions between family members and, acquaintances, and the one between acquaintances and strangers in social relations. One's family is the closest to oneself; outside of families, people who establish permanent regular relations with oneself are acquaintances; other than acquaintances, people who occasionally or never establish contact with oneself are strangers. The pattern of human relationship may be demonstrated by the following figure (Huang 2006: 33).

[1] In the acquaintance society, people often exchange gifts to express their feelings, such as attention, attachment, friendship, responsibility, congratulations, condolences. Therefore, "*renqing*" may be regarded as synonymous to "gifts" and they are sometimes interchangeable.

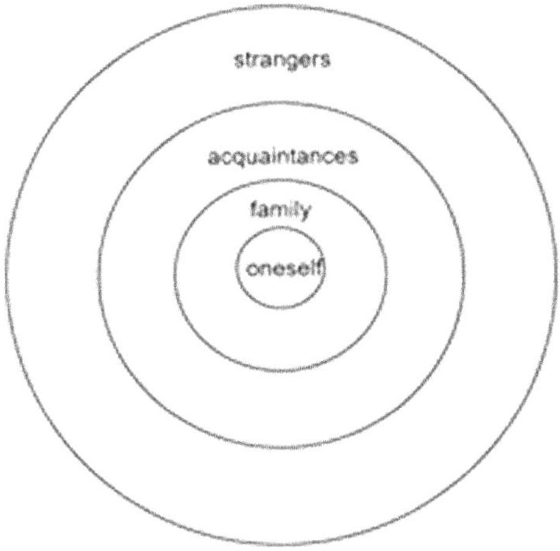

The "human relationships" meaning of "*renqing*" concerns one's family, acquaintances and strangers. To establish public order, cooperation among acquaintances is particularly important. Ideally within the acquaintance society there is not only natural kin and regionalism, more importantly, everybody keeps a relation with others by receiving or giving favors.[2] It is based on such give-and-take relations that the acquaintance society is transformed into an intimate community, or into a society made up of one's "folks". For example, Fei (1992) wrote, "Living closely and intimately with each other, relatives become, over long periods of time, mutually interdependent in many aspects of their lives. There is no way to count the favors given or received. The unity of the intimate group depends on the fact that each member owes countless favors to the other members." "When you owe another person a favor (*renqing*), you have to look for an opportunity to return a bigger favor. By repaying the favor with a bigger favor, you make others owe you more favors in the future. So it goes, back and forth; the continuing reciprocation maintains the cooperation among people in the group. It is impossible for a person not to owe favors in such an intimate group. In fact, people are afraid to square their accounts (*suanzhang*). To settle accounts (*suanzhang*) or to be completely square (*qingsuan*) with somebody means to break off relationships, because if people do not owe something to each other, there will be no need for further contact".

The acquaintance society is characterized by "the owing of favors": everybody owes favors to others or has others owe favors to them. Such reciprocation produces "affections (*qingfen*)"; because everyone is in the network of favors, every member of the acquaintance society becomes one's "own folks", and the society could not have existed if people did not owe favors to each other. That distinguishes rural acquaintance society clearly from urban groups of acquaintances (for example motorist

[2] In reality, things may be different from the ideal state of affairs. Generally speaking, villages in south China are in the ideal state of affairs, whereas those in north China are mainly composed of small kinship groups or several "*renqing*" circles. That's what He (2009b) refers to as "a unit of farmer identity and action". In central China, things may be more complicated as the "*renqing*" circles are not as close.

clubs), whose members are also acquaintances who share formation, but for them, there is nothing like the network of favors. In the acquaintance society, people owe favors in various occasions. People exchange gifts to express their feelings in ceremonies, such as birth celebrations, weddings, housewarmings, funerals, elderly people's birthday celebrations, etc.; and in non-ceremonial occasions, such as drop-offs, visiting patients, paying a New Year call, etc. (Yan 1996). More importantly, favors are created not just by gift-exchanges, but also by mutual assistance in other non-expressive moments of daily life. Favors are involved in any occasion where human relationships play a part.

In the acquaintance society, in addition to owing financial favors to others by receiving gifts, people also owe favors through their social and cultural interactions. An action may lead to "owing favors" on various levels, and "rights" and "obligations" become overlapping and interwoven.[3] Therefore, when it's time to "exchange favors" again, people can only "return one favor", they could not be completely square with someone. In this way, both sides are obliged to carry on their favor-based relationships (*renqing*). People should not be preoccupied with personal gains and losses. After all, gains or losses are not what really matters in a favor-based relationship; instead, "emotions" are what really matters. When others need their help and they are capable of helping others, people feel obliged to give a favor; they do not expect to receive the same number of favors or on the same levels.[4] Such relationships of reciprocity may be established on various levels in various fields. In fact, even if a chance comes up for "squaring accounts", people would not take it and end their relationships. Besides, they could never break off favor-based relationships (*renqing*). As long as life goes on, they would carry on the reciprocal relationships, for in a traditional agrarian society, people often have a strong need

[3] For instance, Villager A asks Villager B to help out at a wedding that is held in A's house. B does not get paid for doing so. This action of A's will cause A to "owe favors" to B on at least 3 levels. First, on the economic level: as B helps A out without being economically compensated, A owes a favor to B, which means that A is obliged to do the same to B when the time comes. Second, on social and cultural level: A's asking favor from B is an indication of A's "good opinions" on B, or B's good interpersonal relations in the acquaintance society. Being needed for favors is an indication of the person's "face". A has asked a favor from B but also gives B face. For that B owes A a favor. In the next encounter, B must show A some face by asking a favor from A. Third, also on social and cultural level, the fact that B does not turn down A's request shows B's willingness to help, which in turn makes A good-looking, but only when B allows so. Therefore, A owes a favor to B for B allows A to look good. In similar occasions, A may not turn down B's request. Owing favors on the above-mentioned levels eventually will lead to "rights" and "obligations" on these levels. First, A is obliged to return the economic favor to B. Second, B may ask A to help out when need be. Third, A may not decline any request from B.

[4] For instance, when two villagers have exchanged favors in major life events such as weddings and funerals, they should not limit their relationship to what they have already established. Instead, they must extend their relationship as needed to any aspect of life that requires mutual assistance and cooperation. If A finds B's lost child, and the child has no other means of getting help, A is obliged to send the child back home. This is not about the happiness brought by helping out others. This is an obligation—the result of interactive micro-powers in people's "*renqing*" relationship. B may feel indebted to A for A's action, but this does not make the obligation any less binding.

for mutual assistance and cooperation, which leads to a high degree of interdependence. In a traditional agrarian society with low productivity and a lack of tools and manpower, people would often borrow and share farm equipment, pitch in for a complete set of tools (called "*datao*" or "*heju*" in different dialects), swap labor (*banggong*), borrow each other's animals for labor, ask others to work (*daigeng*), help out each other (*banggong*) and even set up "mutual assistance teams" (*shushe*) as organizations of labor; therefore, collaborative farming (*nonggeng jiehe*) is bound to happen (Si 2005). Collaboration brings about favor-based relationships. In cases of natural disasters and social crises, people have a stronger need for mutual assistance, for example, in flood control and building defense against thieves and burglars. Besides, in a traditional agrarian society with a fragile environment, risks often come unexpected; coupled with limited ability to handle those risks, people would often expect the occasion for needing help to come up. Therefore, people would like to establish wide connections in daily life. The more favors people give to others, others will feel more indebted to them.

The implications of favor-based relationships go beyond the two parties to across the whole acquaintance society, for their attitude and positions are being reflected with "impression management" implications (Goffman 1959). Even if one can square accounts with others, they would choose not to do so, and would instead, give more favors so that the other side would feel more beholden. Occasionally, when their relationship is characterized by long-term imbalance where one party has no chance to return favors whereas the other keeps giving favors, they would still carry on their relationship. With a long-term view people do not haggle over gains or losses for the present, not even for their lifetime for they have to look out for their future generations. That's also why we would often hear about entanglement that involves generations.

Because of "impression management" implications, favor-based relationships are not about striking a balance on the levels of "rights" or "responsibilities". What matters is not really whether to owe favors or not, but whether one has "tried one's best". When others have a great need for help and one can grant a favor at a very low cost, to give the favor seems to be a thing one must do. A capable person is expected to give more favors, or bluntly, to allow others to benefit from them, and not to be preoccupied with gaining an equal footing on the short term. The requirement for people to try their best has made moral role models. Those who try their best without calculating gains or losses would be widely acclaimed and respected. So "to try one's best" becomes an incentive to everybody, and this incentive system does not involve benefits as the essential element. When productivity remains at a low level, this incentive has helped to form reciprocity as the code of conduct and sustained public order.

In the acquaintance society, owing or giving favors creates power relations between any two villagers (Chen 2010). The acquaintance society would fail to sustain itself as an intimate community without people exchanging favors on different levels. With "favors" as a key element in human relationships, one may assume dominance over another on some issues in some areas or fields. And due to reciprocity of favors, such dominance is also mutual; because people could never square accounts with others, the influence from such mutual dominance endures; because everyone in the acquaintance society has favor-based relationships with others, enduring power relations have been established between any villagers. Here I use the same meaning of power as proposed by Foucault (1995). Since the acquaintance society is in itself a network of favor-based relations, the acquaintance society is also a network of micro-level power relations. Each villager has power over others, and in the meantime, is restricted by other's power. Such power relations are mutual and enduring. It could even be said that the network of micro-level power relations does not have a center, because power is highly diffused, with everybody simply as an intersection in the network. This network of micro-level power relations has served to maintain public order.

Because all people are equal in the relationship network built on micro-powers, their views regarding how to behave in individual cases will be accepted universally as "regional consensus" (He 2006b). When people go through long-term interactions in daily life, regional consensus will be increasingly unquestionable and become what most people will do unconsciously, and accept as the convention for interpersonal relationships. Each member of the acquaintance society will act both as followers and supervisors of such convention, as participants in relationships and in relationship conventions, everybody being restricted by their counterpart's power and by relationship conventions. The restriction from social convention is actually an aggregation of restriction from all the other members of the society. In the above-mentioned example, if villager A unexpectedly did not take care of villager B's child, then he would feel ashamed to see B. In addition to being punished by B in their relations, he will also be punished in the whole relationship network for violating relationship convention. The convention works by people's constant voluntary conformity and self-discipline, which also serves to discipline others. Such mutual discipline turns the society into an intimate community.

Inside the acquaintance society, this mutual discipline mechanism generally sets interpersonal relationships on a virtuous cycle. Everyone tries their best to "give" favors (*geiyu*) to make others feel "indebted" (*kuiqian*) to them. This will also allow them to gain a moral advantage in the relationships and maintain a good image in their networks. They do not mind if others do not respond actively to a voluntary given favor. Acts of calculation may border on squaring accounts, which is not recommended in favor-based relationships. Besides, one should also care to save face for others. In this way, everyone observes the "face" and the "not-going-too-far" principles. Those who have committed errors in relationships are being punished by the person concerned as well as the whole relationship network. They are most likely being punished by public opinion. When someone does now follow the principles for interpersonal relationships (*renqing*), the other side in the reciprocal relationship would express dissatisfaction and encourage other villagers to condemn the rule-breaker with verbal criticism, which is "a common form of punishment" (Zhu 2003: 193). Having been punished in this way over some time, the rule-breakers would almost invariably reform their conduct, set their relationships back on track and observe normal relationship conventions.

Occasionally someone would go too far and public opinion does not work on him. In such cases, villagers would choose to banish him from any social relations, punish him by severing any relations with him, shun and isolate him, or refuse to work with him or help him out in work or life. Many villagers would choose to do so in cases of deviant behaviors. The reach of his social interaction would be greatly constrained, and his life derailed. Such harsh punishment is nothing short of sentencing the person to "social death", because no one could survive in isolation. Surely sometimes the punishment does not take place right away. When someone goes overboard, people would wait in silent resentment until the right moment for punishment comes up. For example, in a village in the south of *Hunan* Province, villagers did not punish a person until when a flood came. They refused to help him out, because his repeated reckless behaviors have jeopardized the village's common interests (Yang 2008a). In cases of disastrous and tragic events, villagers would often attribute those events to deviant behaviors and punish transgressors. This is because Confucianism argues that heaven, earth and all living things are closely connected to human behaviors. Therefore, people believe that disasters and misfortunes occur if someone (the transgressor) goes against the heavenly principles (*renqing* principles).

When the village's interests and reputation are severely damaged, the transgressor will suffer the most severe punishment, i.e. brutal physical punishment, for example, being banished, sunk to the pond bottom, or buried alive. During the Republican Era in *Licun* Village in northern Anhui, members of the *Li* Clan were driven beyond their limits for forbearance when they knew that a member took part in banditry; under the leadership of their clan leader, they beat the man dead in a raid (Hn 2007: 53). I heard about similar tales on my field trips to different villages, with no exceptions. Even in a place like *Jingmen* where social relations are very distant, similar incident happens now and then. In a time when violence is accepted by clan members, transgressors would invariably be subject to violent punishment by other villagers and clan members, even if they may disregard punishment through public opinion and social relations. Occasionally, some transgressors are so powerful that none of the above-mentioned punishments work for them. That creates an abnormality in the acquaintance society, and requires interference from external forces. Examples for such powerful transgressors include "*shengyuan*[5] going rogue" (Qu 2003: 314; Chen 2005: 387), "profiting brokerage" (Duara 1988), "local tyrants and evil gentry" (Mao 1965a). Such abnormality often occurs toward the end of dynasties with lax social control.

To sum up, public order in the acquaintance society is maintained by the mechanism of favor or reciprocity (*renqing*). Based on relations of reciprocity or responsibility, the acquaintance society becomes a nexus of power relations. Conventions for interpersonal relations are followed by each member of the network. In addition to being disciplined by relationship conventions, people are also hands-on participants and supervisors of relationship conventions. Relationship conventions provide

[5] Translator's note: *Shengyuan* (historical) in ancient China refers to a scholar who has passed the entry-level examination to study at a college.

people with guidance by delegating power to individuals and regulating relationships. Under the guidance of those conventions, people are involved in a system of joint and several liabilities, and the acquaintance society is transformed into intimate society characterized by few internal disputes and unity against the external world. Ideally, villagers develop intimate bonds with one another, treat each other with love and respect and will do their best to help one another whenever in need. These fine traditions are passed on from generations to generations.

2.3 Reciprocity and Ritual in the Acquaintance Society

Fei (1992) argues that "Maintaining a rural social order differs, in many respects, from maintaining a modern social order. But that does not mean that rural society is 'without laws and without heaven' or that it 'does not need laws'." "We can say that it is a society 'without law', if we define laws as those regulations maintained by state power. However, the absence of laws does not affect social order, because rural society is ruled by rituals." "The difference between a ritual and a law is the force used to maintain the norms. Laws are enforced through state power. The term state here means political power. Before modern states developed, tribes also held political power. But rituals do not require any concrete structure of political power in order to be effective. Instead, ritual norms are maintained by tradition." "A ritual (*li*) is not something that is carried out by an exterior force. Rituals work through the feeling of respect and of obedience that people themselves have cultivated. People conform to rituals on their own initiative." (Fei 1992) *Fei* explained the difference between a public order ruled by law and one ruled by ritual, and compared their different roles for maintaining order, but he did not describe what constitutes as the foundations for the law in rule of law or the ritual in the rural society. Nor did he clarify the connotations of the law or the ritual. Based on our analysis, we've come to the conclusion that the essential element that maintains public order in the acquaintance society is "reciprocity", and that's also what people's behavior pattern is based on. This seems to suggest some connection between ritual and reciprocity. Therefore, we may consider the foundations and connotations of ritual by investigating into "reciprocity" (*renqing*).

"*Renqing*" contains several meanings, such as emotion, connection, convention, mechanism, but originally it denotes human's natural emotions and temperaments, or in modern psychology terms, instinctive emotions and feelings. *The Conveyance of Rites* (*liyun*) of *The Book of Rites* (*liji*) states "Joy, anger, sorrow, fear, love, hate, and desire; with no need to learn them we have these seven innate abilities". As these primitive and innate feelings have no boundaries, when the society evolves to a certain point, social mechanisms must be established for regulating human feelings so that natural feelings are transformed into social feelings needed on the societal level. In traditional Chinese societies, such transformation is marked by the introduction of "ritual" (*li*). Ritual is established when regulations on human relations have become well-established and well-accepted. *The Conveyance of Rites* of *The Book of Rites* states, "Confucius argues that 'It was by those rules that the ancient kings sought to represent the ways of Heaven, and to regulate the feelings of men. Therefore he who neglects or violates them may be (spoken of) as dead, and he who observes them, as alive.' Therefore, the rules of propriety are for man what the yeast is for liquor. The superior man by his use of them becomes better and greater. The small man by his neglect of them becomes meaner and worse. Therefore, the sage kings cultivated and fashioned the lever of righteousness and the ordering of ceremonial usages, in order to regulate the feelings of men. Those feelings were the field to be cultivated by the sage kings. They fashioned the rules of ceremony to plough it. They set forth the principles of righteousness with which to plant it. They instituted the lessons of the school to weed it." This metaphor, by comparing feelings to field and fashioning of rules to ploughing, vividly reveals the connection between "feelings" and "rules of ceremony (*li*)", i.e., rules are used for regulating feelings. The significance of rules on feelings lies in giving people the freedom to feel at will without breaking rules, or self-restraint and restoration of rites (*keji fuli*). In other words, people are required to follow the rules, exert control on their desires and not articulate feelings without any constraint. Confucianism associates the articulation of feelings with the heaven, earth and all living things. *Zhong Yong*[6] states: "While there are no stirrings of pleasure, anger, sorrow, or joy, the mind may be said to be in the state of Equilibrium. When those feelings have been stirred, and they act in their due degree, there ensues what may be called the state of Harmony. This Equilibrium is the great root from which grow all the human actions in the world, and this Harmony is the universal path which they all should pursue. Let the states of equilibrium and harmony exist in perfection, and a happy order will prevail throughout heaven and earth, and all things will be nourished and flourish."

In a social order built up by Confucianism, "ritual" is used to establish a connection between heaven and humans, for regulating social lives through codes of conduct to achieve a perfect harmony between human and nature. "Ritual" originally refers to offering sacrifices to gods using two pieces of jade in a vessel, and later to all kinds of sacrifice activities. *The Conveyance of Rites* of *The Book of Rites* states: "propriety and righteousness are the great elements for man's character; it is by means of them that his speech is the expression of truth and his intercourse with others the promotion of harmony; they are like the union of the cuticle and cutis, and the binding together of the muscles and bones in strengthening the body. They constitute the great methods by which we nourish the living, bury the dead, and serve the spirits of the departed. They supply the channels by which we can apprehend the ways of Heaven and act as the feelings of men require." The Confucian world view holds that heaven, earth, and all living things are governed by "the Laws" (*li*). Humans are destined to observe natural laws and may not change or disobey them. On the contrary, "feelings" are personal and subject to free will. As the connection between heaven and humans, ritual is established by the sage king to inherit the ways

[6] Translator's note: *Zhong Yong*, or The Doctrine of Mean, is a Confucian classic. The concept of "the mean" is a core idea of Confucianism. It says that in all activities and thoughts one had to adhere to moderation.

of heaven, or regulate human feelings in the name of heaven and let human feelings conform to heavenly laws. But regulation does not equate to not allowing people to have feelings. What is to be eliminated in the later Confucian slogan of "retaining heavenly principles and eliminating human desires" are the human desires that do not conform to heavenly laws. In Confucian views, when human feelings conform to heavenly laws, and when the Mean and Harmony are actualized in the articulation of feelings, Heaven and Earth are in their proper positions, and the myriad things are nourished.

Regulating human feelings with ritual reflects the Confucian thought on the construction of social order. It is much needed in traditional time periods. The traditional Chinese society is an agrarian society with low productivity. People rely solely on hard labor to engage in intensive cultivation on land and to make ends meet. This mode of production and way of living are highly susceptible to natural hazards and social risks. People's livelihood heavily depends on the natural environment, and more on people's cooperation. Stable partnerships or groups are needed. However, in pre-modern times, stable groups or communities of steady relations are mainly formed through blood ties. Therefore, Confucianism heavily emphasizes family ethics. The construction of ritual-based family relationships is essentially a process of ritualizing natural feelings in family relationships. The purpose is to hold communal groups together. Because natural ties have a limited influence and communal groups may not be founded on natural ties alone, social systems are needed for forming a solid foundation for communal groups.

However, in the process of ritualizing natural feelings, Confucianism does not take all kinds of feelings equally, valuing relationships between father and son, between elder and younger brother, but not those between husband and wife. According to Confucianism ethics, "King is graceful and officers are faithful. Parent is loving and children are respectful (filial piety). Elder are friendly and younger respectful. Friends are to be trustworthy." But instead of stressing "love and care between husband and wife", Confucianism argues for "giving different treatments to males and females". In Fei opinion (1992), the reason behind is that "one of the main aims of marriage was to assign responsibilities in child-rearing. Child-rearing was a very complex task that mainly involves economic activities that arise from daily necessities such as firewood, rice, oil and salt. A couple had to handle these activities before they could be qualified to pursue any common interest. In other words, in a society with simple production technologies and low living standards, the couple spent much labor and time on economic activities. People tend to put stress on cooperation in family responsibilities while repressing emotional satisfaction. In other words, only in a society where various available facilities relieve child-rearing burdens and reduce economic labor for them, can couples attach greater importance to their emotional life." However, the Chinese traditional society with a relatively low productivity fails to provide good conditions for stressing emotional attachments between couples. Thus ritual has been used to repress emotions.

The reason behind such unequal emphasis is that the kinship groups that Confucianism attempts to hold together are lineages (*jiazu*) and clans (*zongzu*) rather than families, for only lineages and clans could handle most of the natural hazards and social risks in an agrarian society. With uninterrupted human reproduction, the family gradually expands into corporate forms of kin or a lineage. Lineages are big enough to handle most potential risks in agrarian societies, and are the only social organizations that people could depend on and seek protection from. Undoubtedly, continuity, maintenance and harmony are very important for the sustenance of families and lineages. Natural feelings alone are not enough. Therefore, in order to hold communities together, Confucianism downgrades natural feelings between husband and wife and stresses those between father and son and between elder and younger brother. In Confucian ritual, the ties between father and son and between elder and younger brother are conceptualized as filial piety and ancestor worship. This allows lineages to develop into clans. In villages where people with different family names living together, Confucianism treats kin and non-kin equally, extending consanguinity (*xueyuan*) to regionalism (*diyuan*), a tie to a specific place. Villagers from different clans establish ties similar to father-son relationships or brotherhood ties. Based on regionalism, village community identity has been established to guard against risks in an agrarian society. Just as Yang (2008b) puts it, without clans, the cost for establishing a village community on the basis of "self-interested individuals operating as separate atoms" would have been too much for the acquaintance society.

In the process of establishing social order by Confucianism ritual, another problem is how to construct a unified society (state or *tianxia*[7]) based on small societies (clans) lacking kinship ties or regional ties with one another. Confucianism stresses a sense of identity to be established among small societies, paving the way for visualizing and constructing a large society (state or *tianxia*). First, Confucianism calls for empathy or putting yourself in other's shoes. For example, Confucianism argues for "honoring old people as we do our own aged parents, and caring for other's children as one's own", and that "men did not love their parents only, nor treat as children only their own sons", and for "helping others stand up when one wants to stand up oneself, and helping others to understand things when one wants to understand things oneself; and not imposing on others what one does not desire oneself". In other words, one should feel for others in the same way one feel for his family relatives. Second, analogies are drawn between kinship and political ties, for example, father-son relationships analogous to king-officer relationships, brotherhood ties to colleague relationships, and connections with minority groups were made through marriage (Su 2007). With those analogies, Confucian ritual has actually identified the similarities between family and state. The ties of kinship in kin-based lineages are extended to other social organizations; the kind of internal structure, relationship patterns and code of ethics in lineages are replicated in other social organizations, and that has resulted in family-state isomorphism (*jiaguo tonggou*). However, the family-state isomorphism has been established only in the mind of the elite class (the gentry). It has not become a real-world structure that the general public live in. Instead, the public still lead a life greatly constrained by the acquaintance society. Because their lives are highly

[7] Translator's note: *Tianxia* literally means "land under heaven", often used to refer to a country or a state.

dependent on the acquaintance society, the general public almost have nothing to do with the mighty empire.

As Confucianism establishes social order with ritual, Confucian ritual also regulates people's natural feelings and their reciprocity-based relationships (both may be referred to as "*renqing*"). In other words, the process of establishing social order is in fact one of ritualizing "*renqing*" (incorporating "*renqing*" into the ritual system). Here "*renqing*" refer to both natural human feelings one may have in family relationships and other natural feelings one may have, for example, for land and home. Once incorporated into the ritual system, "*renqing*" means more than just natural feelings. It blends with ritual and becomes a key component of the system. Therefore, a ritual-based society may also be referred to as a *renqing*-based (or reciprocity-based) society. *Renqing* (or reciprocity) becomes deeply embedded in the thinking patterns of villagers in the acquaintance society. Naturally, Confucianism's success of ritualizing "*renqing*" (reciprocity) and embedding it in people's thinking pattern of the acquaintance society must be attributed to the fact that, *renqing* (reciprocity) is necessitated by the acquaintance society in its very nature.

2.4 *Xiangtu Luoji*: Reciprocity-Based Behavior Pattern

In the acquaintance society, people's behaviors are influenced by "*renqing*" (reciprocity); they embrace reciprocity as their code of conduct, which can be summarized as "*xiangtu luoji* (the traditional pattern of behavior in the rural land)". The Master said, "At fifteen, I had my mind bent on learning. At thirty, I stood firm. At forty, I had no doubts. At fifty, I knew the decrees of Heaven. At sixty, my ear was an obedient organ for the reception of truth. At seventy, I could follow what my heart desired, without transgressing what was right. (The Analects · *Weizheng*[8])" In *Fei Xiaotong*'s opinion, "what was right" (*ju*) was ritual and a result of people's learning (Fei 1992). My interpretation for "what was right" is ritualized reciprocity (*renqing*) or the ways of the world in the acquaintance society, that is, the reciprocity-based pattern of behavior. People spend their whole lives familiarizing themselves with or learning such pattern of behavior.

When people have accepted the Confucian doctrines for maintaining social order, they embrace reciprocity (*renqing*) as their code of conduct for different types of social relationships, And when that happens, people's due affection (*qingfen*) and obligations (*qingyi*)[9] in different kinds of social relationships would vary. So would the related friendly sentiments (*qingyi*).[10] Just as Liang (1987: 79) put it, "the affection of intimacy and care springs from familial and blood relationships, and all related persons naturally have affection. Where there is affection, there is obligation. The

[8] Translator's note: *Weizheng* is the second Chapter of *The Analects*. It means "the practice of the government".

[9] "*Qingyi*" is a combination of affection (*qing*) and obligation (*yi*).

[10] "*Qingyi*" is a combination of affection (*qing*) and friendship (*yi*).

obligation of the father is kindness, the obligation of the son is filial piety, the obligation of the elder brother is friendship, and the obligation of younger brother is respect. The husband and wife, and friends, and all related people naturally have obligations toward each other." As "there is no afterlife to love the family", one must fulfill one's obligations toward family members in the present life; as "one cannot get away from village folks", one must have affection towards acquaintances and give them due respect; as one has no particular affection towards strangers or just happen to have met them by chance, one is bound by nothing.

To sum up, the reciprocity-based behavior pattern may be summarized into at least the following 4 principles: first, the reciprocity principle (underpinning the other principles); second, the "not going too far" principle; third, the "discriminatory" principle (for dealing with strangers); forth, the "attachment" principle.

First, the reciprocity principle. In the acquaintance society where people live from birth to death in the same place, people seldom go from place to place and are born familiar with one another. Intimacy and affection (*qingfen*) have thus arisen from such familiarity. They must provide help and financial assistance to one another and avoid constant disputes. As they cannot get away from their folks, people must have affection towards acquaintances and give others due respect. Familiarity and intimacy have generated the reciprocity principle, making it the basic principle in interpersonal relationships. It requires people to be sensitive, respectful, impartial, reasonable and tolerant in the way they conduct themselves in relation to others. Interpersonal relationships are greatly influenced by "reciprocity (*renqing*)" and "face". Being inevitably involved in face-saving and reciprocation, acquaintances form a rough understanding of their "rights and obligations" based on the "give and take" that has happened. They seem to have kept an "account" of such happenings.

Because people all care to show consideration and respect, the "give" and "take", "rights" and "obligations" become inconspicuous under the cover of reciprocation and face-saving acts, but a balance must be established. This may be different from the balance between rights and obligations required by a Gesellschaft. Instead of striking an exact balance in every encounter, it requires obtaining mutual benefits in engagements over a long term. That's how the reciprocity principle works. This means there will not be unilateral favors or concessions. Though most people are used to giving way out of consideration or respect for others, every imbalance caused by any yielding act will be recorded in their "account books". The books gradually gather information, and when imbalances build up and become intolerable or hopeless, when there is no point in having considerations for feelings or faces, people may come to an open break in their relationships. This is not about judging a case as it stands. Instead, people will have to settle accounts and be square with the person. Here are a few things suggested by the reciprocity principle—one will be looked down upon if he is indifferent and calculating; or if he caves in every time and never speaks up when things are not fair for him; either way one will end up leading a miserable life. Therefore, in village life, people should not seek revenge for the smallest grievance or comply with others' requests and never speak up. Instead, people should keep the eye on the long run, have mutual respect and receive mutual benefits.

Second, the "not going too far" principle. When Hiroaki (1998) investigated land disputes during the *Qing* period, he contended that dispute resolution rested upon the ethics of mutual concession, because traditional China lacked a mechanism for determining the belonging of benefits. Such an understanding is very enlightening, but on a deeper level, the lacking of such a mechanism all boils down to the functioning of the traditional behavior pattern of the rural land (*xiangtu luoji*). As one of the basic principles of "*xiangtu luoji*", the "not-going-too-far" principle plays a pivotal role in rural life and fits in well with rural life. All people's fates are interconnected. People have a constant demand for cooperation in the fields of hydraulic engineering and public security. That's precisely the reason why no one can live independently from others and why people must reciprocate compromises and avoid going overboard. The reciprocity principle has generated the "not-going-too-far" principle. In addition to requiring reciprocation, this means "not stubbornly clinging to reason" when there is a conflict between reason (*li*) and feelings (*qing*). Someone who persistently appeals to reason without any consideration for others' feelings would be regarded as unreasonable or being incapable of having feelings. One does not go overboard if he is understanding and reasonable. So people should make a decision while taking feeling into consideration. When there is conflict between feeling and reason (or law), people should put feelings first and find ways around reason or law. Otherwise, one would still be judged to be unreasonable even if he has a justified reason to do so. In Confucian views, being unreasonable makes one almost inhumane. Acquaintance society seems to hold onto the same belief.

One nurtured by the acquaintance society would live up to Confucian ideals and act with benevolence toward others; he is understanding, reasonable, easy-going, and self-restrained; and he would put feelings before reason, keep reason in check with an emphasis on feelings, refrain from stubbornly clinging to reason or going overboard. Lin (2006) states, "To a westerner, an idea is acceptable if it is rational. To a Chinese, being rational may not be enough, because it has to also conform to feeling. In fact, *he qing* (conforming to feeling) seems to be even more important than *he li* (being rational)." Being rational means being consistent with "reason". "Reason" is highly valued in the acquaintance society, because "one can go anywhere if what one wants stands to reason." That being said, reason is still being restricted by feeling. If one completely disregards feeling, he will be judged irrational no matter how many reasons he has to do so. Because reason is always limited by feeling in the acquaintance society, resolution of disputes often typically factors in people's feelings. As disputes (especially family disputes) often have arised from minor disagreements in long-term communal living, the line between right and wrong may often be blurred. When the elder villagers are invited to sort things out, they would factor in feeling rather than reason things out. When the judiciary settles a lawsuit, it also has to consider human feeling under specific circumstances. All principles of judiciary resolution—such as "*zhunqing zhuoli*" (consider reason on the ground of feeling), "*heqing heli*" (conform to feeling and reason), "*ruqing ruli*" (agree with feeling and reason), "*tongqing dali*" (acknowledge feeling and stand to reason) and "*zhuoqing chuli*" (feed feeling to discretion)—put feeling before reason.

In daily life, not going too far is not only required by Confucian teachings, but also a common-sense choice. People often say "one will live a better life if one allows others a good life." So people choose to be tolerant as much as possible and would not push others to limits even if their propositions stand to reason. In the acquaintance society where people live from birth to death in the same place, people expect to live there for the long run, so people live in the same physical world that holds up the same ethical principles. In reality, a family may not keep its advantage forever, because "wealthy families lose their fortunes in three generations", "the first generation may be heroes, the second fools, and the third losers". To winners, their future generations may find themselves in the same disadvantage for their opponents today. So in interpersonal relationships, they would act with benevolence with others, show others respect, and leave themselves a way out; and never haggle over temporary losses of benefits or take extreme measures. What's lost today may be retrieved tomorrow (Chen 2009a).

People must follow the not-going-too-far principle, because the acquaintance society lacks a clear regime (e.g. land tenure) for regulating interpersonal relationships, and people have to get along in the village community for a long term. Maintaining harmony makes it easier for them to work together to handle natural and social risks. Discontinuity in cooperation and mutual relationships may make life hard to carry on. Besides, no matter how justified one's rights and reason are, the other side cannot be entirely wrong. Ethically nothing is absolutely legitimate. Right and wrong may even switch sides. What's more, going too far may bring uncalled-for harm to people's daily life. Therefore, even when having gained an advantage, people would not give full play to such a position and recklessly push the others to a corner.

In traditional time periods, rural order at grassroots market towns are generally controlled by elite groups. In other words, the acquaintance society's order has naturally extended to grassroots market towns. At the market town level, though people from different villages do not know each other well, they would still be somehow connected by marriage, trade, security, drought relief and flood control efforts, lineage, and religious beliefs. In these interpersonal relationships, people should also be understanding and respectful, and refrain from an uncompromising mindset and avoid going overboard. Disagreements are bound to take place among people from different villages. The resolution of these disagreements relies on the face and not-going-too-far principles, though these principles do not apply directly to unfamiliar people from different villages, but between local elites or between elites and villagers. Local elites would often act as the go-between to mediate disputes or conflicts between people from different villages, which may be proved by local social studies (Liu 2004: 146). When elites act as mediators between people from different villages, they are bound by the "face" principle; when elites act as mediators between people from the same village, they would invariably settle the disputes by asking the two sides to "do me a favor". In other words, the "face" principle works mainly by its application to elite-villager relations. Therefore, the grassroots market town is a natural continuation of the acquaintance society and the "face" and "not-going-too-far" principles also apply to interpersonal relations that occur at market towns.

Third, the "discrimination" principle. Different groups in the acquaintance society have different social positions. Men inherit from their ancestors and fathers by birth the qualification to live in the village; women acquire their qualification through dependence on men, i.e. their fathers, husbands or sons (Yang, 20). As members in a complicated network of human relationships, villagers form a circle of "their own folks" that clearly distinguish themselves from "strangers". Strangers are those that sojourn or settle among villagers, including: (1) new members who have joined the kinship ties—such as called-in son-in-law (*shangmen nvxu*), son adopted from one brother to another within the family or lineage (*guoji* form of adoption) and girl adopted into a family as future daughter-in-law (*tongyangxi*); (2) the "temporary strangers"—villagers with other surnames who are connected to the village through regional proximity, and (3) "the obvious strangers" such as Buddhist monk, Taoist priest, craftsman, vagrant, and youth sent down to rural areas (*zhiqing*). With members of the acquaintance society, these outsiders have not established the kind of relations needed for gaining a foothold in the society.

People make a clear distinction between relationships with acquaintances and strangers and treat them differently. The reciprocity principle applies to relationships between acquaintances. Doing business is considered a "non-feeling" area, and business must be done at the market, where people would temporarily put aside their relationships and make deals as strangers. Settling accounts face to face is what strangers would do when other social relationships are not taken into account (Fei 1992). In stark contrast to the reciprocity principle that applies to acquaintances, strangers are being treated in a ruthless manner. They believe it is equitable to discriminate against strangers; violence or threats of violence may even be used as means of negotiation; it is also equitable to disregard interests of strangers and favor their acquaintances and the local folks (Chen 2006). Such a ruthless manner not only applies to the "obvious strangers" from outside the acquaintance society, but also to the "temporary strangers" with special backgrounds, accents or experiences. When the "obvious strangers" have been wronged or treated unfairly in the village, it would be much harder for them to get assistance, be it financial assistance or opinion support. The same may be said if they need assistnace outside the village. The ruthless manner for handling strangers is also a reflection of the reciprocity principle. In other words, the discriminatory principle is generated by the reciprocity principle.

Although Confucian teachings favor "honoring old people as we do our own aged parents, and caring for other's children as one's own", "men did not love their parents only, nor treat as children only their own sons", "Among the four seas, all men are brothers", rural people seem to be unaffected by such teachings. They seem to have no feelings for strangers; their acts towards strangers are dominated by the discriminatory principle. This is because in the acquaintance society supposedly people have the same ethics and provide mutual assistance. In traditional time periods, state power does not penetrate to the local level to influence every aspect of village life, and fails to deal with public affairs at the village level. At the same time, the family also fails to provide public goods needed for irrigation, mutual assistance and social security, or the other goods that are directly related to farmer life and production. That's when the acquaintance society, characterized by both consanguinity and regionalism, becomes an important social unit for unifying ethics and ensuring mutual assistance. It ensures close connections among people through establishing hard laws,such as family and lineage rules, village regulations and agreements, and soft laws, such as Confucian ethics and public opinion. Because people need assistance from others, people must act by shared code of ethics and would inevitably treat their own folks and outsiders differently. What outsiders must deal with are not individual villagers, but the whole village group, or the community of acquaintances with shared life experience and code of ethics. Members of the community are connected through kin, benefit and mutual assistance.

Of course, how outsiders and strangers are treated also depends on their influence. In *Fei Xiaotong*'s opinion, the concept of "neighborhood" (*jiefang*) in a village is not a permanent group, but a physical area. The size of it depends on the influence of the resident villagers. The area of a neighborhood with influential resident villagers may extend to the whole village, whereas one where the poor live may include just two or three families (Fei 1992). Compared with the acquaintance society members, outsiders and strangers have a much smaller influence and fewer available resources. They are often not qualified to be admitted into "mutually beneficial and fair" favor-based cliques of acquaintance, nor do they have what it takes to establish a competitive clique of their own. Therefore, when treated unfairly, they can hardly retaliate; when their interests have been harmed, those who have caused harm may be well protected by their influential families and acquaintance cliques. Imagine an influential person coming to the village as an outsider or stranger. He may be in a better situation because with his influence, he may gradually build up a clique of his own and eventually fit his clique in the greater villager network.

Forth, the "attachment" principle. In the acquaintance society, micro-powers in interpersonal relationships not only exist in the daily occurrences inside the society; their influence also extends to other places where interpersonal relationship occurs. The most typical example would be outbound elites who are still under the control of interpersonal relationship conventions. For example, in traditional time periods, the gentry are highly respected in social etiquette, and enjoy privileges in legal, economic and political affairs; they have close relations with bureaucrats and have access to important political resources; they have access to cultural resources such as government documents and litigation techniques; though not present at the villages, they still play a dominant role in public affairs in the rural society. Such dominance over local society is shared by powerful and wealthy local elites with access to land, lineage connections, special professions and military forces (Hilary 1979; Keith 1982; Zhou 2000). However, the precondition to their dominance is that their behaviors are still bound by their relationships with village members and relationship conventions. If this system is no longer binding for them, for instance, if they refuse to help out villagers, work for the public good or allow fellow villagers to "benefit from" them, they would no longer be respected. In fact, the more attention they give to the village, the higher their social status is and the better their reputation; also they would gain more respect and appreciation from fellow villagers (Luo 2002; Lv 2006b). Those outbound elites for whom relationship conventions no longer work would have to give up their chance to get re-established in the acquaintance society.

In this way, the definition of "relationships" is broadened to also include the relationship between people and homeland, which may be described as "attachment to the land" (*xiang qing*). No matter where they are, people feel their roots in the land and have deep and enduring attachment to the rural land (*xiang*). Those who have left the land, the gentry included, still have a close connection with the village. They may take examinations, join the bureaucracy or go into business, but their roots are still rural; they still have their mind set on "ploughing and studying to carry on the family name" (*geng du chuan jia*). Sentimentally attached to their homeland, they have a lifelong ambition of achieving eminence and influence (*xianda*) in the homeland. They look forward to spending their final years at home for a sense of belonging tied up to it. Wherever they are, they have their personal honor and grace closely linked to the land. They are emotionally connected to the land through homesick feelings, home accents and customs, and would often seek solace in homesickness. They all desire to return to their hometown in embroidered silken robes (*yijin huanxiang*), unless—as a popular saying goes—"they are unable to face the elders of the South (*wuyan jian jiangdong fulao*)", because "if they do not return home when they've become wealthy, their wealth would be as inconspicuous as wearing embroidered silken robes at night". When they die, they have to transport their coffins back to the hometown and be buried together with their ancestors. This "fallen leaves return to their roots" sentiment has arisen from having reciprocity-based relations in the rural land. What's behind these "going back to the hometown for a sense of belonging" customs is the attachment principle, an extension of the reciprocity principle. Being attached to rural land, people would always come back after leaving. This ongoing cycle establishes a healthy connection between the urban and rural societies.

The attachment principle encapsulates the clan's function as the foundation for the values that farmers live by. This principle is highly related to their ancestor worship and clan identity, both essentially a reflection of the "root" sentiment. Generally speaking, the gentry and outbound "travelers" (*youzi*) have bigger aims in life than ordinary farmers. They are interested in topics like what makes one's death "heavy as Mount *Tai* or light as a feather"; they are looking to establish virtue (*lide*), establish meritorious achievement (*ligong*) and establish words (*liyan*); their ambitions are to "bring stable government and order to the world" (*zhiguo pingtianxia*). However, in a structure of family-state isomorphism, they would also seek a sense of belonging or the significance of life by returning to their clan or hometown. They also need to worship ancestors and feel included within their family and clan. In addition to bringing worldly honors and achievements, this sense of belonging or inclusion may allay their loneliness, fear and anxiety caused by the shortness and limitation of life, give them a sense of security and a sense of meaning in life, and allow them to be respected and influential in the clan and rural areas.

2.5 The Impact of Collectivization on the Village

This dissertation focuses on the period since the 1980s. Previous analysis on the ideal model of the acquaintance society has drawn empirical evidence from Chinese villages in traditional time periods, therefore, now I must discuss how rural China has been influenced by the collectivization period from the 50 s to the 70 s, or most importantly, to what extent has the villages remained the acquaintance society in nature. In this section I will briefly talk about special policies implemented during the commune era and their impact.

Communes (embracing scores of villages) are grassroots rural organizations intended to serve political, economic and cultural purposes. They are often made up of one natural village or several neighboring villages. The political and social restructuring process during the commune era may be summed up as "a change of fortunes (*fanshen*)" (Hinton 1966). The real meaning of *fanshen* (literally means to turn the body over) is a change of identity for elite groups and shift in grouping standards. By the new standards, the rich are grouped as the exploiters while the poor the exploited; what must be accomplished after a change of fortunes is to overthrow the exploiters and reveal truth in history. The change of fortunes is done through class struggle, i.e. a revolution weapon used against country elites of the old times, including the gentry, landlords, rich farmers and *baojia zhang* (heads of the *baojia* [a rural administration system based on households]). An important result of class struggle is a fundamental change in the elite class. By changing the elite class, the state rebuilds rural order and integrates the rural society into the bigger unified political community. In the new era, with a highly ideological identity, the new elites take on revolution as their important political task. As revolution is, by definition, related to poverty, the qualifications for rural elites are no longer education and wealth, but poverty and revolutionary involvement. Of course, those who are qualified to join the elite class are just a minority of the poor population, and the Party organization of the commune era is mainly formed by the village's elite class (Wu 2002: 79; Student Tan 2007: 206).

The basic policy during the commune era is "ownership by a three-level structure (communes, production brigades, and production teams), with production teams as the grassroots units". Production teams, often of the same size with natural villages, are the units for collective farming, accounting, and income distribution, which has caused them to become acquaintance society marked by frequent interactions among members. Production brigades (consisting of several production teams), as the main administrative units for issuing orders, delegating agricultural production tasks, are also the acquaintance society in nature, with a lower frequency in human interaction. Production brigades and teams are both marked by distinct boundaries: naturally by geographical boundary and socially by villager identity. Inside these acquaintance societies, each villager's past and present are well understood by others; the reciprocity principle also applies. At the same time, villages are independent, isolated and dispersed. The structure of the rural society resembles that of a "beehive". Compared with villages in traditional time periods, the basic social unit of rural society during the commune era is characterized by a higher level of independence and isolation. In such structures people would more distinctively use the discriminatory principle against outsiders. Even in the 1980s, we may still very strongly feel it at play for interactions between the acquaintance society and the external world.

A look into daily life in villages will reveal that collectivization has to some extent created an ideal world of "small country with few people" (*xiaoguo guamin*). People from neighboring villages may "live so close to each other that they can hear the roosters and dogs but may not socialize with each other". Because people have access to resources and social networks in their production brigades and teams, they do not need to come into contact with people from other villages. All that has been impossible before collectivization. Therefore, during the commune era, people still behave in similar patterns based on Confucian thoughts and village ethics; the reciprocity principle and discriminatory principle still apply in dealing with different social relationships (with acquaintances or strangers). To some extent it may be said that collectivization has set a clearer natural and socializing boundary for the acquaintance society, and thus has heightened the effects of the face and discriminatory principles. As the natural and social boundaries become clear, the effect of the not-going-too-far principle also gets heightened. This is because during the commune era the acquaintance society (village group) has a stricter control on social resources, and the transgressors may face more severe punishment and are less likely to escape punishment. Transgressors may also be punished in the name of "revolution" against the backdrop of national movements. That was what happened to *Li Yuting*, the evil landlord who had killed his fellow villager. He got burned to death by someone during the Cultural Revolution, but no one investigated what happened because villagers believed it served him right (Han 2007: 121). In the 1950s, *Ma Wenhong* and *Huang Chongdao* from *Xiaocun*, *Yunnan* Province, among others, were punished in the mass movement to "clean up bandits and oppose local despots" (*qingfei fanba*) for what they had done in the 1930s (Zhu 2003: 102).

Despite a dramatic shift in the identity of social elites, no fundamental change has taken place to their pattern of behavior towards ordinary villagers. Compared with the traditional time periods, the new elites exhibit more "commonness", because people could no longer join the elite class through economic prosperity, clan connections or educational success; the only way is being accepted by the Party institution. However, the elites still observe the reciprocity and attachment principles. Looking forward to living in the acquaintance society for many more years, they must still give others due respect, and when handling administrative affairs, strike a balance between representing the state power and being a member of the village. Though during the commune era the state power, not the village or locality, is constantly being emphasized, deep down the outbound elites are still deeply attached to their homeland; they still look forward to returning home in embroidered silken robes (*yijin huanxiang*) and work to the benefit of their hometown. Though ancestor worship is regarded as superstitious and regionalism as narrow-minded and both as outdated compared with revolutionary ideas, these values are still secretly embraced by many of the elites. In the 1980s when the state relaxes its control on the people, the elites soon become important figures backing the revival of old ethical values. For them, acting in line with revolutionary values is just an expedient during the commune era. Deep down they are still loyal followers of the face and attachment principles.

Generally speaking, the four "*xiangtu luoji*" principles remain largely unchanged. They are still the code of behavior for the general public. Though the traditional pattern of behavior (*xiangtu luoji*) sometimes has to give way to code of behavior required by the revolution, when it comes to daily routines, people still follow the old behavior pattern. In the same way, the order mechanism and the network of reciprocity-based micro-powers have remained largely the same. What's more, due to more institutionalized power at the village level, those behaviors that go against the common behavior pattern and interpersonal relationships conventions may suffer more timely punishment.

To sum up, while introducing revolutionary ideas, the commune era has kept the traditional pattern of behavior (*xiangtu luoji*) alive. The acquaintance society has maintained its moral order. The attachment principle still applies, while the effects of the other three principles have even been heightened. All in all, contrary to common belief, the traditional pattern of behavior (*xiangtu luoji*) have remained intact. It has coexisted with revolutionary ideas, possibly because the villages have remained the same in form and nature. Actually, when the Party and government organizations have penetrated to the grassroots level or into the acquaintance society, and carried out collectivization practices, they have served as the cornerstone for the continuation of the village's nature as the acquaintance society.

Chapter 3
The Rise of Gangs and the Acquaintance Society

My fieldwork has started in the 1980s, a time period that may be regarded as epoch-making in the development of the Chinese society. In the three decades since, all the changes that are taking place to rural China have something to do with this time period. What happened in the 1980s—the retreat of revolutionary ideals, rapid development of rural economy, greater social mobility, and transformation of urban economic development—has had a major impact on rural China today.

In the 1980s, under the guidance of the central government, the whole country step by step gives up the political commune system and establishes the contract responsibility system. Land is owned by the state, but is contracted (*cheng bao*)[1] to rural families. To some extent, this system is a return to the traditional small-scale farming. It also ends the era of villagers having their time and space being entirely tied to land. Villagers are granted some freedom. Instead of having inflexible time arrangements under the commune system, they are allowed to use their time freely, which leads to some physical mobility. Though free flow is still restricted by the household registration system, their mobility is not influenced if they intend to move within certain areas. As revolutionary ideals retreat from national high-level politics, they also begin to withdraw from rural China, which loosens up the close relationship between individual villagers and village community. Under revolutionary idealism, villager's personal work and fate are once associated with the commune and further with the national destiny and the bright future of Communism. These ideals that used to be a great encouragement to several generations of villagers seem to have disappeared altogether overnight. As people give up their revolutionary ideals, their personal freedom and benefits are increasingly being recognized. The control from communes starts to diminish.

[1] Translator's note: The *Chengbao* system refers to the private or individual contracted operation of public assets such as bus lines, hospitals, and schools. These operators pay a fee as well as a percentage of profit generated to the state. It is regarded as a form of entrepreneurship. ("*chengbao* system", *Academic*, 20,211,215, https://en-academic.com/dic.nsf/enwiki/11526038).

B. Chen, *The Underworld of Rural China*,
https://doi.org/10.1007/978-981-19-8710-6_3

Well-controlled physical mobility, a retreat from revolutionary idealism and loosened commune control all serve to bring a new order at rural China. Rural residents have exhibited a pronounced increase of deviant behaviors, partly in defiance of the invariably repressive Commune order and partly being at a loss with the retreat of revolutionary idealism, occasionally also in pursuit of personal gain. Such behaviors exist among people of different age groups and genders. But the youth group has exhibited distinctive patterns that are different from previous or later time periods or from other groups in the same time period. This Chapter will describe various deviant behaviors by the young people in the 1980s. With that, the author is interested in revealing the special rural underworld of the 1980s, investigating the attitude and reactions from the rural society and state power towards deviant behaviors, and analyzing the relationship between the rural underworld and the acquaintance society.

3.1 The Idle Youth

In interviews with rural residents, I asked many of them about their adolescence life during the 1980s and their views on that time period. I often got them into deep thinking. Most of the times I got the same answer: "that was a boring and strange age." Many of them find their own behaviors very interesting, in fact, so interesting that they almost border on the incomprehensible. Some others said that they looked forward to the future but did not know how to make it happen. *Li Junfa* from *Chujiang* described to me his life then in this way:

> In high school I liked goofing around. Many people of my age stayed at home after graduating from junior high. Though I was in high school, I had no hope for going to college because of low admission rates, so all I did was hanging out with friends. We would make a flag out of a red undershirt, pretended to be traffic cops and pull over cars for a ride into the city. We wanted to play but we had no money, so we chose to hitchhike. But we had to get up early, for when it was still dark, drivers couldn't see clearly and would mistake us for real traffic cops. Once they had stopped their car and we were already on board, they would have no other choice but to drive us. Back then there were not as many road bandits (chefei luba) and drivers were less vigilant. Later with worse public order, it was much harder to do the same.

In his account of what he did back then, he often used the word "boring" to describe it for he did not know how to kill time. At that time people have just come out of the commune system. They no longer have to go to work on time like factory workers. Nor do they have to obtain "work points" (*gongfen*: the basis for output redistribution during the commune era) by working in the fields. Once land is contracted (*cheng bao*) to rural households, they no longer cultivate the land for the communes. Being more motivated, they no longer slack off at work, which greatly shortens their working time in the fields. Though constant increase of labor input in farming and sideline production is still required by the traditional "involuted agriculture development mode" (Philip 1990), rural labor is still in oversupply—idle labor thus becomes a common problem.

In the three decades after the founding of P.R.C., with the agricultural sector less susceptible to natural risks after collectivization, grain production has increased tremendously, keeping the rural families "warm and fed" (*wenbao*) and making it possible to support a large population. At the same time, higher quality of healthcare and the prevalence of barefoot doctor scheme in the wide rural areas have granted rural residents low-cost health care. Both these two factors have served to increase birth rate, decrease mortality rate and increase people's life expectancy. By the end of the 1980s, large drought-relief, flood control and water resources projects have been completed, and mountain forests and lakes have been effectively developed. Therefore, agricultural development does not require additional input of labor. The contradictory relationship between people and land becomes more glaring than before liberation, not in terms of having difficulty in keeping the population "warm and fed", but of labor oversupply. Existing land does not need more people for cultivation, but the population keeps growing.

After continuous efforts by the whole country for 30 years, our country has established a relatively complete modern industrial system, but the size of these industries and related service industries are still quite limited, thus unable to transfer excessive rural labor to urban areas, which keeps urbanization at a slower rate than the rate of population growth in rural areas. That has exacerbated the excessive supply of labor. Even with agricultural involution, villagers still have too much free time. The oversupply of labor in the 1980s quickly turns into an idle villager issue, which in turn becomes a youth problem and then an adolescence problem. My interviews and analysis have highlighted two relevant issues: night life and adolescence. Both are youth-related issues.

In traditional time periods, villagers had no night life. But during the commune era, under the influence of revolutionary ideas, people start to have a night life with unique Chinese characteristics. People go to work in the fields during day; during night they summarize their work, calculate their "work points" and make arrangements for the next day; they go to study at night schools to acquire literacy; they have class struggle and rectification movement meetings. People gradually get used to this lifestyle of having all their time arranged by the communes and party organizations. Work time is for collective farming, while non-work time is for making collective arrangements and political study. Their whole lives are associated with revolutionary visions and practices. In the 1980s, all that changes: no more collective farming during the day or political life time arrangements during the night; no more class struggle or political study. With technological advancement and the introduction of technologies in rural areas, the commune night life is superseded by a new film-watching fashion. The introduction of films to every corner of the country by the Party and central government in the 1980s marks the beginning of a film-watching age, which does not end until it is superseded by television ten years later. Looking back, many people are quite excited about film-watching. Back then, villagers from 6 or 7 villages in the same standard market town (Skinner 1998) would gather to see a film. Sometimes the whole village goes to see the same film. The stories that are being told at the open-air theatres—romance, entanglements in the Martial Arts world and also violence of course—all feature the young as the leading characters.

In the 1980s, the young people have gained much more freedom. With an oversupply of labor and limited land for each family to cultivate, people do not have to labor at an early age. That's nothing like the commune era when families need the young people to participate in collective farming for "work points", and families with more people would have had a better life. Things change when land is allocated to families. In families with more people, surplus labor becomes a burden. During the off seasons, aside from helping out in the fields, the young people almost have nothing to do, so they gang up with their playmates. In the beginning they gang up with people from the same village teams, but later with people from outside the teams too. As new friends join the gangs, subgroups are formed, and the gangs start to stratify and reorganize. The young from big families with many brothers have more free time because they have no work to do and thus are more likely to hang out at faraway places. The following is an extract from a police record for public security investigations of *Linjiang* county during the 1980s.

> Some day in late April, I went from Shawan to Dongfengqiao to hang out in a barber shop owned by Zhang Xiuli, my senior alumna. Another young man nicknamed "Lakouzi" (literally a veteran in diverse professions) from Shiping Village was also there in the shop, and we came to know him in chitchat. Past 10 o'clock in the morning, he and another person nicknamed "Wanzi" (literally meat ball) from the same village invited me to hang out in their village, so I went with them to Shiping Village and stayed for a day. That night we did not find a place to crash in until over 4 o'clock in the early morning. There "Laokouzi" and I stayed up till dawn. Then we had a bowl of dumplings and hung around for another day at Shiping. On the third day "Laokouzi" asked me to go to the city with him. The two of us, together with "Wanzi" and another person whose name I didn't know, went to the bus station to look for someone nicknamed "Caoshangfei" (literally flying across the lawn). We had lunch there. In the afternoon at around six the man came with a woman. He scolded "Laokouzi" for bringing too many people. That night he found us a place to stay at the travel agency. But because he was mad at us, we rented a car and drove to Li Dongping's mother-in-law's and hung out there for two more days…

We could see from this record that the young people have leisure time that they could spend on their own will, so they have time for going to places. But they seem to be really "bored", and what they do is not so "fun". They wander around like aimless wild horses, and grab a drink and snack as they go. This may be the result of not knowing what to do with the new-found freedom, or they would like to simply relish the freedom to wander. Of course, many young people do not have the chance to wander across counties. They may only hang out in the standard market town close to their villages. They have relatively limited freedom, but they are just as "bored". They do not have the opportunity to continue studying like the young people in urban areas; nor do they follow in their father's footsteps and work as a blue-collar worker; they do not have the opportunity to make money by working in the cities. They have to find something fun to do.

To sum up, on one hand, rural China has marched into an era of film-viewing nightlife; on the other hand, the young people have nothing to do and have to find something fun to do. A combination of these two aspects gives the young people a special living environment, in which going to the theatre and having leisure time to kill have become important components of their lives. Under such circumstances, they develop their own life expectations and aspirations. For instance, they look up to distorted hero images, and they start to pursue their needs and interests, some of which have become legitimate just recently.

3.2 A Period of Idleness: The Formation of Regional Gangs

3.2.1 Regional Gangs in the Rural Underworld

In the 1980s, in the process of fun-seeking, the idle youth form rural gangs that gradually develop into a special gangland of that era. Everything about the young people of that era has something to do with the gangland. Some become active members or "celebrities", while others are ordinary or peripheral members; some members have moved away from the rural land and only come back for occasional visits, while others seem to be suffering the misfortune of being tied up to the land for their whole lives like their father's generations or "having their face pressed against the land and their back against the sky". Everything they do has something to do with the rural gangs, including their visions of heroic acts, deviant behaviors and criminal acts. The rural underworld has become an integral part of the youth life.

The rural underworld is initially an association of interactive cliques that are organized by young people from the same village teams. Therefore, rural gangs come in the form of regional gangs with relatively permanent members. Many who have lived in the 1980s told me, "In the early 1980s, villages may be viewed as a whole unit. Young people often hang out with other members from the same village team. Being of a similar age, being a member of the same village team and hanging out together, people gradually develop a collective identity." Whether such collective identity is developed out of regionalism or cordiality established by collective activity is unimportant. In fact regionalism and cordiality may have been mutually reinforcing in the formation of regional gangs.

There are no clear leaders in regional gangs. When they fight or hang out, some may have more influence on others, but none are official leaders. Not do regional gangs have names or symbols. By referring to them as regional gangs, we are actually making a retrospective generalization. They have no other intentions than just hang out together. Nor do they intend to form a clique or association in a strictly technical sense. People may easily associate the concept of "clique" or "association" with organized crime groups. But that may be a misunderstanding on regional gangs. Of course, when gang members experience further personal developments in later periods, their forms of organization may no longer be defined as regional groups.

Members of regional gangs in different localities have their unique ways of "hanging out". At some villages with wastelands, gang members often gather there to have discussions on life and plans, as if they were exchanging "martial arts secrets". At villages with a cattle pasture, they would meet up there on cattle-grazing trips and play cards for fun or get up to some mischief. The purpose of their gatherings may have been checking out a special person or something special that could provide some fun for all. Mr. *Lu* from *Linjiang* County gave us an account of his years in the regional gangs.

> My home became a place for gathering because we had a cassette recorder, which was one of the first in our village. We did not have a special venue then and everybody was simply attracted by the recorder. Dancing to the music played by the recorder became the latest trend. We often had one or two dozen of people, sometimes 3 or 4 dozens.

Besides "hanging out" together, the young people would also stir up trouble, especially by stealing, described by the villagers as "stealing dogs and chickens" (*tou ji mo gou*). "Stealing dogs and chickens" is very common among the young. Almost each of them is directly or indirectly involved, or at least has "benefited" from the "fruits". However, if they had taken it to the next level, things would have been very different. Here is the second half of *Lu*'s accounts.

> We would pick up trends in films. For example, we would learn from gangsters in the films to put on some kind of "air" (pai). We all looked up to Chow Yun-Fat who has a special "air" about himself. When we were done dancing, we would wander about in the village and inevitably would cause some trouble. A typical example would be to steal some chickens and cook them. In the beginning we were just up to some mischief, stealing one chicken or two. Later some became criminals, for example, involved in blatant attempts of robbery. Many cliques were involved in criminal activities. But our clique did not go that far. Because of my reputation in the village team (cundui),[2] I was able to protect my members. Villagers hated us, but could do nothing about us. Naturally we were very selective in only stealing from those timid villagers. We dared not to provoke the tough ones.

The bored youth in the early 1980s to some extent resemble the street-corner gangs formed by Western European immigrants in the early days of the American history. Gangs are formed by the idle youth who have nothing to do and no way of climbing up the social ladder (Riis 1892: 236). As for those young people who have something particular to do, they are mostly peripheral members of the gangs and only participate in gang activities occasionally. They are the minorities who have a particular way of social climbing, for example by continuing their study, or being sent by their parents to learn craftsmanship for becoming a blacksmith, a carpenter or a bricklayer. Having their time occupied with these things keeps them from participating in gang activities. However, in contrast to street-corner youth gangs in the U.S. that only accounted for a very small share of the American population, joining regional gangs is a common lifestyle for the whole generation of rural youth of the 1980s. Frederic Thrasher (1927: 37) wrote, "Gangs represent the spontaneous effort of boys to create a society for themselves where none adequate to their needs exists". His view may also provide an explanation for the forming of regional gangs in rural China and for the universality of the gangs. Regional gangs represent the

---

[2] Translator's note: Village teams are more formally referred to as "production teams". A production team was formerly the basic accounting and farm production unit in the people's commune system in People's Republic of China from 1958 to 1984. Production teams were largely disbanded during the 1982–1985 agricultural reforms. Since 1984 production teams have been replaced by "village groups" (*cunmin xiaozu*).

spontaneous effort of the rural youth to build up a stage of their own. The venue for common activity just happens to be open-air theatres.

When people have limited means of having fun, films possess irreplaceable charm to the excitement-craving youth. They would get very excited just hearing about a film to be played at some village and all the youth in a village may travel to another village just to see the film. Thus, the theatre becomes a place for gatherings of the young people and encounters between members of different gangs, and a lot of stories happen during the film-viewing time. We may regard the theatre as their stage where members from various gangs flaunt their talents and act out their fantasies. Therefore, the theatre offers us a window into the rural underworld. What's most likely to happen at the theatre is fight or brawl. A case in point would be fighting between youths from *Zengling* Village and *Dingxiang* Village of *Linjiang* County. The following two passages were extracted from police records at a police station in *Linjiang* County. They narrate fights during film-viewing that happened for two days in a row.

> After ten last night, the film ended and we were all heading back. On our way back, I was leading the group; Huang Falong and Feng Dejun were the last. When we came to the west of Henan village group, we ran into a group of young men from Dingxiang Village. One of them came up to us, took Feng by the arm and said: "I know you. Me and you, let's square up!" With that he punched Feng in the nose; Feng gave him two kicks on the feet and then started to run towards home. Just then the man took out a dagger and thrust it in the air as he chased after Feng. Once he caught up with Feng, he gave Feng another punch. At that, Huang Wengao in our gang yelled, "Take him to the police!" The man with the dagger fled, leaving the others behind. I said, "now that you've beaten up Feng, what the heck are you still doing here?" One of them said, "None of your business! You are just lucky you didn't get beaten." I replied, "Go on! I dare you!" So he jumped on me and I knocked him down with a punch. Then another two men ran to us. The man rose from the ground and drew a knife out of him and swung it in front of me. One of them grasped me from behind and another two punched me from front. Just then one of our men Huang Falong came and they ran away. At the theater they tried to hit me with a stool but didn't get me. Huang Wengao yelled again, "Take him to the police!" Then they all ran away.

> At around 1 at noon, I was goofing around at the main base of our production brigade, where I ran into Pan Zhanlin. He asked me, "There is a film on at Zengling Village today. You come?" I said, "Yeah. I have no plans anyway." So I went to his place at the Fifth village group. After dinner, we went to the area of the Forth Group and heard some young men asking about whether there was a film that night. Pan said he had to go to Zenglin Village just to be sure. On the way we ran into a pupil who assured us about the film. So we went back to the Forth Group and told others the news. Yao Xin said, "We should all go to the film theatre. Find some more people to go with us. Bing'er (literally the son of a soldier) got injured in a fight last night. If they are not being nice, we should fight them back tonight." So we went to find more people separately and managed to find altogether 25 people. … When we walked past a store, we say a young man, and heard Pan say, "This is him! From last night! He injured Bing'er pretty bad! Screw this son of bitch!" I said, "If we pick a fight right in front of the store, people would accuse us of robbing. Better not do it here or we may get stood up!" After a while, the man felt something was wrong and ran out of the store. Qu Pei chased after him and tried to kick him but did not get him. He got away. After a while, he brought back 5 or 6 people. I picked up a stick from the ground and ran up to the river bank on the west. As I turned around, I saw a man at my heels, and I gave him two hits. He lashed me with a belt on the head and held my stick. I put down the stick and tried to run but got stopped by 5 or 6 more people. One of them charged at me and hit me on the arm. I

ran into the field and got caught after running across two ditches. Then they drove me to the police station.

3.2.2 Collectivism: The Formation of Regional Gangs

Fighting seems to be the main activity for many rural regional gangs in the 1980s. Street-corner gangs in urban areas fight to distinguish between members from opposing gangs and promote internal unity. For regional gangs of rural China, fighting does not seem to serve the same function, because who belongs to which gangs seems to be plain as day, though fighting may certainly improve internal relationships among members.

According to Thrasher (1927: 46), "the gang is an interstitial group originally formed spontaneously and then integrated through conflict". He divided its development into three stages. During the first stage, the gang is a diffuse and loosely organized group with no leaders and not likely to exist for long; some gangs enter the second stage with enhanced unity. Conflict with other gangs plays a very important role: it marks the gang distinctness from other groups, strengthens the bonds between members, and fosters solidarity against external members. In the final stage, the gang becomes conventionalized or incorporated in some way into the structure of the community, and the gang disintegrates; or if it does not become conventionalized, it drifts into habitual crime and becomes completely delinquent (1927: 54–55). A sharp difference between American urban gangs discussed by Thrasher and regional gangs in rural China is that the latter is characterized by a very high level of collective identity, whereas such identity is lacking for the former. Regional gangs in China are formed by members from the same village teams. They develop a high level of familiarity out of common experience of growing up and schooling. They have a collective sense of honor and share everything in the village teams (acquaintance society in nature). When they meet at open-air theaters, they would naturally unite against outsiders in conflicts. All is united against one man's enemy—their collective sense of honor brings about common action. In other words, troublemakers would have to deal with the whole village in its entirety. This may be reminiscent of what the rural residents have said, that is, "The village may be viewed as a whole unit".

As has been discussed, the young fight out of a collective sense of honor to protect the village, and conversely, because the village is a whole unit, the young are also protected by the village. This mutual reinforcement process may enhance unity within regional gangs and people's sense of honor for the village. This may firstly be proven by the fact that the young would mostly engage in group fights out of a collective sense of honor. The above-mentioned fighting incident between young people from the two villages of *Linjiang* County was the result of individual action mixed with collective action. Back then, gang members may keep in touch most effectively by going to the film theatre. Trouble is often made by young people from the same village team gathering to see a film. Members of regional gangs at *Shaqiao* Village of *Chujiang* City are the most formidable fighters among the neighboring villages. The youth from *Xinwang*, *Wangji* and *Linsha* Village are all afraid of them. But those from *Liji* Village are unconvinced and thus have got into many group fights. Back then anything might be associated with the honor of the village, and escalate into gang wars. For example, even when different gang members play games on cattle pastures, they would associate the winning with the honor of their villages, which often lead to reckless behaviors. Some would prompt others to do things by saying, "if you dare not to do it, you would disgrace your gang!" Conflicts are even more commonplace in competitions like dragon boat races. In many parts of *Hubei* Province, every year dragon boat races lead to group fights, and the races are repeatedly canceled and resumed. Of course, the young would also fight alone, but also for the group's "name", often because "the village team's name is at stake".

The mutual reinforcement between unity within regional gangs and people's sense of honor for the village is also proven by the fact that the young not only fight out of a collective sense of honor, but also to protect the collective interest of the village. Sometimes nothing big is involved, but as long as the village mainstream regards it as affecting the village's honor, the young people, mostly the core members of regional gangs, would fight for it. This suggests that the young people in regional gangs and the village mainstream share the same sense of honor and values. They would even fight over the boundaries of commune land or water conflicts between village teams. They would also actively participate in armed confrontations between clans. A typical example is included in the following archive information.

> On March 26, a group of more than 300 people (over 40 with the surname Li from Xianan Production Brigade of XX Commune, over 50 from Tianxin and Xinhe Production Brigade of XX Commune, others from Shicao Production Brigade of XX Commune of XX City) carried with them their clan flag, dragon lanterns and weapons such as harrows, iron spears, batons and sticks, pounding the gongs and drums as they marched by. All of them were clearly marked by rolling up the right sleeves. In an well-organized group, these people planned, organized and implemented this riot at a cemetery at Heli Production Brigade of XX Commune of XX County. They smashed 32 housings, chopped down almost 10 trees, trampled the crops, and injured 4 people (1 seriously injured), causing a great deal of damage.
>
> On March 27, about 100 people (with the surname Duan from Xianan, Daxi Production Brigade of XX Commune, Xinqiao Production Brigade of XX Commune, and Tangwan Production Brigade of XX Commune) gathered at Lifu Production Brigade of XX Commune of XX City, ready to put up an riot. When their intentions were revealed, most of them were convinced into returning, but over 20 people at Tangwan refused to be persuaded. Instead, they gathered with some other 800 people, also with the surname Duan, carried 6 dragon lanterns and several clan flags and marched while pounding the gongs and drums to Liuyi Production Brigade of XX Commune of XX County. They refused to be stopped and persuaded by officials of criminal justice agencies or by community officials. They tore down an four-room office building at Liuyi Production Brigade and smashed all the furniture in it, including storage units, tables, chairs and tableware.
>
> These two riots of our county were participated by not just ordinary villagers but also leaders of production brigades, such as Li Zhifan, former secretary to Xinhe Production Brigade of XX Commune, and Duan Taihong, former secretary to Xianan Production Brigade of XX Commune. Some leaders of production brigades allowed the riots to take place. After knowing the intentions of the mob, some leaders did not try to stop the mob or report to the higher Party committees.

> We also came to know about the intentions of other groups to settle disputes or put up riots at cemeteries. We must remain vigilant.[3]

Both public security system's archives and my interviews indicate that such incidents are very common in the 1980s. Another record mentions, in the first 4 months of 1983, at *Shanxiang* County of *Hunan* Province, 41 clan conflicts killed 4 people and injured 87, mostly disputes over cemeteries and deaths.[4] Such group conflict is common not just in villages with a strong sense of clan identity like *Shanxiang* but also in Double-lake Plains with compromised clan identity, where as far as we know, rural residents are operating as "separate atoms" (He 2007e). The records did not reveal who are the main culprits behind those incidents, but through interviews the author has come to know that the spearheads in group wars are often the young rebellious generation, who also constitutes the backbone of the regional gangs. Some people say those young people are so naive that they get taken advantage of by the clans. We cannot say for sure whether that is true, but one thing is for sure: they are indoctrinated with clan identity and collectivist values. Perhaps from today's point of view, there are many things they should not have done, but then they believe otherwise. So do the rural residents. They firmly believe in fighting for their village teams and clans. In this sense, they are similar to what Joan Moore describes as "gangs in their early stage". As Joan Moore (1991: 31) puts it, the gangs start out as friendship groups with a more or less clearly defined territory; they are committed to defending one another. The same is true with regional gangs in rural China during the 1980s.

Because the young people and regional gangs fight out of a collective sense of honor to protect the village interest, village officials would intercede for them if they run into trouble. Village officials would let the police officers know that those young men have always been good except for occasional scuffles and that they are forgivable and deserve leniency. For the police and the state, the active members of regional gang are the main factors jeopardizing social order, but from the perspective of rural residents and village officials, the gangs have engaged in fights to defend village teams' common interests and honor. Therefore, instead of being rejected, they have gained approval from rural residents and officials.

About what has led to their bellicosity, former gang members, villagers and officials seem to have reached a consensus—their bellicosity is put down to susceptibility to the educated youth (*zhiqing*) sent down to the countryside during the Cultural Revolution. Cliques are often formed by those "sent-down youth" who have come from the same work unit and often have ended up in the same village. Across neighboring village, different cliques may have been formed by sent-down youth from different work units, and often do not see eye to eye with each other. Therefore, petty disagreements between cliques from neighboring villages would easily escalate into group wars. The local youth come under their influence by hanging out with the sent-down youth. As an old official once put it, "good people may benefit from the company they

---

[3] Public Security Bureau of *Shanxiang* County: "A Report of Social Affairs in March", the 6th Issue of *Brief Reports on Public Security*, April 1 1980.

[4] Public Security Bureau of *Shanxiang* County: "An Overview of Public Security", the 6th Issue of *Brief Reports on Public Security*, May 18 1983.

keep, whereas bad people may be led astray." In my opinion, we cannot overlook the influence from sent-down youth, but more importantly, the prevalent sense of honor and collectivism among the general public constitutes the social foundation for their susceptibility.

Later regional gang members have very different personal development. With the development of the market economy, most members have achieved social standing in the mainstream society and thus withdrawn from the gangs. Some members become involved in a larger gang, or seek to make known their names in a "relationship-based gang"—I will discuss this type of gangs in the next section. Some gangs even develop into organized crime groups, which will also be discussed later. Still some others are incorporated into legitimate organizations. This is also worthy of our attention. Take Mr. *Lu* from *Linjiang* County for example.

Though Mr. *Lu* and his fellows were once also involved in behaviors that disrupt public order, such as petty thefts and brawls, their behavioral problems are by nature adolescence issues. Because of his reputation in his gang, he was later elected as secretary of village Youth League branch, or incorporated into the government system. This fundamentally changed his path of personal development. After being elected, he relied on his influence in regional gangs for encouraging the young people to participate in legitimate and positive cultural activities, such as playing chess and performance rehearsals. His influence brought positive results—one of the "villains" that people dislike asked to participate in Youth League activities. Such positive results made *Lu*'s village the epitome of youth work in the county. *Lu* attributes the young's deviant behaviors to adolescents' propensity to make mistakes, and a lack of positive guidance from the state or the commune. He still holds the same opinion today. Therefore, good guidance could have kept the social order under control. He also believes that the youth today still lack proper guidance, or that the Youth Leagues are loosely organized. He has many ideas for reforming the young people's life today, but he seems much less passionate in carrying them out.

3.3 Heroism: The Formation of Relationship Gangs

3.3.1 Heroism in the Rural Underworld

Now that twenty years have passed, rural gangs of the 1980s seem very distant to us. Looking back, many researchers, me included, all find the incidents very confusing. In fact, "the gangland on-lookers" (people who have witnessed the changing gangland but are not directly involved in it) also find what has happened hard to understand. Even the main gang members also describe that time period as absurd and incomprehensible. When compared with the present, that decade from the not too distant past becomes even more mysterious.

There is nothing special about *Zhang Chuanrong*, an ordinary farmer now. When I saw him, I could not believe he used to be one of the most notorious "naughty boys" at *Shaqiao* Village of *Chujiang* City. His name was known to all and feared by all in the vicinity of his village. Many people still vividly remember his "reputation". But now he is very well-behaved. I intended to interview him about his adolescent years as a "naughty boy". However, out of my expectation, the interview could not possibly go as planned. I could not engage him in a conversation, because he was so indignant at the "naughty boy" years and how things are today. The reasons behind his indignation also deserve further investigation. I did not see it coming because that was contrary to what I came to know about him before the interview. His wife comes from the neighboring *Zengji* Township. In the 1980s when people got married, a bride token was required, but he did not give it by virtue of his "reputation". He threatened to tear down his parents-in-law-to-be's house if they refused to marry their daughter to him. They were intimidated into consenting to the marriage. In the later crackdown campaign, he did not get punished or put into jail, whereas some other equally notorious "naughty boys" in neighboring villages got sentenced to jail time. The last time he came into contact with the police was in 1992, when he beat up a villager and seriously injured the man, but again he got away with that. Despite all that, he still harbors resentment and dissatisfaction.

His father divorced his mother a long time ago, and had to raise six children alone. Life was very hard for them. In 1978 after graduating from high school, he went to buy watches at *Guangzhou* and sell them once they came back. He went into the watch business because another fellow villager called *Yang Guang* who used to sell turtles at *Guangzhou* happened to know something about the watch business. *Zhang* went to *Guangzhou* with *Yang*, bringing with them some silver Chinese coins he'd collected, traded the coins for some watches from dealers and brought the watches back for sale. Back then there were few watch sellers. They could sell one for 60 *Yuan*. In his words, the "bad things" they did, was nothing but hanging out with his peers, taking up cudgels for those being bullied, or making a pass at a girl. He never did anything evil, not to mention having any abhorrent behaviors. He said:

> We came to like Kong fu when we saw it on the films. We learned some moves from a Kong fu teacher who came to our village. We were all looking to make a name for ourselves. If we heard about someone being a bad ass, we would have to fight and defeat him. I was never afraid of bad ass. I've beaten up so many people that I don't remember how many times I got caught in police stations or detention centers. I also had to compensate others out of my own pocket. In 1983, I came back from Guangzhou, I stabbed a guy who was trying to take advantage of a girl in our village team. I had to give him 300 Yuan for stabbing him. That was quite a lot, for a whole pig only cost something around 60.
>
> Whatever I did, I was trying to behave myself. I did not want to be "crowned with a tall paper hat" (gaomaozi) and paraded through town. The government prefers such public humiliation than giving a fine. In our days "naughty boys" like us could not even find a wife. But now, if one had done time, that's even better than serving in the army. Now a bad ass can make money.

In 1982 when *Zhang* got married, over a hundred "naughty boys" from all the neighboring villages and even neighboring townships came to congratulate him. That says something about his reputation and the expansion of his network back then. He could have harnessed his network to make a big profit if he still had it now. From here we may come to understand that his indignation may have been caused by the stark contrast between his time and how things are for the gangs now. Even though he had had a big network, it had not brought him any financial gain. He could not have possibly done that because the gangland of the 1980s has been a world of "heroes". Unlike his father's generation, the young people of the rural gangland have not been indoctrinated by the Commune system with collectivist values and idealism on the future of the nation. Instead, they believe in a special type of heroism unique to their time, and that's probably why the rural gangland values reputation more than anything else. Therefore, their heroic acts seem incomprehensible to the people today. Even "former rural gangs" regret having done something that seems to be quite "silly" from today's point of view.

In the rural gangland characterized by heroism, reputation matters more than anything else. How to build up reputation is closely related to the social trends and youth subculture. Trendy look is the precondition to reputation and fighting the main means to its establishment. Besides, loyalty and camaraderie are also indispensable for building up reputation. These are mainstream values in the youth subculture. Such a climate has served to spread romantic stories of "beauty falling in love with hero". During the 1983 crackdown campaign, one of the "naughty boy" who has seriously injured others made an escape; several girls from neighboring villages wanted to run away with him and marry him.

Back then, to establish reputation, one has to dress up in a style that fit in with the gangland ideals. Flower patterned shirts and bell-bottomed or skinny pants are ideal for the youth in the gang; sun glasses, army caps and sunhats are the accessories they'd love to possess. They have a special way of wearing clothes and accessories. This is especially true with wearing an army cap. Instead of being worn uptight, it should be tilted to form a crooked style. Not all young men have a trendy look, but they all want to look in this way. Some cannot afford to look this way; others fear being judged by other rural residents. *Zhang* told me that despite his reputation, he was being very careful and refrained from bold dressing when at village. Though he liked bell-bottomed pants, he often carried them to the road and did not put them on until he could not be seen by other rural residents. Mr. *Lu* from *Linjiang* County stated:

> People in the 1980s started to be more conscious about living in the moment, not in terms of having very good food, but more of dressing up to look good. The same consciousness spread to ordinary rural residents too. When they had money, they started to build houses, and then cared to dress up nicely. The young people went for style and swagger that came with stylish clothes, such as bell-bottomed pants, skinny pants, sunglasses, sunhats and army caps. All the young people would love to own these items, but only the rich ones could do that. In the eyes of ordinary rural residents, these items spoke of indecency. The elder generation opposed to this way of dressing, but the young would not listen to them. Once the household responsibility system has been established, the elder generation could no longer tell the young what to do. The Commune system used to regulate people's behaviors. When the system disintegrated, rules and regulations no longer applied. Looking back, I could say that the same was true to the period after the war of liberation. But back then, because the young were required to practice filial piety (for example to promote the public name of their family), it was unlikely for them to pick up this habit of dressing nicely. Besides, when social

order was not restored yet, there was no social foundation for one to give attention to one's looks.

In the 1980s, when people do not have to worry about food and clothing, and are enjoying a period of social stability, they start to express their desires that have once been suppressed during the Commune era. With a good social foundation for expressing their desires, building new houses and wearing new clothes becomes trendy. But the younger and the elder generations have different attitudes towards these new trends. The old regard the dressing up behavior as indecent, while the young all look up to the "deviant" dressing style. To build up a name, some of them want to grab public attention by trying out new styles, or to be chic. Chic allows them to gain approval, admiration and compliment from the young, and then help them building up a name, but is rejected and oppressed by mainstream culture. In other words, the youth subculture is being rejected. That being said, the elder generation are losing grip over the young. The young people start to have their own minds and standards, and the "naughty boys" are the predecessors and pioneers in applying those standards. They dress up to look different from ordinary young people. Having good looks is the precondition to establishing reputation.

A more noticeable way to express themselves is to build up a tough guy image. They may try to do that by being involved in many things: "I will steal what you dare not steal, I will rob what you dare not rob, and I will fight who you dare not fight." Thus, the troublemakers gain reputation and social status among their peers. Many former gang members told me that many of them were not bad guys; they all had their mind set on joining the People's Liberation Army or did not have any intention to become a bad guy. They said they were just eager to express themselves and in the self-expression process, they involuntarily acquired some bad guy qualities. The "naughty boys" are indeed very different compared with today's organized crime groups. They are nothing more than some local bullies. Their actions, though having led to disruption of rural order, are mostly unorganized and unplanned. They tend to take chances with whatever comes in their way. They form gangs when embroiled in fights. What the villagers detest is they often cause damage to public property and order, but what they do is nothing but goofing around. Their efforts to build up the tough guy image may seem incomprehensible today, but are very commonplace back then. The following are two typical occurrences of the time.

In the evening of June 18th, Zhang Mingcan and 6 other people from XXX Village went to Yanjiang Village to drop in at someone. As they walked past a field owned by the second village group, a young man Yu Shengtian who was trying to catch frogs there accidentally flashed his torch at them. Without asking why he had done so, they immediately pressed Yu against the field and beat him up. As Yu called out for help, 6 other young men from Yu's village group heard his calling and arrived in time to witness their fellow countryman being beaten up, so they got into the fight. The free fight between the two sides injured 5 of them.

In the evening of June 20th, Chen Hongxin and 6 other people of the Sixth village group of Taishan Village XXX Township were hanging out in the county. On their way they ran into a young man named Liu Daorong. Liu nicely greeted them, "where are you from?" Chen replied, "Why the hell do you want to know?" The group charged at him and bashed him up.[5]

Sometimes the young engage in a fight not in pursuit of financial interests, but simply out of boredom and impulsion. If they get together and one of them dares the others to beat up someone, they would all be infected with the impulse to engage in a fight. No one wants to be left behind. Apart from fighting, this belligerent attitude may also be seen in many other aspects. Records of a few young people's uncalled-for aggressive behaviors are kept by the following excerpt from a report of a school in *Chujiang* City to the township government.

> Four shifty-eyed young men with straw hats lopsided, shirts draped on their shoulders, sticks in their hands, stormed into the school and swaggered past classrooms and faculty dormitory. Mr. Tang tried to stop them, but they did not listen, and even barged in on a preschool class. Seeing them lying down on the desks, the students were all scared away. Later they used the sticks to beat at the desks, chairs, doors, windows and railings. The bell went off and teachers bringing bowls and chopsticks were walking into the cafeteria and getting ready for a meal when the young men rushed in. One of them who looked like their chief ordered, "We eat first. You wait up!" And then they tried to grab the teachers' bowls and chopsticks and were going to eat before the teachers did. They did not stop until Mr. Zhou promised to cook for them later. The teachers were worried about them injuring the students and had to ring the bell earlier to start classes sooner. They mistook the bell for a sign of the school teachers not cooking for them, so they stormed down the staircase and were going to leave. When they got to the playground, they pointed their fingers at the teachers and told them spitefully to "watch themselves!"

Judging from what happened, we may conclude that the young people's constant fights have indeed disrupted the school order. This is also proven by a document jointly issued by the *Hubei* Provincial Department of Public Security and Department of Education. The document states: "Unauthorized personnel are not allowed to enter schools. There should be no excuses for unreasonable behaviors such as affray, binge drinking, gambling, picking quarrels, provoking troubles, insulting or assaulting faculty, staff or students, molesting female faculty, staff or students, robbery, and disrupting public order.".[6] Field interviews reveal a high frequency of such incidents at that time.

Public security organs tend to record serious events that have caused grave consequences and leave out those "tough guy contests" that have not led to major consequences. My interviews reveal that their bellicosity is more commonly seen in behaviors of forcing others to change shoes or belts with them, or grabbing others' army hats. Such behaviors are very common in all of my fieldwork destinations of the Double-lake Plains. It's trendy to wear these things, and when they see someone wearing the things they don't have or in a better condition, they would grab the things from others or force a trade. If they get rejected—mostly by disobedient

---

[5] Public Security Bureau of *Linjiang* County: *The Re-occurrence of Hooliganism, Brawl and Plunder Deserves Attention and Intensified Efforts by Local Authorities*, published on the 47th issue of "Reports of Current Events", July 7 1986.

[6] *Hubei* Provincial Department of Public Security and Department of Education: *Announcement on Maintaining Public Order at Schools*, issued on April 25 1983.

young men—they would engage in a fight. If the trade is a success—mostly because the men they are dealing with would like to avoid confrontation out of timidity—their reputation in the gangland would certainly grow. The "tough-guy contests" are a vicious circle—their brutality and fierceness are reinforced by the competitive atmosphere. Once some of them successfully make a name, they gradually develop their own "influence". They are feared not only by the other young men in the gang, but also by other rural residents and even village officials. The following information is a typical example of such influence.

> In October 1982, when a wedding banquet was held for Yao Shougui's younger brother, five people led by Wang Weiwu [known under the nickname "Wu Ye" (literally a highly respected sir who knows martial arts)], broke into Yao's to pick a quarrel. Wang took off the sunglasses worn by a male young guest and said, "What a show off! Give me your sunglasses!" The young man said, "There is no need for that. Come on! After all, we are all young people, aren't we?" No sooner had he finished his words than Wang and his gang beat up the guy. Yao's mother kneeled down and begged for mercy. She said, "Wu Ye, please stop beating. It's not easy for us to hold this banquet." Wang and his gang did not leave until they have smashed windows and torn down some roof tiles.[7]

3.3.2 Relationship Gangs in the Rural Underworld

In the process of competing to look better and be a tough guy, the young people gradually expand their socializing network to beyond the village level. Some who have made a name for themselves extend their influence to the township level or even across several townships. At *Shaqiao* Village and its neighboring villages, different areas are dominated by a few toughest guys nicknamed "Lord of the East Side" (*dongbatian*), "Lord of the West Side" (*xibatian*) or "Lord of the North Side" (*beibatian*). At *Shaqiao* Village, one of the rural residents told me about an incident about *Zhang Chuanrong*. His brother-in-law's army hat got snatched by a "naughty boy" from a neighboring village. When the boy's companion told the boy about his brother-in-law's relationship with *Zhang*, the boy gave the hat back immediately. *Zhang* still felt uncomfortable when he came to know that, so he went straight to give the naughty boy a good lesson. This incident may provide us with evidence for *Zhang Chuanrong*'s growing influence. As the notorious "naughty boys" extend their influence to out of their own villages, their social network expands in the same way. In the process of building up their reputation, they gradually come into touch with other "naughty boys" from neighboring villages or townships, and make friends with them. A new type of gang emerges, which the author refers to as "relationship gangs".

The appearance of relationship gangs in the 1980s marks a main threshold in the development of rural gangland. Relationship gang functions in a different way compared with regional gangs. It is worth noting that the two types have coexisted.

---

[7] Office under County-level Committee of the Central Political and Legal Affairs Commission of CPC County Committee of *Linjiang* County: *XXX Commune Uncovered Six Gangs Involved in Robbery, Street Fights and Theft*, published on the 15th issue of "Political and Legal Affairs Brief", Sept. 1, 1983.

Every "naughty boy" has started their gang life in a regional gang. As he makes more contact with the outside world, he joins or establishes a relationship gang. But that does not mean he has quit the regional gang, or even though he does quit, the regional gang still continues to exist. Besides, gangs in different villages of the same county may have started and developed to various stages. Therefore, regional and relationship gangs coexist in the same county or township.

There are two types of relationship gangs: temporary or permanent. Technically temporary relationship gangs may not be regarded as an actual gang. But when governments or public security departments compile statistics for crime and public security issues, they regard a crime involving several suspects simply as a crime committed by a gang. Truth is these suspects may or may not have steady long-term relationships. Those gangs that are temporarily formed are referred to as "temporary relationship gangs", which are not uncommon in the 1980s. However, because the public security departments do not distinguish different types of gangs or have any statistics thereof, it is impossible for us to precisely estimate its prevalence. Of course, temporary relationship gangs are capable of whatever the permanent ones could do. Temporary gangs may also develop into permanent ones, for occasional deviant behaviors may serve to strengthen relationships and turn temporary ones to permanent long-standing ones.

The relationship gangs I discuss in this book are mainly permanent relationship gangs. They are formed by the young men looking to present a tough guy image and build up their reputation. They have maintained relatively steady relationships for a long time. They become acquainted either "from an exchange of blows", or because "like knows like". Just like regional gangs, most relationship gangs in the early stage lack organizational structures or leaders, but as they get to know each other better and the disparity in influence between members grows wider, some become leaders and others ordinary members. In some relationship gangs, clear standings are given at the beginning, such as boss (*laoda*) and underboss (*laoer*). Some learn from movies to take an oath and become sworn brothers. Some tattoo the same patterns or signs on their body to show membership to the gang or club. Some give nicknames to members based on the person's quality, for example, a good swimmer may be called "Walker on Water" (*shuishangpiao*), a good fighter "Iron Hand" (*tiezhangwan*). Other names include "*Ke Daxia*" (a hero under the surname of *Ke*), "Seven Freaks of the South" (*Jiangnan qiguai*), etc. Such practices may have been influenced by martial arts films seen on television or at open-air theaters or by popular martial arts novels. They like to imitate organized crime groups and *Jianghu* (the world of martial artists) sects or schools.

Jianghu customs are pervasive in relationship gangs. From what they see in films, "Naughty boys" carried out certain practices of organized crime groups and *Jianghu* sects. They are formed with the same organizational structure; they also pick up the nicknaming habit and a strong sense of honor towards friends (*yiqi*). A sense of honor towards friends can't hurt. Nothing bad would have happened if that has been all that there is. However, a sense of honor combined with their bellicosity would invariably lead to many problems. As mentioned previously, the bellicose tendency (competing to be the tough guy) is shared by all the young people and the "naughty boys" of their time; whereas for members of relationship gangs, the *Jianghu* sense of honor has become a valuable element in their respected character, and thus the main means of reputation building. Inevitably both have become important values for regional gangs. If a member gets defeated in a fight, other members would invariably help him out. Otherwise, they would not be living up to the "tough guy" criterion or the value of honor. The following serious "beating, smashing, looting" incident may be evidence of them upholding the *Jianghu* sense of honor.

> At around 11 p.m. of June 12th, Zhang Guangfu, Zhang Guanghu, Zhang Guangbao, three brothers of an eight men group nicknamed "the Eight Powerful Flying Tigers Squad", from Wuchang Production Brigade of XXX Commune, together with 24 others from Baolian, Chenwan, Zhuqiao, Wuchang, Huaqiao and Wusan Production Brigades of the same Commune, drove a tractor borrowed from Baolian Production Brigade into a kiln owned by Yuejin Production Brigade. They held lethal weapons such as wood bats, iron bats, shovels, sabres, and started beating, smashing and looting there, leaving 12 injured, 4 seriously injured. The value of broken or looted items was estimated to be over 250 Yuan.
>
> A preliminary investigation revealed the reason behind: At 1:30 earlier the same day, Zhang Guangfu was selling ice pops at the kiln. Six people at the kiln, Tang Jisong, Tang Jiping, Chen Xiaojin, Tang Luowen, Li Zijiang, Liu Weimin, wanted to buy ice pops, but three of them (Tang Jisong, Chen Xiaojin, and Liu Weimin) ran away with the ice pops without paying. Zhang Guangfu asked Li Zijiang to also pay for his companions. They got into a quarrel, which later turned into a fight. Tang Jiping gave Zhang Guangfu three punches. Li Zijiang gave him a slap in the face. Zhang Guangfu was indignant and he promised to come back for them at night. When he went back, ...[8]

This is a major case caused by people sticking out for their gang companion who had got bullied. Members of relationship gangs would not only stick out for their friends, but also for friends' friends. Unlike the gangs today for whom only financial gain would work, they do so mainly out of a sense of honor for friendship. On one hand, "one could not possibly ask for money for everybody was poor". On the other hand, they honor friendship and heroism. The above-mentioned incident was one with serious consequences. Similar incidents were very common, except that the result might not always be as bad. Another typical example of sticking out for friends is to help friends break off an engagement. Members of relationship gangs would go to the fiancé's and demand a break-off. If they get rejected, they would smash something; if the other side refuses to return the bride token, they would threaten to take away some cows or pigs. However, gang members would never intervene in conflicts between residents of the same village. They would not offend residents of their own village to honor gang friends, because they value their relationships with acquaintances and expect to live in the village for a long time.

The "leader" in a relationship gang is often the most notorious "naughty boy". He is often better off and fashionably dressed. He is also a loyal friend, a gallant fighter and a tough guy. His followers respect his leadership mainly because he is known for having all these qualities. Once he assumes the leader's position, he must maintain his status and safeguard his reputation by doing more than others. This may include

---

[8] Public Security Sub-bureau at *Linjiang* County: *A Mob Engaged in Beating, Smashing, Looting at ABC Commune*, published on *Zhongda Qingkunag Zhaibao* (Major Incidents News Briefs), 11th issue, 15 June 1981.

the following aspects. First, he must fight harder than others. As the "expertise" for gang members, fighting is something "the leader" must be good at, or otherwise his followers would refuse to obey him. Second, he must be smart and resourceful. Reckless efforts may lead to failure and in turn cause disobedience. Third, he must be capable of protecting his followers or managing to get the face back, especially when they have been treated unfairly. When his followers get into trouble with the police, he should also try to keep them out of trouble. Forth, he must be generous to his followers. He must be ready to give them some charity or assistance. When they pay him a visit, he must treat them to nice meals. When they have no money, he must give them some. Fifth, he must be trendy. He must keep up with the times with what he eats and wears. He must try out new foods, dressing styles and new ways of having fun before anybody else does to maintain the status and reputation of being "the leader". In a word, the leader must constantly prove him worthy of the job; otherwise, the gang would not sustain itself. Either a new leader takes his place, or the gang would have to disband.

The relationship between the leader and fellows in relationship gangs of rural China in the 1980s is somewhat similar to that in street corner societies of the U.S. in the 1930s. In street corner societies described by William (1993), the leader's obligations are closely related to his status. "The leader is depended on by all the members to meet his personal obligations. He cannot fail to do so without causing confusion and endangering his position." In relationship gangs, the leader first grabs others' attention with their outstanding personality. When he becomes the leader, he must maintain his position by doing everything a bit more than others. That sends him further down the deviant road. In the process of keeping his position, the leader often moves the whole gang towards an organized crime group which further disrupts public order. Honoring friendship is no sin, nor is keeping up with fashions or trying to be the tough guys. But as the leader takes everything to the next level, the gang slides further into the abyss. They would become unruly mobs that ordinary residents detest. The gang evolves into a crime group. Few gangs may break out of this pattern.

3.4 Dark Corners in the Rural Underworld

The rural gangland of the 1980s is a product of the youth's adolescence issues. During an age of idleness, they are prompted by boredom and emptiness to form their own gangs which suit their bellicose nature and the need to establish reputation. They are characterized by a kind of distorted heroism peculiar to that age. Their heroic acts make them seem even a little lovely. But that's only one side of the story. The other side suggests they are the main challengers of public order. The harm they have inflicted upon other young people and the other well-behaved rural residents has been enormous.

A report on criminal and public security cases at *Linjiang* County shows, in 1985, 655 criminals have been arrested, with only 11 aged 50 and above, and 98.3% aged below 50. The number of adolescents under the age of 26 was 494, accounting for

75%; rural residents 411, 62.7%; unemployed idlers 97, 14.8%. In public security cases, 4132 were arrested, with 49 aged below 16, 595 aged 16 to 18, 1511 aged 19 to 25, 1602 aged 26 to 50, and 555 aged above 50. This report shows that adolescents account for 3/4 of the criminals, and rural residents account for a much larger proportion than urban residents. The proportion of juvenile delinquents in public security cases is a little smaller than that in criminal cases. But that does not mean public security is less affected by adolescents. The explanation for this may be that public security authorities have mainly targeted gambling, which mainly involves middle-aged men. The effect of gambling itself on village order is not much too significant, but an excessive focus on gambling may have statistically understated the influence of adolescents on public security. If taking this factor into account, we may conclude that adolescents, in both criminal and public security cases, have become the main roles threatening public order. Another source of information states, in the first half of 1986, 243 criminals were arrested in *Linjiang* County, with 189 adolescents accounting for 77.7%. Similar results are also offered by other sources.

Due to the transformation or relaxation of the overall social control system during the 1980s, people are exhibiting much more deviant behaviors compared with previous historical periods. This is true to people of different age groups and social classes. A former public security official divides crime and public security problems into the following categories. First, sabotage attempts or retaliation against government officials by those who resent socialism. When social control becomes relaxed, such attempts are made by traumatized victims of the revolutionary period. Second, criminal offenses or acts undermining public security by itinerant criminals. They have criminal records or past experience of undermining public security in their hometown and have fled to get away from repression. Third, thefts or robberies by locals. People are still trying to get "warm and fed" (*wenbao*), so they steal or rob to address their basic needs, which is entirely different from stealing and robbing that have occurred in later periods. Forth, crimes committed by young people when goofing around. They are still too young to realize serious consequences of their acts. Fifth, regular cases that are still regarded as criminal ones today. It may be seen from his categorization that he sympathizes with the third and fourth groups. Other than that, this way of categorization is basically the same with the way in published documents or briefings by public security authorities. As far as I can judge from my interviews and the information I have consulted, the five categories may well summarize the crime and public security issues, except for the lack of a universal categorizing standard and some overlap between categories. The second, fourth and fifth categories all involve the young people in rural gangs.

When recalling "the naughty boys" who have later become criminals, former gang members and the *lao shih* (law-abiding villagers) have expressed both sympathy and regret. However, criminal records suggest otherwise: rural residents detest gangs. Such distinctly different attitudes towards rural gangs may be explained by the fact that the same behaviors that are intolerable to rural residents back then are nothing special to rural residents today. However, one thing is for sure—in the 1980 they have inflicted serious harm on the rural residents and disrupted rural order.

First, the disruption of rural order is caused mainly by property crimes, theft being the most common one, followed by blackmail and robbery. Earlier, I explained that the young people in relationships gangs would "steal some dogs and chickens" out of boredom; they also steal and rob to live up to the "tough guy" image or even simply to save "face" or establish reputation. The gang leader also needs some finance for maintaining his position and reputation. Theft and robbery are undoubtedly the shortcut to being financed if they cannot get enough money from their families. In the process of building reputation and parading heroic acts, the gang members are eventually led into the dark corners. We could not possibly imagine the kind of terror the young people have inflicted upon the others. A country woman told us about her childhood in the 1980s: when she and her mother dropped in on their relatives, her mother would insist on going home before dark and never spend the night. That used to disappoint her a lot. Her mother had to decline their relative's offer for them to stay by explaining that their chickens and dogs would have been stolen if they had not gone home soon.

Theft and robbery are not just done by young people working on their own, but also by organized groups resorting to brutal violence. In addition to property damage, this has caused both physical harm and psychological terror to the rural residents. This may be proven by the following record.

> In the process of case investigation, ABC Commune decided to dig deep and thoroughly investigate criminal groups. By August 30, they have uncovered 6 gangs involving 33 people, 1 of them with 7 people involved in robbery, 3 with 20 people in street fights and 2 with 6 people in theft. ... The seven-man robbery gang led by Fan Xiaoyin (21 years old) has committed 4 robberies at Chujiang City, Luochang Village, and Youth Groups in the Commune organizations. On the night of August 15, the seven men in the gang took a boat to Wen Guangxin's duck shed at Chujiang's Gangkou (port) Production Brigade and tried to rob him. Wen kneeled and said he had no money on him. The gang belted him and took away a tape recorder, a pair of liberation shoes, two straw hats, six packs of Changde cigarettes, and smashed his pans, bowls, stove, and thermo bottle.[9]

When being robbed by the young, the old people are often terrorized into kneeling down, being already overwhelmed by the rising rates of public disorder and violent crime. Such theft and robbery cases may no longer be simply explained by the young men's boredom. Instead, the social order is at stake. Apart from robbery, the rural residents are also terrorized by a special way of blackmailing, then referred to as "writing letters to demand money" (*diuzi hankuan*), as is shown in the following record.

> On the night of June 25, at the Fifth Village Group of Shima Village of XXX Township, a letter was found pinned down by a soil brick on the threshold of Pi Jiasong's home. The letter was addressed to Pi, "We are tight on money. Deliver 500 Yuan in cash to Sun Guangzu's toilet at the Fourth Village Group of Shima Village by midnight three days from today. You must come alone and leave right after you drop the money. If you fail to do it, we will kill

[9] Office under County-level Committee of the Central Political and Legal Affairs Commission of CPC County Committee of *Linjiang* County: *ABC Commune Uncovered Six Gangs Involved in Robbery, Street Fights and Theft*, published on the 15th issue of "Political and Legal Affairs Brief", Sept. 1, 1983.

> your only son and set your house on fire. You know what they say: better an open robbery than a backstabbing. Watch it! Don't do things you will regret later." On May 15, a similar blackmail letter was given to Yan Liangquan, from the Fourth Village Group of the same village. The letter asked him to deliver 300 Yuan to under the west side of the west levee that crosses the Jinjiahu Lake.[10]

Records kept by public security authorities show that "writing letters to demand money" blackmail is quite common towards the end of the 1980s. Investigation reveals that blackmail is mostly done by relationship gangs who have been brought in through their connection with the local "naughty boys". Terror is exacerbated by the sneakiness of the threat. In addition, it poses threat to both property and personal safety.

Second, the disruption of rural order is caused by sexual harassment and assault, which are both very common throughout the 1980s. Sexual harassment and assault cases may be found in almost each record in classified information held by the government agencies, for example "Public Security Briefs". People's highly universal lifestyle during the Commune era has almost numbed their sensitivity for sexual repression. They have not become aware of their urges that have been repressed until they are granted more freedom by reform and opening-up. Their sexual repression may be reflected by the following information extracted from a case file at a police station of *Linhu* City.

> When I drove my cattle close to Zhang Chang'e, she was reading a novel I gave her. I went up to her and said, "Giv'me the book. I'm heading back." She said she wanted to read some more, so I took the chance to get closer to her. I held her left arm and body from behind her and touched her tits with my left hand. She said, "Don't do that." I said, "I'm still single. But I have my needs. Why don't you help me out (meaning have sexual intercourse)?" She tried to break away from me, but I held her by the waist with one hand, and touched her down there with the other. I tried to pin her down on the ground. The more she struggled, the hard I held her. I pressed my lower body against her butt and thrust my hip forward (like during sex). I didn't let her go until I've come.

The man confessed that he was aware that his behavior was illegal and that his recklessness was driven by his sex impulse. The high rates of sexual harassment and rape may be explained by the lack of normal means for young people to express their sexuality. When they are granted freedom, the young people are under the illusion that they may get away with all sorts of illegal and criminal activities. In fact, as my field research reveals, they do have a big chance of getting away (Chen 2005). When the young people gather in their gangs, they seem to feel their sexual urges more strongly Being a member of the gangs also emboldens them. As a result, more of the young people become perpetrators of sexual harassment and assault. Their boldness may be proven by the following incident.

> On the night of June 25, Liu Honggao, Chen Jinglong, Chen Gonghu from the Third Village Group of Jiati Village ABC Township, and Zhou Renjun from the Second Village Group of Guantai Village saw a film at Guantai Village. After that, they stalked five young women,

[10] Public Security Bureau of *Linjiang* County: *Two Blackmail Cases Solved*, published on the 10th issue of "Selected Reports on Major Events", June 30, 1989.

> including Xiao Hongmei from the Third Village Group of Guantai Village. When they came to about 500 meters away, they charged at Xiao, grabbed her, and pressed her on the ground, unzipped her pants, and covered her mouth. Some touched her breasts, and some rubbed her lower body. They were just about to rape her when they got dispersed by an incoming crowd.[11]

Once violence is used in sexual assault, the result would be very serious. Needless to say, violence is inflicted upon the victims. Violence is also used on rural residents who try to stop them, which in turn reinforces the people's impression on compromised public security and order. This may be proven by the following record.

> On the night of August 12, 1983, Hou Yijie and 18 friends of his gathered on the road next to Shuanghe Production Brigade. When Zhou Jinzhi and five other women passed by, they surrounded the women and attempted to fondle, kiss, and touch the women all over the body. The women were scared and screamed for help. Fan Qingming from the Sixth Production Team of Shuanghe Village saw what they were doing and tried to stop them. He called out to them, "Stop it! That's what the Kuomingtang[12] would do!" Hou snarled at him, "If you say we are the Kuomingtang, so be it! Better watch your mouth!" They continued with making sexual advances at the women. The women were desperately trying to get away when Zheng Qixiang and some other young men from the First Production Team passed by and said, "What the hell are you doing to these innocent women?" Hou and his friends charged at Zheng, pin him down and beat him up. They didn't stop until Zheng was badly bruised and had big lumps in his head.[13]

Third, the disruption of rural order is caused, most typically and seriously, by violent crimes. The young people are in a constant contest to build up a tough guy image and rely on such an image to build up reputation. Therefore, violence is not only pervasive in the rural gangland; it also affects the *lao shih*. They are often under attack for no obvious reason and even more so at open-air theatres, where the young gangsters are emboldened by the large crowd and the dark environment. It is very hard for people to see faces. It is also very hard for victims to state clearly what has happened, or for the police to collect evidence. The following record may show you the high rate of uncalled-for violence at open-air theatres.

> From late June to Early July, the high rate of hooliganism in this township has caused the general public to panic. For just seven days, from June 22 to July 8, a total of 28 brawls took place at open-air theatres. 30 people were injured, 7 of them in serious condition. Some hooligan leaders encouraged their fellows publicly by saying, "Fights at open-air theatres require quick response and quick withdrawal."...

---

[11] Public Security Bureau of *Linjiang* County: *Attention to and Tightened Measures on Reoccurrences of Sexual Harassment, Brawl and Looting*, published on the 47th issue of "Reports of Current Events", July 7 1986.

[12] Translator's note: *Kuomingtang*, or the Chinese Nationalist Party, is a major political party in the Republic of China. Corruption was rampant under the *Kuomintang* regime, and contributed to its fall in 1949. The party retreated from the mainland to Taiwan on 7 December 1949, following its defeat in the Chinese Civil War.

[13] Office under County-level Committee of the Central Political and Legal Affairs Commission of CPC County Committee of *Linjiang* County: *ABC Police Station Uncovered Three Gangs Involved in Theft and an Analysis of their Behaviors*, published on the 34th issue of "Political and Legal Affairs Brief", Sept 26 1983.

> On the night of July 8, at Taiping (literally peace) Village's theatre, a hooligan gang led by Shen Xianbing with over 20 people and another gang of hooligans engaged in 5 consecutive bouts of violence, referred to by the rural residents as "3D films"—violence on and off stage (martial arts films were on). When the film was over, they chased after passers-by and people who were swimming in the river. People joked about what happened by saying "there is nothing peaceful about Peace Village."[14]

To sum up, the gangland of the 1980s has two opposing sides. On one hand, gangs would honor friendship regardless of their personal interests. This may be put down to regionalism deeply engraved in their mind, and their constant need to flaunt bravery and to seek recognition with heroic acts. On the other hand, they are also on a slippery slope into the abyss of bringing endless harm on the rural residents. Such harm is caused by crimes against people's properties, sexual assault, and uncalled-for violence on anyone that comes in their way. Rural residents are suffering from an undermined sense of security almost to an unbearable point.

3.5 State Intervention and the Decline of Gangs

Measures have been taken to deal with the disruption of rural order during the 1980s. Initially the government attempts to restore public order with a combination of "legal education classes" and "help and education programs", but the result has been disappointing. As rural order continues to deteriorate, in 1983 the central government launches a "crackdown campaign against illegal and criminal activities", which eventually led to the decline of rural gangs.

Before launching the "crackdown" campaign, the government holds two forces accountable for the disruption of rural order. One is counter-revolutionaries and those anti-socialists who attempt to launch a vindictive counterattack (*fangong daosuan*); the other is juvenile delinquents. This is actually a continuation of the same revolutionary standpoint, which puts down juvenile delinquency to their incorrect worldviews that have been adversely influenced by the decadent capitalism. For the former, the government carries on its strict crackdown as it has done so since the founding of the nation. For the latter, the government takes different measures against different groups. For the criminals, the state will undoubtedly give them severe punishment, but for those whose behaviors fall short of crimes but still pose major threats to social order, the state chooses to use moral education. Before the launch of the crackdown, treating juvenile delinquency is one of the government priorities in its efforts to restore rural order. The main means is to give them "legal education class", reforming them through moral and legal education and redirecting them to "the right path".

> Legal education classes allow minors to learn about the legal system by studying "the Criminal Law" and "Regulations of the People's Republic of China on Administrative Penalties for

[14] Public Security Bureau of *Linjiang* County: *Huge Success in ABC Township's Special Campaign Against Brawls by Hooligan Gangs*, published on the 59th issue of "Reports of Current Events", Oct 13 1988.

Public Security". Under the general guidelines of "making clear reports, setting moral standards and enhancing legal awareness", these classes encourage minors to gradually enhance their moral and legal awareness by: voluntarily reporting their own issues, reporting companions, making restitution, making a clear break with their past, and turning over a new leaf. Legal education classes do not limit their freedom or torture them by binding and lifting them for beating. The classes work on their beliefs and worldviews by teaching them about policies and laws and teaching them to build a career and plan for the future.

Li Liangkun, juvenile delinquent from the Seventh Production Team of Tongxin Village, together with his companions, carried out 3 thefts from January to May this year. When he was told to take legal education classes, he ran away out of fear; even his parents kept him from taking the class. Officials of the Party branch went to his home to reassure his family by informing them of the Party's policies, legal knowledge and clarifying the purposes and methods of these classes. Once his parents were convinced that their son would not get beaten up there, they went to get him back and send him to the classes. After a few days of study there, he clearly reported his own issues, realized his mistakes, and even reported his companions. He turned a new leaf by trimming his curly hair, changing his bell-bottomed pants, and wrote a letter providing assurance that the same mistakes would not be repeated.[15]

For those juvenile delinquents who have broken some rules but not involved in criminal activities, public security authorities categorize them into "conspicuous (serious) offenders" and "regular (minor) offenders". Serious offenders are re educated by township legal education class about the legal system; they are required to report their own issues and show a willingness to reform themselves; after some time, they are educated by support teams back at their village groups. Minor offenders are also educated by legal education classes, but after that they are mainly educated by the general public, and thus no need for assigning support teams for them. The combination of "legal education class" and "support team" has produced good practical results, which received government attention. It is later widely advertised by the government, as shown by the following text.

Intensified efforts for the past several years to help and educate juvenile delinquents by the general Party branch of ABC Management Unit (the same as production brigade) at ABC Commune's have produced positive results. In the Unit there were 26 serious offenders whose unlawful behaviors have not come to the point of being arrested. Through help and education programs, all of them have been redirected to the right path: three took up roles of management in the production brigade, one joined the Communist Youth League, and the rest of them stopped offending. Because of such good results in the efforts to help and educate juvenile delinquents, crime rates in the Unit declined by 50% on a year on year basis, resulting in a clear improvement in public security.

Due to a lack of attention on juvenile delinquency from authorities at ABC Production Brigade, minors developed habits of brawling and stealing, disrupting public order, normal production and rural living. For the recent two years, with increased attention on the issue, the Party branch organized 16 successive groups of juvenile offenders to study the socialist legal system and offered them patient guidance and education. Three brawlers corrected their

[15] Public Security Bureau of *Linjiang* County: *Improve Public Security, Stability and Solidarity*, published on the 42th issue of "Reports of Current Events", August 25 1982.

mistakes, and 13 stealing minors returned stolen items and stopped stealing. Public security greatly improved.[16]

The same record contains information for *Tang Xiao'en*, from the Third Production Brigade of *Xinfa* Village, a 21-year-old who started to engage in brawls with his "brothers" after graduating from junior high in 1977. He has engaged in over 50 fights within 3 years. Later he was admitted to "legal education class". With the help and education from *Xiao Desheng*, Director of Public Security Division of the Production Brigade's Party branch, he was redirected to "the right path" and honorably joined the Communist Youth League. When I interviewed him, he told me that it was a disgrace to his family that he got admitted in legal education class and he felt like a second-class citizen. But once he got admitted, the Director would often drop in at his place and talk to him like a close friend, because the Director wanted to "get to know the changes in his state of mind, help him out when he was in trouble and encourage him when he made progress". That also put much pressure on his family by drawing much neighbor attention—there was some eavesdropping and peeking. As "an official", the Director was a very nice person. The Director gave him some legal papers and magazines and even found him a job at the kiln of the local production brigade. With all these ongoing efforts, he "would have been totally embarrassed if he had not taken the righteous path".

Perhaps "legal education class" or "help and educate program" works not by directly enhancing the juvenile delinquents' moral awareness, but by applying pressure on them and their families, in the way *Tang Xiao'en* has experienced. The young people who spend their day mostly in the gangs may not feel the pressure as strongly as their parents who have to live side by side with other rural residents. If both the minors and their families feel being judged, the young would face greater pressure and refrain from being willful. Therefore, the success of "legal education class" and "help and educate program" largely relies upon the fact that the rural society of China is still the acquaintance society in nature. The young people of the gangs are still bound by the acquaintance society, so despite their strong subculture, they may not evade facing the acquaintance society. At least they are still bound by it through their family ties. So, they may not choose to be indifferent to mainstream opinion and judgment, and therefore may not take their deviant behaviors and collective rebellion very far. Nor would they be able to alter mainstream judgment. The acquaintance society does not exert much pressure on the young people when they seem to be very assertive in the pursuit of their ambitions for their gangs, but once when they must face the real life in the village, they may not get away from such pressure.

Wu Changhua from the Second Production Brigade of Wusi Village was a persistent offender of fighting and brawling. Having stabbed someone at an open-air theatre, he became notorious among his "fellows" as "the tyrant". His family didn't know what to do about him, and nor did the production brigade. But the Party branch and the help and education group did not give up on him. After getting to know him, they came to the conclusion that he had no respect

[16] Public Security Bureau of *Linjiang* County: *Persistent and Painstaking Efforts to Redirect the Prodigals to the Right Path*, published on the 39th issue of "Reports of Current Events", August 7 1982.

for the law or discipline mainly because he had been deeply influenced by anarchism. So on one hand, they talked to him, told him about his mistakes and the causes behind, and analyzed the consequences and encouraged him to correct them and reform himself; on the other hand, they gave him an assignment to do—guarding the watermelon field. He was deeply touched by their earnest efforts and things started to take a turn for the better. Later he got acquainted with Xie Jiamei, the sister of the Party branch secretary. When Xie's parents disapproved of their relationship, he was filled with deeper remorse for his past mistakes and felt self-contempt. Being aware of his feelings, the Party branch secretary comforted him by asking him to make a clean break with his past, and in the mean time tried to get Xie's parents to come around. Xie Jiamei expressed her willingness to be his lifetime partner if he promised to reform himself. Their faith in him deeply touched him and gave him the resolution and confidence in changing his behaviors.

I do not intend to discuss the role of ideology here. Instead, let's focus on the interaction between the young people's subculture and the mainstream of rural society. We may clearly see that the young people have their own subculture: they rely on their bravery to establish reputation and romantic relationships. But they may not make their own decisions when it comes to marriage. Instead, they must gain approval from their parents. And that means accepting the elder generation's values, making a clean break with the rural gangs, being redirected to the right path, living a life that the elder generation approves and conforming to mainstream norms in rural life. In this process, the younger and elder generations would have a falling-out, which often ends up with triumph from the elder generation. Instead of caving in, the young may also choose to elope, which may allow them to save both their romantic relationships and subculture values, but eventually often lead to loss on both sides. Therefore, when the society remains the acquaintance society in nature, subculture could not possibly win over mainstream culture.

Despite continuous efforts by public security authorities to restore order, in the first few years right after reform and opening-up, the level of public security in rural China still continually declines, with increased rates of theft, robbery, "group brawling, smashing and looting", sexual insult, and rape. People feel very strongly that "bad guys do not fear the law, good people fear bad guys". Compared with the five years before 1980, crime rates after reform and opening-up (1980) increased dramatically to higher than 7 per 10,000 people and remained at high levels for years. Specific information is given in the following table:

Year	1974	1975	1976	1977	1978	1979	1980	1981	1982	1983
Crime rate	5.7	5.2	5.2	5.8	5.6	6.6	7.7	8.9	7.4	6.0

(*Unit* rate per 10,000 people. Crime rates statistics from Wang 1989: 461)

The direct result of undermined public security is the launch of the crackdown campaign in 1983. On July 19, 1983, Deng (1993: 33) pointed out that "the number of crimes, including serious ones, has increased substantially, and the people are very disturbed about it. Over the past few years, far from being checked, the tendency has grown. Why is that? Chiefly because we have hesitated to take prompt and stern action to combat criminals and have given them very light sentences. This is true of both economic crimes and violent crimes such as robbery and murder". He suggested "why not organize a relentless campaign against crime—or two or three campaigns? … In every campaign we should crack down on a large number of criminals." "Serious offenders, including, for example, murderers, robbers, members of criminal gangs, instigators of crime, habitual criminals who continue to pass on their criminal skills to others while being reformed or educated through labor, traders in human beings and proprietors of brothels, should be arrested and prosecuted without fail, reformed through labor or severely punished according to law. A number of criminals should be executed according to law, and some others should be put behind bars for a long time. We should keep cracking down on criminals, arresting them whenever they surface. Otherwise, they will have nothing to fear, and 10 or 20 years from now the problem will still not have been solved." On August 25, 1983, the Central Committee of the Communist Party of China released "*the Decision to Take Severe Measures Against Illegal and Criminal Activities*", kick-starting a three-year-long campaign.

An old official who had personally experienced the campaign told me that, "In the climate of political revolution built up by the campaign, a legal crime and a moral offense are treated indiscriminately. For example, chasing after a girl may be charged with hooliganism. Past behaviors of fooling around may be hyperbolized into criminal activity and one may end up in jail for trivial matters." Mr. *Hou* from *Linjiang* County is a case in point. He was once sentenced to labor camp for theft; after being released he became used to acting in an overbearing manner. He once ruthlessly assaulted innocent people and even police officers. On one occasion, with the presence of on-duty police officers, he threatened to "tear the ligaments and break the bones" of someone and even dared the officers to fight with him. When people tried to stop him, he threatened to beat them up and burn down their houses if they dared to intervene. But all he did was calling some names, so the police had insufficient evidence for charging him. However, the public security officials believed he had aroused silent resentment from the public, and that it would have gone against the people's will if they had done nothing about him, which would have "emboldened the bad guys and angered the good people", and would not have done any good to public order. So, they organized a "special investigation team" and very soon convicted him. Another recorded case may be more illustrative of such hyperbole. *Xu Changzhen*, nicknamed "female hooligan", got arrested and judged guilty[17] by public security authorities based on her long-term unemployment and hooligan nature—she was a bad influence as she kept hanging out with 17 male hooligans. Her case should not have been counted as legally indictable crime. Her records suggested nothing more than an issue of moral concern.

The above-mentioned incidents may be explained by the rules of game in "the crackdown". On one hand, the rules stipulate dealing a heavy blow to all the real criminals to create tension among them; on the other hand, the rules also target those not so serious "troublemakers" who have notoriously threatened social security and safety, even though they may not have committed any crime. The latter is essentially

[17] Office under County-level Committee of the Central Political and Legal Affairs Commission of CPC County Committee of *Linjiang* County: *The County Judged Cases Involving Another 453 Criminals*, published in the 39th issue of "Political and Legal Affairs Brief", Oct 15 1983.

a "punish to set an example" practice. According to an old official who went through the crackdown, besides these general rules, the crackdown further pinpoints three groups of people to be arrested and convicted: first, recent law-breakers; second, those who have broken the law after recently released from labor camps or punished in other forms; third, those most notorious and conspicuous local bullies who have angered the public. The police collected archive information on these people, verify information by comparing it with their actual behaviors, and put them into different categories. While applying the principle of "harsh and speedy punishments", they called for achieving precise hits with a "one check, four combines, three reports" method, i.e., checking present behaviors against past behaviors; combining the efforts of villagers, township authorities, help and educate personnel and police officers; and reporting criminal information and criminal facts, and regular reporting.

Apparently, the "punish to set an example" practice may be explained by the extended case method. Legal anthropologists initially use the extended case method to sum up local methods for investigating and resolving disputes in the Filipino and African tribal societies. It reveals "facts" and identifies the "nature" of incidents in contexts and resolves disputes accordingly. "Facts" are considered in the socio-cultural context. To determine the nature of a fact, one has to connect the fact to the past in anticipation of the future consequences (Zhu 2004, 2007). The same method has been adopted in the "punish to set an example" practice in the crackdown campaign. In determining the nature of the cases, the methods of combining archive information and real-life information, and that of "one value, four combines, three reports" both requires analyzing the interrelation of past and present behaviors, and that of official's and people's opinions on the concerned person. Therefore, for cases of the same nature, persistent and first offenders may be subject to entirely different punishments. Similarly, for the same behaviors, people having different state of mind may end up with different results. Apparently, the extended case method is not consistent with the rule of law principle of the modern legal system. It only works in a small community and the acquaintance society. Only in the acquaintance society would people look beyond the case information, consider past behaviors, and form a consistent judgment or identical opinion on the people involved. A holistic view on case "fact" and "nature" outside the legal framework only exists in the acquaintance society. Therefore, we may infer that the social nature of the acquaintance society may have served as the foundation for the practice of "punish to set an example".

As the crackdown campaign carries on, the state often holds public trials, after which criminals are paraded in the streets. Many such incidents have been recorded by archives. Here is an example:

> To generate publicity, educate the public, deter crime, boost people's morale and suppress criminal aggressiveness, in the afternoon of August 20, our county transported in 42 vehicles 375 criminals arrested in XXX Township, XXX farm, XXX Commune and XXX Commune. Escorted by over 200 police officers and armed militia, they were transported to XXX City and paraded in front of the public. Crowd applauded as the parade was approaching.[18]

[18] Office under County-level Committee of the Central Political and Legal Affairs Commission of CPC County Committee of *Linjiang* County: *People Applauded Efforts to Eradicate Social Evils*, published on the 9th issue of "Political and Legal Affairs Brief", on August 22 1983.

Public trials and parades aim to "suppress criminal aggressiveness, boost people's morale, educate the general public on legal policies with vivid examples, and reform criminals and their families through education and persuasion":

> Shan Yongkang (convicted with hooliganism) from XXX Farm was brought to trial. After hearing the judgment, his father Shan Qingdao deeply regretted having spoiled his son, which may have led him astray. He said, "The law will punish all criminals, including my own son! I strongly support the government's decision!"
>
> Xu Chenglong from the Sixth Production Brigade of Shagang Village said, "The last trial executed a group of criminals; they got a few more this time. Neatly done! What a great relief! Now we all feel safe and may focus on tilling the land." Government officials commented: the crackdown allows us to stand up straight (a big confidence boost). Keeping up the efforts will motivate us even more.[19]

Though people's comments may have been embellished for a dramatic effect, they are actually backed up by my interviews and thus are trustworthy. A rural resident told me that all the 3 people were executed by shooting during the 1983 crackdown because all of them had been convicted of murder. He said, immediate execution and parading would enhance public security, reveal "karma" to the people, and educate "the naughty boys". He even believed that the present day still needed this manner of "doing visible justice. One of the government officials said keeping up coercive measures may enhance social security. Apart from gaining positive comments from the general public (please refer to footnote 19), the campaign has also remarkably lowered crime rate.

Year	1979	1980	1981	1982	1983	1984	1985	1986	1987	1988
Crime rate	6.6	7.7	8.9	7.4	6.0	5.0	5.2	5.2	5.4	7.7

(*Unit* rate per 10,000 people. Crime rates statistics from *The Law Yearbook of China*)

Our interviews reveal that the three-year-long "crackdown" has led to the fall of the rural underworld: gangs of different types seemed to have vanished. Most of the active members or the leading roles are put in jail. A few that have got away have given up their old business and quit their gangs. Most of the ordinary members are deterred and educated by "the crackdown" and gradually stop gang activities. After 2 or 3 years, when their adolescence ends, they start a law-abiding life as ordinary villagers. That's perhaps what people say about "quit the gang when coming of age". The lucky ones get a job inside the system (*tizhinei*: government organs, public institutes or state-owned enterprises). They would often work very hard because they want to seize the opportunity to climb up the social ladder, as it has not come by easily. When the gang leaders get released from prison, everything has changed. The right climate for rural underworld to prosper does not exist anymore. The old brothers in the gangs have long begun a settled life. The rural society has developed a new mode of association: people have customs, life styles and ideals that are entirely

[19] Office under County-level Committee of the Central Political and Legal Affairs Commission of CPC County Committee of *Linjiang* County: *Another 453 Criminals Got Convicted in our County*, published in the 39th issue of "Political and Legal Affairs Brief", on Oct 15 1983.

different from before. Looking back nostalgically to their past youthful days, they know they have to take everything as it is. They have no other choice but to try to make a living in the new environment, tilling the land or doing some business, but no matter what they do, they will have to do hard work and engage in lawful activities. For a very limited number of them who still plan to carry on the old business, they could do nothing more than bossing people around or petty theft. They may never be accepted as an orthodox group or become the mainstream of the rural society as they once have been.

3.6 The Rural Underworld Embedded in the Acquaintance Society

The rural underworld of the 1980s, either rising or falling, has always been bound by social norms in the acquaintance society. "Naughty boys" in both regional gangs and relationship gangs build their lives based on the basic rules or pattern of life of the acquaintance society. The state also relies on "local resources" in the acquaintance society (Su 1996) for managing rural gangs. The "legal education class" and "help and educate program" in the early 1980s force the "naughty boys" to quit the gangs by placing them under pressure of public opinion. The functioning of the later "crackdown" relies on the "extended case method", which works well within the acquaintance society. Likewise, the legal punishment imposed on the "naughty boys" cannot deviate from mainstream values of the acquaintance society.

Despite harming the rural order, the "naughty boys" and the rural gangs have not changed the basic nature of rural China as the acquaintance society. Nor have they altered its nexus of micro-level power relations and reciprocity-based behavior pattern or the moral order of rural China. Everything that has happened in the rural underworld, including what the "naughty boys" or gangs have done and the country's management on the underworld, has not broken out of the acquaintance society norms. The reciprocity-based behavior pattern still plays a key role in rural China and also in the rural gangs. In some occasions it may have undergone some readjustments, but these changes are nothing radical. It may be safely concluded that the rural gangs of the 1980s are well embedded in the acquaintance society.

Chapter 4
The Renaissance of Rural Gangs and the Acquaintance Society

Having gone through the 1983 three-year "crackdown", the rural underworld collapses. However, no time period may be void of deviant behaviors and disobedient people or those who tend to disrupt established order. During times when coercive measures are taken, these behaviors are kept in check, and a "repressive" order is established (Nonet and Selznick 2001). During transitional times with "collapse of the etiquette and ruin of the music" (*libeng yuehuai*) when the old social control system is breaking up, people become emboldened to challenge the established order. The rise of deviant behaviors develops into a social trend. Just as a rural resident has described, "every generation produces its own great characters, and each exerts impact for decades". This expression adapted from a famous poem[1] is a true reflection of the rural gangland. The introduction of reform and opening-up ushered in a period of "collapsed etiquette and ruined music", with rural gangs as its prelude. The 1983 "crackdown" may have led to the fall of the rural underworld, but very soon it experiences renaissance and accomplishes its transformation. After that, rural gangsters again become the dominant roles in rural life.

The rural underworld of the 1980s seems basically the same in different counties of the Double-lake Plain, but after the 1983 "crackdown", gangs in different localities are not developing in sync. This is mainly because with the renaissance and transformation of gangs in the 1990s, they have shifted from going after "fame" to "real gain". Other reasons may be summed up as differences in local resources, transportation and economic development. That being said, their development trend is basically the same: they have gone through or are going through the same stages of development. It is worth noting that, instead of regarding each decade as a phase of development, this book looks at the 1980s and the period since the start of the 1990s as two phases, based on the changing characteristics of the rural underworld. This book does not look at the period after 2000 as a separate phase, because it is

[1] Translator's note: the original poem was written by *Zhao Yi* of *Qing* Dynasty. The original poem means "Each age bring forth new genius on this noble land, each will rule its own domain for hundreds of years to come".

B. Chen, *The Underworld of Rural China*,
https://doi.org/10.1007/978-981-19-8710-6_4

only a natural continuation of the 1990s with no significant changes in the gang characteristics.

4.1 Renaissance and Transgressions

With some "naughty boys" put in prison by the "crackdown" and some others getting married and quitting the gang, the rural underworld goes through a phase of decline, but this phase does not last for long. A few years later, a new generation of young people become active and reform gangs in the 1990s. Most of the "naughty boys" in the underworld of the 1980s were born in the 1960s, while gang leaders in the 1990s were mostly born in the 1970s, with an age gap of not more than 10 years. With the new generation growing up, the rural underworld starts its renaissance. More deviant behaviors appear.

Since 1988, crime rate has started to go up again. It reached a high level in 1989 and has remained at high levels since then, with occasional insignificant drops. After the 1983 "crackdown", the state enforced a second across-the-board crackdown and several special campaigns, but none of them were as effective as the 1983 effort. This may be proven by crime rates and rates of administrative penalty as revealed by the table below.

Year	1989	1990	1991	1992	1993	1994	1995	1996
Crime rate	18.1	20.1	20.9	13.7	14.2	14.3	14.4	13.5
Rate of Administrative Penalty	17.0	17.8	21.4	25.9	29.1	28.4	28.0	28.4
Year	1997	1998	1999	2000	2001	2002	2003	2004
Crime rate	13.4	16.5						
Rate of Administrative Penalty	26.8	26.8	27.6	36.2	46.2	51.7	47.6	52.7

(*Unit* rate per 10,000 people. Crime statistics come from *Zhongdong* 2004: 2; statistics for administrative penalty come from *The Law Yearbook of China*)

Since changes are under way to social control in urban and rural areas at around the same time, and the forms of social control are generally the same, the increased crime rates and rates of administrative penalty may be indicative of social order in both urban and rural areas.

Interestingly, instead of referring to gang leaders as the "naughty boys" in the 1980s, people call them "gangsters" (*hunhun*). This change of appellation deserves our thoughtful deliberation. The appellation "naughty boys" indicates that being a member in the gang is the result of adolescent issues, implying that all the transgressions are nothing but some deviations during the growing up phase. The appellation "gangster" does not carry the same implication. As a matter of fact, in the 1990s, the gangsters do not quit when coming of age. Instead, they will stay in the gang as long as they get to keep their positions in the gang. They are not just "boys", either. Many are married, some middle-aged. This is quite a different story compared with the rural underworld of the 1980s. The difference in appellation reveals a basic difference between the two phases.

The 1983 "crackdown" has decreased crime rate, hit hard the gang leaders, and to some extent, is conducive to character building among gang members of the 1990s. The "crackdown" is a warning to the young people: reputation has no practical significance. They've come to realize that acts of flaunting their bravery will create nothing but an illusion of reputation, and have not brought the "naughty boys" any real gain. Acts of flaunting bravery often lead to very serious legal consequences as the "naughty boys" have been severely punished by the crackdown. Momentary audacity may lead to years of life spent behind bars. Drawing lessons from what their predecessors have experienced, the successors have come to understand the rules for being in a gang and become more "mature", shifting from earning "reputation" to seeking "financial gain". The end of the 1980s witnesses a reappearance of deviant behaviors, but none for earning fame.

In the early 1990s, unlawful and criminal acts are still marked by serious violence. But instead of using violence to flaunt bravery, violence is used to seek financial gain, most commonly through theft and robbery. They often steal chickens, dogs, power lines and other electric power infrastructure, and slaughter and steal cattle. The "naughty boys" of the 1980s used to steal chickens and dogs too, but only to get something nice to eat—most of the stolen goods are shared among the members of regional or relationship gangs. This is quite different from the "steal to sell" practice in the early 1990s. With stolen goods being sold, the harm of theft is taken up to the next level. Unlike the random acts of stealing in the 1980s—since they could not eat that much of food, acts of "stealing to sell" in the 1990s are much greater in terms of frequency and number of goods involved. In a single theft case, during the time period of over a month, three people ganged up 22 times, stealing over 130 chickens.[2]

In early 1990s, it is a common phenomenon for gangs of thieves to seek financial gain through "steal to sell", which even changes the livestock-breeding habits of rural residents. At *Xianning* of *Hubei* Province—the author's hometown, due to frequent acts of drugging and stealing dogs, dogs often get stolen shortly after being brought home. Starting from 1993, in townships adjacent to *Chencun* Village, rural residents gradually have stopped keeping dogs. Take the author's childhood experience of keeping dogs. We kept a dog from 1972 until it got stolen in 1987. Later, we got another dog and kept it from 1987 to 1992, when the second dog got stolen again. The following two dogs we kept both got stolen in less than a year. So, we simply stopped keeping dogs. Besides chickens and dogs, cattle for ploughing often get stolen and sold. Stealing cattle is a more difficult job, which often requires cooperation between internal and external gangsters, sometimes even between external gangs and local discontented rural residents. The following record may testify to the commonness of stealing cattle.

---

[2] Public Security Bureau of *Linjiang* County: *Chicken Thieves Arrested*, published on the 24th issue of "Brief Reports of *Linjiang* Public Security Bureau", on March 12 1991.

> After having solved 8 cattle-theft cases in mid-March, XXX Public Security Bureau solved another 12 such cases in the first half of this month, recovering all stolen cattle and financial damage of over 14,000 Yuan. … Shen Deying confessed that since 1988, he and his associate Xie Shouyin has stolen 12 cattle, which were disposed of at over 10,000 Yuan.[3]

Apart from stealing, another common way for gangs to seek financial benefits is robbery. The most typical perpetrators are "rogue robbers" at village boundaries, or "bandits" on highways. A 1992 report by *Linjiang* County states that public order in key areas along highways and (or) village boundaries have been compromised, because they are favorite haunts of highwaymen, with 153 robbery cases in less than 9 months, accounting for 59.5% of cases in the whole county during the same period.[4] In the whole county, over 9 month, the number of robbery cases added up to an alarming number of 257. There are mainly two kinds of robbery. One is to rob drivers or cargo owners on non-local trucks on the highways, with an example recorded as follows:

> At around 11 p.m. of July 21, 9 criminals (16 to 23 years old) carrying butcher's knives, nine-section whips, and clubs stopped a Dongfeng truck at 143 miles from the segment start on XXX Segment of XXX Road, beat up the driver, stabbed the cargo owner and took away 480 Yuan in cash and two watches.

Another kind is to board a coach as passengers and rob the passengers once the coach drives to a remote place. People refer to such robbery as "clean-out" (*xijie*), implying all the passengers are robbed indiscriminately and much chaos has been created. The following record may show us the unscrupulous ferocity of robbers.

> Under interrogation, Suspect Yang Chunmei confessed: at 8 o'clock in the morning, at one end of Wancheng Bridge, Yang and Hu Dingxing, Lei Zehong boarded a coach bound for Wuhan. When the coach drove to XX market town, Yang sneaked a handbag from the luggage rack, got off the coach and ran away. A few moments later, Lei, who'd been staying in the coach, gave the knife he carried in his belt a slap and shouted: "Giv'me your money! Party members first!" Nobody made a sound. So, Lei started a reckless search for money in bags on the rack. A tea cup fell on the head of a male passenger but he did not dare to make a single sound. After having collected over 700 Yuan in cash, Lei and his companions took out a package of San Wu cigarettes, handed out the cigarettes to the passengers, one for each, said "thanks for working with us" to them and left. Since April this year, the group of six, including Yang Chunhai, Hu Dingxing, Lei Zehong, Tian Chunying and "Donkey", has committed 17 robberies or thefts, involving property worth more than 70,000 Yuan.[5]

"Rogue robbery" at peripheral areas or "banditry" on main roads often also involves theft. Sometimes their strategy is to "steal whenever they can and rob when

---

[3] Public Security Bureau of *Linjiang* County: *XXX Solved 12 Cattle Theft Cases*, published on the 51th issue of "Brief Reports of *Linjiang* Public Security Bureau", on June 15 1991.

[4] Public Security Bureau of *Linjiang* County: XXX Solved a Major Robbery Case, published on the 64th issue of "Brief Reports of *Linjia*ng Public Security Bureau", on July 25 1991.

[5] Public Security Bureau of *Linjiang* County: *XXX Public Security Bureau Achieved Remarkable Results in Efforts to Combat Highwaymen*, published on the 53rd issue of "Brief Reports of *Linjiang* Public Security Bureau" on July 1 1991; *Linjiang* Security Bureau of *Linjiang* County: *XXX Public Security Bureau Made Progress in Efforts to Combat Highwaymen*, published on the 67th of "Brief Reports of *Linjiang* Public Security Bureau" on August 2 1991.

stealing is not an option"; some other times they simply decide to steal or to rob based on their mood; or some steal and others rob. The above record is a combination of robbery and theft. The same record also states that, in a special campaign to combat highwaymen, after one month of tightened efforts, one of the townships destroyed 2 highwayman groups that involved 57 criminal suspects, arrested 26 people, and uncovered over 150 cases (19 of them major or one of the biggest cases), involving stolen or robbed property worth over 200,000 Yuan. These numbers are good evidence for social chaos and the gravity of property crimes. At peripheral areas and main roads, they often choose to rob non-locals and steal from locals, because they dread being recognized by their own folks or being easily identified in a police lineup. That's something they don't have to worry about if they are dealing with non-locals. At *Linjiang* County, handbag theft along the highways is also very common in the 1990s. A report states that on the second day of the special campaign against handbag theft, plain-clothes police have succeeded in successive coach trips along two national highways; 7 criminals were caught stealing and got arrested on the spot.[6]

In addition to harming their property, frequent theft and robbery cases also hold great terror for the rural residents. A lack of sense of security over their property has led to worry over social order and even fear, especially when they have to go around. Many rural residents at *Huchang* Village, *Linjiang* County told me that to keep their money from being stolen, they would never keep cash in their pockets when shopping in town; if they travel to faraway places, to guard against searches or robberies that often happen during coach clean-outs, they would put money in secret places, such as in socks, under insoles, or in underwear pockets, and even in bras (female). A young man from *Qiaotou* Village of *Linhu* City, who was still a college student in 1996, told me that on his trip to college—his first long trip ever—his family were very worried because of the chaos caused by a large number of highwaymen, and after racking their brains thinking about where to keep the money safely, they decided to use an insulated water bottle for carrying cash on the trip.

The rural gangs of the 1990s have come up with various ways of getting money. Pretending to be police officers in a gambling bust must have been the result of "thinking out of the box".

> A gang of 11 people, including Li Xinmin, Wan Debin and Li Houping, pretended to be officers from Public Security Bureau of Linjiang County, officers from Civilian Alliance led by Public Security Bureau of Linjiang County, policemen from XX and XX Police Stations. Equipped with a pair of walkie-talkies, two pairs of handcuffs, dressed in police or armed police uniforms, they committed 20 robberies on the pretext of gambling busts at Minzu Road, Renmin Road, Zhonggulou (bell and drum towers), Yueshan Village, Yincheng Village, XX Township, XX Tree Farm, XX Town, and XX Township. The amount of gambling money they have seized and the amount of "fine" totaled almost 6000 Yuan.
>
> In the process of committing a crime, they often intentionally exposed handcuffs and daggers, and used walkie-talkies for communication. They threatened, searched, bound up, and beat

[6] Public Security Bureau of *Linjiang* County: *Plain-clothes Police Succeeded in Catching Handbag Thieves on Coach Rides Along Two Important Highways*, published on the supplement to the 35th issue of "Brief Reports of *Linjiang* Public Security Bureau", on Oct 18 1990.

up the rural residents, and confiscated their properties. At XX Longzhou Village, they gave each gambler a 70-Yuan fine, and when some of the gamblers had no money for paying the fine, they took off the three gamblers' clothes, tied two of them up with ropes, handcuffed the other one, and punished them by making them kneel down on the ground. They did not ask them to stop kneeling until the family members of the gamblers handed over a total of 210 Yuan. At Yangsi Village of XX Township, also on the pretext of gambling bust, they grabbed 1660 Yuan in cash and over 500 Yuan of gambling money from someone who had just returned from a business trip and did nothing but watching a gambling game.[7]

The reason why the gangs can get their way has something to do with the way laws are being enforced by public security organs, or to use an expression with political overtones, the reason may be that "China's legal system still has room for improvement." Unlike the present time with prevalent slogans about legal rights and procedures and legal rights and procedures well understood by the general public, people in the 1980s and the early 1990s would be intimidated at the sight of police or police uniforms. That's even more so if some unlawful activity is going on. The gambling-bust novelty was enabled by such permeating fear of the police. The chance of it happening today is next to zero.

In the 1990s, to gain illegal financial benefits, it is a common practice for the gangs to use some "trickery" on the payment bills at the grain stations[8] (*liangzhan*). At *Shaqiao* Village of *Chujiang* City, several gangsters got their "first pot of gold" using such underhanded methods. Some coerce grain station staff into forging the amount due to farmers when grain or cotton are delivered; some collude with the staff and share with them the profits from the forged amount; other fraudulent methods include directly changing the due amount. The following record is an example of such a scam.

People reported to the local police stations that Zhao Yi, being unemployed and having no lawful income, had recently purchased a motorcycle with money of an unknown source. ... The police investigated and collected evidence from the general public... During cotton harvest and delivery season, Zhao used the method of changing the due amount in five scams and obtained 4900 Yuan, which he used for purchasing a 90 cc motorcycle.[9]

Despite high crime rates in the early 1990s, most of the crimes are done against property, a completely different type of crime compared with the 1980s with brawling as the main type. A report in 1998 states that in the first half of that year *Linjiang*

---

[7] Headquarters Office of Restoring Social Order of *Linjiang* County: *We Must Make Thorough Investigations to Eradicate Evil Forces*, published on the 10th special edition on restoring social order of "Reports of Current Events", on April 24 1989.

[8] Translator's note: Grain stations were part of the grain rationing system imposed by the People's Republic of China in the 1950s to control the production of food and to boost industrialization. People were given grain coupons with which they could buy a certain amount of grain at a low, state-imposed price, which varied somewhat with age, profession and location. Any quantity exceeding the coupon quota would have to be purchased at market price. In the 1950s, communes were established in rural areas and the government imposed 'procurement planning', which required peasants to deliver a certain amount of food to the state at state-imposed prices.

[9] Public Security Bureau of *Linjiang* County: *The General Public Assisted in Catching a Con Man*, published on the 13th issue of "Brief Reports of *Linjiang* Public Security Bureau", on January 8 1991.

County filed a total of 141 criminal cases, including 61 theft cases, 38 robberies, 2 murders, 12 injuries, 2 scams, 2 rapes, and 27 others. 108 crimes against property, theft or robbery, accounted for 76.6% of the case total.[10] A comparison of crime statistics between the 1980s and 1990s will reveal considerable changes, the most radical one being that of crime types, from crimes against person to those against property. Financial gain is being highly valued in the 1990s. The 1980s gangsters look forward to becoming heroes and building a good future, though they may have failed to find the right way of doing it, but in the 1990s, gangsters care about nothing but material gain.

Economic growth in the 1990s brings more job opportunities and more means of climbing up the social ladder. This keeps people from "boredom", which in turn stops new regional gangs from forming. Instead, all the gangs are organized in pursuit of illegal gain and may be divided into two types.

One is well-organized gangs, similar to organized crime groups. They are very small in number. They are formed mainly by unsophisticated young people who are more affected by televisions and novels than having any real experience in the gangs or understanding the rules of gangland. They take everything very seriously. For example, they would develop their own gang names, elect gang leaders, tattoo same symbols on members and etc. They also aim for financial gain, but they often resort to more brutal and violent means. Take "the axe gang" from *Xianning, Hubei* Province. It is a well-organized gang of young perpetrators of murder and robbery. But as time goes by, especially when it comes to the end of the 1990s, such gangs have become much fewer in number.

The other type is loosely structured temporary gangs that lack a common emblem or symbol and aim for short-term common interests. Most of the gangsters are in an acquaintanceship network where only loose connections are maintained. They have no clear purpose in life. They may form short-lived groups with others in pursuit of some common illegal gain, but very soon, as common gain ceases to exist, their relationship ends. They may reform a group when common gain appears again. The longer common gain holds up, the longer they stay in an illegal gang. They choose financial benefits over honor towards friendship (*yiqi*). They do not look forward to becoming heroes—they may seem unimaginative in this respect. Even the formation of gangs changes with what they are after. Therefore, they do not feel the strong impulsion to form a universal gang, or to build up their reputation among the gangs, not to mention becoming a leader of the whole underworld. As far as they are concerned, if it does not bring financial benefits, reputation does not make much of a difference.

The underworld (*Jianghu*) of the early 1990s is completely immersed in the pursuit of financial benefits. There are no more gangs formed out of boredom or to kill time, but just crime groups going after real gain. The ongoing emergence of gangsters and gangs in the underworld are like bubbles surfacing on the lake (the Chinese term for the underworld "*Jianghu*" literally means rivers and lakes). Some are organized

[10] Public Security Bureau of *Linjiang* County: *Analysis of Social Order in the First Half of 1998*, published on the 25th issue of "Brief Reports of *Linjiang* Public Security Bureau", on July 14 1998.

crime groups, but larger in number are successive property crime groups that are formed and suppressed. This is a very confused age. The whole gangland bears no honor for collectivism or heroism as the gangs of the 1980s. Nor does it have the kind of structure as the gangs will do at the turn of the century. Though unorganized as loose sand, the underworld has produced many heroes. Just as "a wide sea allows fish to leap about, a vast sky allows birds to fly", the gangland of this age knows no boundaries or limits, and for its members, it is an age granting "the indulgence to go after material gain".

4.2 Opportunity and Adaptation

4.2.1 Second-Generation Gangsters in Townships

Fieldwork in the counties and cities of Double-lake Plains suggests there are four generations of gangs since reform and opening-up. The "naughty boys" born around the year of 1960 and active in the 1980s are the first generation; those born around 1970 and joining the gangs in the early 1990s are the second generation; those born around 1980 and joining the gangs at the end of the 1990s or the beginning of the century are the third generation; those born around 1990 and joining the gangs in the present day are the fourth generation. Interestingly the rural residents vividly remember the second-generation gangs, including details like their birth, personality, hobbies and notorious records. Most of the stories they told me concerns the second generation. Their generalization of the gang characteristics is mostly based on their experience with the second generation. What's so special about them? We must deal with this question from the perspective of the chances they'd encountered as they grow up. These chances allow them to stay in the gangs since the onset of rural gangs. They have never left. They are still active in rural gangs, exerting a big influence on the life of rural residents. The table shows information on the most active gangsters in *Shaqiao* Village of *Chujiang* City.

Status	Name	Education	Year of birth	Social relations and "influence"	Notorious records	Personality	Current sources of income
At village	*Wan Qing*	High school graduate	1971	3 brothers and 19 cousins, some of whom working as the section head in a bank, Party secretary of the village, or the town mayor	Stealing chickens and dogs; drinking; brawling; blackmailing	Loyal to friends	Having liquor and lobster monopoly in the village; Contracting to cultivate 20 *mu* of undeveloped land in mountains

(continued)

(continued)

Status	Name	Education	Year of birth	Social relations and "influence"	Notorious records	Personality	Current sources of income
At village	*Wan Min*	Primary school dropout	1970	4 siblings, 3 being brothers; having a cousin working at the public security bureau of the city	A brutal fighter, stealing chickens and dogs, working as a thug for hire	Ruthless, having no respect for friendship	Undertaking ground works in the township; running a transportation and a well-drilling business; contracting to manage 70 *mu* of fish pond and 15 *mu* of undeveloped land in mountains
At village	*Wan Rong*	Secondary school graduate	1972	2 brothers, the elder brother being a section chief in traffic brigade of the city; having a cousin being a gang leader in the city	Stealing chickens and dogs; brawling; theft; gambling	Cranky; fond of gambling; womanizing	Contracting to manage *Huang Zongdang* reservoir in another village; renting a crab farm
At village	*Wan Ming*	Secondary school dropout	1972	2 sisters, *Wan Ming* being the only son; having a relative being the leader of a steel company at *Guizhou*	Stealing chickens and dogs; brawling; swindling; theft	Loyal to friends; ruthless to others	Currently in jail; previously relying on stealing for squandering money
At village	*Wan Shan*	Secondary school dropout	1969	Having no special social relations to rely on	Beating up neighbors; stealing chickens and dogs; adultery	Cranky; domineering	Mainly farming; making extra gains through bullying
At town	*Wan Liu*	Secondary school graduate	1968	Having 2 brothers	Theft		Running a transportation business and a Mark six lottery outlet
At town	*Liu Lin*	Secondary school dropout	1974	Having a brother being the deputy Party secretary of the township	Stealing chickens and dogs; blackmailing; trouble making; gambling; swindling	Aggressive; a roughneck	Charging sanitation fee in the name of urban management bureau (*chengguan*); running a taxi business and a Mark six lottery outlet

(continued)

(continued)

Status	Name	Education	Year of birth	Social relations and "influence"	Notorious records	Personality	Current sources of income
At town	*Wan Jin*	Secondary school graduate	1972	Having many cousins; one of them working as director of the county Family Planning Office; three of them doing business away from home	Stealing chickens and dogs	Fond of gambling; womanizing; smart	Undertaking ground works in the township but failing to make ends meet, owing several hundred thousand of debt
At town	*Wan Sheng*	Secondary school graduate	1977	Having a sister and being the only son		Lazy	Charging sanitation fees in the name of urban management bureau (*chengguan*); barely making ends meet
At town	*Wan Xin*	Primary school dropout	1972	Living alone; having run away from home after his parents died in his early childhood and because he couldn't stand being abused by his sister-in-law		Unscrupulous	Contracting to manage 400 *mu* of cotton field and a small reservoir in the village
At town	*Xu Shan*		1970	Having 3 brothers, having moved to *Shacun* Village when his mother remarried	Blackmailing; gambling	Aggressive	Unemployed; making a living by blackmailing and gambling; rural residents bothered by his occasional visits
In city	*Wan Hui*	College graduate	1970	His elder brother working away from home	Organized blackmail	Calculating	Forming a clique of their own; running a cell phone repair shop and an automobile repair shop

(continued)

(continued)

Status	Name	Education	Year of birth	Social relations and "influence"	Notorious records	Personality	Current sources of income
In city	*Wan Shuai*	Secondary school graduate	1972	Orphan; cousin being the commander of Major Case Squad in municipal Public Security Bureau	Blackmailing; rape	Ruthless	Charging protection money from urban builders
In city	*Wan Liang*	Secondary school graduate	1976	Having received professional martial arts training			Working as a "thug for hire" at a illegal factory at *Shenzhen*

These people are second-generation gangsters. They are about the same age: 34 to 40 years old and born around the year 1970, with only a few exceptions. Some personal qualities are common to all of them. First, they are mostly local. Though having different sphere of influence, they are active mainly in local places. Those with a smaller sphere of influence are active in the village. Others extend their influence to the town, the county or city, or even developed areas. Second, they have begun their gang life early in their adolescence. When they stop going to school, most of them experience a period of "idleness", when they goof around all day having nothing important to do. Initially they make a living from stealing chickens and dogs, theft, blackmail and robbery. For such behaviors, some have received administrative penalties or sentence. Many of them are slackers. Some have developed bad habits of gambling, prostitution and drug abuse. Third, most of them come from powerful local lineages bearing powerful names. They are supported by influential lineages and often have many brothers and cousin brothers. At *Shaqiao* Village, the most populated surname is "*Wan*". People bearing this surname account for only 40% of the total population, whereas gangsters bearing this surname account for about 90% of the total number of gangsters.

The second-generation gangs are active mainly in townships from the end of 1980 to the mid and late 90s. That's when the gangland plunges into a state of chaos and "a multitude of heroes" are produced. As described by the last section, at the beginning of the 1990s, most of the gangsters mainly rely on coercion and intimidation for committing crimes against property, such as theft, robbery and scam. With varied backgrounds, temperaments, and fortune, they have ended up with contrasting futures. If they had a powerful lineage that comes with many brothers, they would have more people to ask help from when in trouble and have less difficulty in solving whatever problems they may encounter in the gang. For those who enjoy no support from a powerful lineage, their acts of coercion or intimidation are hardly effective in local places, and may have a bigger chance of being caught by public security authorities. Therefore, after being identified, some are immediately arrested and jailed, and others use their social connections to evade punishment; some squander the stolen money or property on enjoyment, and others rack their brains to accumulate wealth and turn stolen money or property into legal assets.

The reform and economic and social development at the end of the 1990s have brought all the gangsters great opportunities. First, county and township governments need gangsters for collecting taxes. After the mid-1990s, county and township governments encounter greater difficulty in governance, especially in the collection of taxes and public dues. This includes agricultural tax and township management fees. Second, "the economic growth comes first" strategy allows gangs to shift from gaining through property crime to gaining through shady (*huise*: literally gray) dealing. Since the mid to late 90s, the government has put "economic performance" on the top of their agenda, and thus has started all kinds of projects. This has provided great opportunities for those resourceful gangsters who have connections in the government. Third, the development of rural markets and allocation of national and communal resources by the market mechanism give gangsters a chance for "primitive capital accumulation". Those who have built up wealth may shift to legal means for further pursuit of gain. Forth, businesses need gangsters for "protection". For instance, entrepreneurs need gangsters to ensure their personal and property safety, or with the help of the gangs, they may seek greater profits using unethical methods.

Gangs in different places often face different opportunities. These opportunities, big or small, are often determined by the distinguishing characteristics of the local economy. Of course, not every gangster can seize them. The minority who can seize the opportunities often meet the following requirements. First, they have a good sense of propriety. Instead of being reckless, they are both courageous and prudent. Second, out of experience, luck, or with reliable connections, their criminal acts have not been exposed or they have managed to evade punishment. Third, they do not blindly seek pleasure and gradually manage to accomplish "primitive wealth accumulation". Of course, to grasp different opportunities, gangsters have to meet up with different requirements, and have to reform themselves. When it comes to seizing the chance, their competence, or the lacking of it, has resulted in stratification in the gangs.

When faced with opportunities that have arisen in late 1990s, with different coping approaches, gangsters at *Shaqiao* (at village or at town) have ended up with contrasting futures.

Among at-village gangsters, *Wan Qing* has a very powerful family with many of his family members working as government officials. At the beginning of the 1990s, he has been arrested many times but always managed to get out. In 1995, he formed a group of over 40 people and tried to create a monopoly in the wholesale and retail market of eel in the town, but failed. In a monopoly dispute, *Wan Qing* was arrested after he and his accomplices physically assaulted and caused ruptured spleen in a town market eel seller, but soon he managed to "be saved" and nobody seemed to be able to do anything about what he had done. So, all of a sudden *Wan Qing* became an "awe-inspiring household name" in the town. Since 2000 he has started a liquor factory and created a monopoly in liquor sales to small shops. During lobster seasons, he and several other gangsters in town created a monopoly in lobster sales. Besides, he also contracted to cultivate 20 mu[11] of barren hills.

When he was still young, *Wan Min* was apprenticed to a master in town and learned about becoming a professional thief. Being responsible for procurement and marketing in the town cement plant, his master had plenty of illegal gain. In 1993, being envious of his master, *Wan Min* asked a few gangsters to beat up his master and blackmailed him for 5000 Yuan. He used that money for buying a wolf dog and a shotgun and started some dodgy business. Very soon he managed to make 10,000 Yuan by falsifying receipts for inbound goods at a town grain station (*liangzhan*) and get away with that. In 1995, he cottoned up to the secretary of the management area (analogous to the township level) by using his sister-in-law's good looks and managed to get a loan of 10,000 Yuan. With that money, he built ten single-story houses along the provincial highway and started a restaurant there. He also started to come into contact with town officials and gradually made a fortune through undertaking town government projects. Currently *Wan Min* mainly lives off using his truck for transporting farm produce from hilly areas or selling local rice to hilly areas. He also bought a machine to be rented for drilling wells. Besides, he also contracted to develop 70 *mu* of fish pond and 15 *mu* of barren hills.

Wan Rong has also always managed to evade punishment after being arrested. In 1994, he started a small-scale edible oil refinery, but the refinery went bankrupt in less than two years because he was both a gambler and a womanizer. Currently he is under a contract to develop *Huangdanghu* Reservoir and also rented a crab farm. But because old habits die hard, he now owes 100,000 Yuan of high interest rate debts, which is more than what he could ever pay back. Gangsters from the city often seek him out to collect debt.

Wan Ming got arrested many times. In 1998, he was sentenced to ten years for theft, and is currently still in prison. *Wan Shan* hangs out in a smaller area—just within his village group. He is very cranky and domineering. He often beats up his neighbors or steals one thing or two from them. He also works with external gangsters when they are up to something in his village.

Among the above 5 at-village gangsters of *Shaqiao* Village, *Wan Qing* impresses the rural residents as having respect for camaraderie. He intends to run a lawful business, except that he relies upon intimidating others for primitive capital accumulation to build up a foundation for his business. *Wan Min* is known for his brutality and lack of respect for camaraderie. The future does not look bright for him as he is used to "resting on his laurels". *Wan Rong* has a strong sense of honor for friendship, but no principle seems to apply for *him.* He is both a gambler and a womanizer, and therefore, the future looks bleak for him. *Wan Ming* is a generous person when he has money, but his bellicosity may cause fatal damage—bad things are doomed to happen to him one day if he goes too far. *Wan Shan* impresses the others as being

[11] Translator's note: *Mu* is the Chinese unit of land measurement. One *mu* is commonly 806.65 square yards (0.165 acre, or 666.5 square meters).

short-sighted but also generous. As he does not get involved with serious criminal activities, he may still maintain his position in the village.

In the long term, as at-village gangsters like *Wan Min, Wan Rong and Wan Ming* waste away their vigor, their influence on the village will diminish. Some are too extravagant and hence not capable of "maintaining a respectable life"; others resort to such brutal violence that they cannot keep a friend circle, which is crucial for mutual support and personal development. Still some others, with their reckless violence, may easily break the law and therefore be hit hard by the state.

In contrast, gangsters like *Wan Qing* and *Wan Shan* have a bigger chance of "making it" or maintaining a respectable life. The reasons may be summarized as follows. First, they are mainly involved in the gray area of the law. As they avoid engagement in serious crimes, they will not "cross the line". They may constantly make "minor" mistakes, but they refrain from making "major" mistakes. Therefore, the country will not interfere with their activities, or when it does, it does not intend to destroy their influence altogether. Second, they honor camaraderie and respect friendship. So they will easily win over gangsters of the same kind, who may back them up when in need and foster further "development" in their common "cause". Third, their unethical methods are used under the cover of a legal profession. This allows them to quickly accomplish wealth accumulation, live a "respectable life", and make it very hard for the country to crack down upon them.

Among at-town gangsters, *Wan Liu* was sentenced to 5 years in prison for theft. After being released, he "picked up his old business", but did not get caught again. Rumor has it that during the reform-through-labor period, he took advice from "a master" and improved on his "skills". Having saved some money from stealing, he bought a house in town. In 2001, he bought a van and started a transportation business.

Liu Lin used to make a living by gambling. Rumor has it that he earned over 200,000 Yuan from gambling scams in 2003. Because one of his brothers works as the deputy secretary of the town Party committee, he was given a job at the town urban administration and law enforcement office. His domineering and unscrupulous temperament helped him in collecting environment sanitation and management fees. As fee collection is a seasonal job, he bought a car for running a taxi business. When underground "Mark Six" (lottery game) became prevalent in the Double-lake Plains, he grabbed the chance and made a fortune.

With an extensive network of connections provided by his family, *Wan Jin* also almost always has managed to get out after being arrested. He bought a bulldozer, and having established a cordial relationship with government officials, he would often undertake earthworks in the town and has made some money over the years. But his fondness of gambling and women cost him even more, and now he is burdened with debt of several dozen thousand Yuan. This has deprived him of any position in the gangland.

Wan Xin has a stubborn and domineering personality. He does not back down until he gets what he wants. No one dares to irritate him. Now he owns a small shop in a ranch close to the town, 400 *mu* of cotton land of the ranch and a small reservoir for growing fragrant lotus flowers. With a shrewd finance manager and a husband manager as his wife, he now leads a peaceful life.

Xu Shan is not officially employed. He lives off blackmailing others and gambling.

Most of the "at-town gangsters" from *Shaqiao* Village have a legal livelihood that allows them to make ends meet and live a "respectable" life. Most of them have got married, and their wives are often full-time housewives who dress up fashionably. Before they get married, they used to form cliques, taking care of each other in need or simply engaging in a "common cause". The formation of such cliques serves as a foundation for further expansion of their groups, which allows them to have what it takes for "handling things" by themselves. The *lao shih* would of course look at at-town gangsters as belonging to "the bad guys", and would deliberately keep a distance from them and avoid provoking them. Either to "make ends meet" or "make a fortune", they mostly stay out of the village, but they would also cause disturbances when they come back. They would come back when two things happen. First, when a public project is carried out in the village, they would come back to "get a slice of the cake" or extort money from project contractors. It is more likely for them to use extortion if the contractor is not a local one. Second, when the interests of their close relatives are at stake, or when their close relatives become embroiled in disputes with other villagers, they would interfere and help out their own folks. People are often intimidated by their coercive forces. They can also be very unscrupulous and unreasonable, for they do not back down until they get what they want.

For some of those who have indeed grasped opportunities provided by the times, their financial gain is no longer made through property crimes as in the early 1990s. They have either started lawful business operations with the help of their primitive wealth accumulation, or rely on violence or threats of violence to engage in more significant unlawful gain. But either way, they have shifted from committing crimes to engagement in shady deals for gaining. Without breaking the law, they are faced with a much lower risk. They have successfully transformed their methods of dealing. During this process, those who have failed to transform themselves would not only have to deal with suppression from the public security agencies, but will have to face the gangland after its restructuring. Undoubtedly their risk of staying in the gangs has greatly increased. At the turn of the century, having gone through its transformation, the rural gangland is no longer a place with "endless possibilities". The time for "a multitude of heroes" to prosper is over.

4.2.2 Third-Generation Gangsters that Have Entered the City

When the third-generation (born around 1980) are about to join the gangs at the end of the 1990s, the second-generation members are going through a period of restructuring. They have taken up important positions in the gangs or have held onto important resources. After almost ten years of struggling in the gangland, the second generation has become better-off veterans. Being born just a few years later, the third generation finds it very difficult to make their way in because they have missed out on the golden period for "establishing a sphere of influence by showcasing one's abilities". *Wan Sheng* from *Shaqiao* Village is a case in point. Born in 1977, he intended to join a gang in town right after graduating from junior high, but he has never done anything bad to rural residents in his village. His experience stands in stark contrast to the second-generation members. He is also much younger than other members who are still active in the gangs. People of his age from *Shaqiao* Village are either migrant workers or members of urban gangs—very few stay in the village. Therefore, when *Wan Sheng* joined the town gang, he had no peers or fellows, so he had to work as "a help" (*ma zai*, literally means foal) to second-generation members. Later he had to scrape by on collecting sanitation and management fees for *chengguan* (the town urban administration and law enforcement office). In recent years, with minimum annual income, he could not possibly make "primitive wealth accumulation". He even had to live off his parents' farming income. As he could not carry on his life as a gang member, he had to quit the gang and get married, move back to the village, and go back to being a farmer.

As the second generation holds onto profitable resources, the third-generation members have no luck. But that's not necessarily a bad thing for them—as second generation have vested interests in rural China, the third generation decides to try their luck in the city or to "fight their way out". Of course, another reason for them to go to the city could be that their "appetite" may not be satisfied by what's left in the rural society.

Those with special connections among the second generation are the first group that has gone to the city. Take *Wan Hui* and *Wan Shuai* from *Shaqiao* Village. *Wan Hui* has been active in the city for many years. Now he and some other members co-established a blackmail gang with 20 or 30 members. He is working as the third boss in rank. Both the first and second leaders are brutal killers. A few years ago, he opened an automobile repair shop with the support from the gang. Rumor has it that he pestered government officials into having their vehicles for official use repaired at his shop. Another person with special connections is *Wan Shuai*, a ruthless and cruel person who was once put in prison for rape. But his brother is working as the chief of Major Case Squad of the city's Public Security Bureau. He used to work as undercover for his brother, which helped his brother solve some difficult cases. With his brother covering his back, he later starts to charge protection fee from urban builders.

It seems like the third-generation gangsters who were born around the year 1980 are bound to establish their presence in urban areas. They have started their gang life early, when they are still studying in secondary technical schools or vocational schools in the city. Their positions have been earned by hard efforts, and anyone who has succeeded already have a group of followers. Though turf-wars are inevitable, the "state of affairs" has become stabilized. The most influential gangs in *Chujiang* City are "*Linsha* Gang" and *Hougang*[12] Gang". The main members in *Linsha* Gang are *from Shaqiao* Village, *Linsha*[13] Village and their neighboring villages. Also coexisting with them are many less-known gangs. When the third generation attempt

---

[12] "*Hou*" literally means "behind", and "*gang*" means "port".

[13] "*Lin*" literally means "next to" and "*Linsha* Village" means a village that is next to *Shaqiao Village*.

to "conquer the world" (*dajiangshan*), state-owned enterprises have gone bankrupt, where gangsters from the same factory or mine would not work together as a single unit. In contrast, those who have migrated from rural areas would unite together because of their regional closeness or consanguinity (many are brothers or cousins), and thus would have an advantage over their urban counterparts in the battle to "conquer the world". *Wan Hui* and *Wan Shuai* are both quite influential in "*Linsha* Gang", acting as the rock for several dozen gangsters who have migrated to the urban area of *Chujiang* City from *Shaqiao* Village, *Linsha* Village, and their neighboring villages. In the process of "conquering the world", the third-generation gangs have engaged in savage killings, being in no way different from organized crime groups. They certainly have broken the law, but they are lucky to have evaded punishment. When things are settled in the gangland, they no longer have to gain through criminal offenses. Instead, they may do so through shady dealings, such as running a "brothel" described as tea houses, bars or nightclubs.

As is described earlier, the third generation establishes a presence in urban areas for the abundant profitable resources there. But in rural areas that still offer such resources, the third generation may also choose to "do business" in the rural areas. Take *Linhu* City as an example. It is profitable to occupy the lakes of the rural area. The lake area has been taken by the second generation, and the third generation chooses to stay to be the help or "thug for hire" for the second generation, to safeguard their vested interests. Local villagers commented by saying "*Peng Batian* (literally Heaven's Devourer) is back!" At *Jingmen* City where coalmines are concentrated, the mine owners need gangsters for protecting their interests and personal safety, and for deterring competitors. Thus many gangsters become their "help" or "thug for hire". *Xiao Qing* is the chief thug for hire to the owner of *Shengli* (literally "victory") Coalmine. *Xiao* is asked by the owner to settle disputes at the coalmine. Because of this, *Xiao* also obtained shares in the mine, taking a cut in the mine earnings.

Like the second generation, the third generation also goes for nothing but financial gain. Apart from that, reputation and camaraderie are both acts of expediency. When the third generation establishes their presence in the gangland, the "gaining is the top rule" setup has already taken shape. "The masters" of calculation will prosper, while those who do not follow this rule will be banished to the periphery. No other rule applies. What used to be fashions in the 1980s has become the laughing stock to the gangsters. When those young men born in the 1990s want to establish their presence as the fourth generation, they would have to "pay tributes to former members" first, or they have no way of surviving. This is no longer a world that gives free entries and exits to "heroes".

4.3 Gang Alliance and Mobility

4.3.1 Loose Alliance Based on a Hierarchical Nexus of Connections

At the turn of the century, almost all of the gangsters have established some kind of connection with others. Having established a connection with one another, gangsters gradually form an invisible hierarchical nexus, with gangsters of the same kind belonging to the same stratum. This nexus of connections keeps those in different strata in a loose alliance.

That is true to over 20 gangsters from *Shaqiao* Village. As their "capacity" or territory differs, the size of their networking circles can be big or small. They maintain a good personal relationship with their "peers" (of the same stratum). They also do so with some members of "the upper stratum", and try to win over a group of members in "the lower stratum" as their "help". The leaders of the strata are "the best performers", while others act as their subordinating "soldiers". Those relatively independent gangsters who have stayed at town or at village are "landlords" with their own "territory" of different sizes.

In their gang life, the hierarchical nexus of connections acts as their "talisman" that mitigates the risk of gang involvement. As they mainly rely on violence or threats of violence for carrying out their activities, inevitably they are subject to some risks. They've come to understand the limit of a single man's power and "the strength of unity". They need to stick together with others to form a network or a "clique" for taking care of one another when in need. This requires them to honor camaraderie by "being a good friend". Only in this way may they win over their peers for helping them out and "developing their careers". Surely some would also choose to work alone, but those are the minority. Such loners are often known in several adjacent villages and towns for their brutality or recklessness. Cautious residents would never offend them; even some non-local wealthy business owners would bear with them because "the mighty dragon cannot beat a snake in its old haunts". But in the end, they do not have what it takes to join the upper stratum of the nexus. They would often "have their dignity undermined" by those who dare to stand up against them, and would find it very difficult to handle members of a gang.

Rural gangsters would generally maintain a good relationship with big bosses of urban gangs—mostly one of moderate dependence. They would never set themselves against the big bosses. If they have offended the big bosses, especially those who have migrated from their villages or townships, they would be more easily under attack. Big bosses do not even have to do it by themselves, but would easily cause trouble for them by asking the dependent subordinates to do it. Therefore, rural gangsters are often intimidated by the big bosses, for their power seems meager compared with the latter, and they would never be as cruel. Compared with their unethical shady dealings, the urban gangs often resort to criminal methods. Just as local villagers have described, "Those punks are just chicken compared with their urban counterparts, who are armed with real guns." In a land dispute between *Wan Min* and some village folks, initially *Wan* insisted on getting his way and refused to make concession in any manner. When some relatives of his opponents sent for urban gangsters who came with shotguns and long knives, *Wan* suddenly changed his "arrogant" attitude. He not only dropped his claim, but even apologized by treating his opponents to dinner.

The hierarchical nexus of connections has formed in many ways: (1) based on natural ties such as blood ties and regional closeness, for example, ties between brothers, cousins, or fellow-villagers, etc.; (2) based on kin-like ties deliberately established through marriage, sworn brotherhood or parenthood; (3) based on ties between comrades or classmates; (4) simply because they happen to get along. In a word, when these relationship networks become interconnected, they become an "organization" for the rural gangsters. In sociological terms, organizations are distinguishable social units with their own functions that are made up with an arrangement of individuals. The hierarchical nexus of connections is technically also a social organization, similar to structurally cohesive social groups or institutions. It keeps rural gangs from being hit hard by the government. In my fieldwork I've found—which is also common knowledge—that the real "bad asses" are those who are good at harnessing the power of the nexus, and that only the green hands or innocent newbies would pay an undue serious attention to things like drafting articles of association, forming an organization and declaring organizational aims. Such members are more susceptible to government crackdown.

Unlike an official organization with clearly defined articles of association or one that specifies eligibility, the rural gangster's hierarchical nexus of connections does not even distinguish members from non-members. It offers nothing more than vague customs, indirectly-suggested rules and exchanges of privilege or benefit. Both the forming and functioning of strata rely on reciprocity or the exchange of "favor" (*renqing*). It is a popular practice in the gangs to "exchange gifts" (*gan renqing*) on occasions of marriage, birth, birthday or parents' birthday. At *Jingshan* County, one exchange of "*renqing*" costs at least 200 Yuan. *Zhang Fei* from *Dengwan* Village told me that he spent about 5000 Yuan annually on gift-exchanging. In 2007, when his brother got admitted by a college, over 30 gangsters came voluntarily to give him gifts. If it had been a celebration for himself, he would have had to send an invitation and possibly 70 or 80 people would have come. Gift-exchanging is an important way for gangsters to keep in touch. They establish an "acquaintance relationship" with each other and act in the way acquaintances do.

However, big bosses and their subordinates are in a state of disequilibrium when it comes to gift-exchanging. Gifts given by big bosses cost at least 500 Yuan each, which is much higher than those required for their subordinates. They must be generous not only in gift-giving but also daily life. *Wan Hui* from *Shaqiao* Village runs an auto repair shop and he hires many of his subordinates to work for him. He treats them very well. He often treats them to dinner or buys them drinks; in addition to paying them he also gives them some when they have run out of money. In this way, he gains control over them. They would engage in a fight if he tells them to do so. They even take pride in being picked by him for a fight. The rural residents are convinced that if *Wan* attempts to settle a matter with someone, he is capable of doing it "without leaving any evidence behind", because he does not even have to do it by himself. The relationship between the big bosses and their subordinates is essentially one of social exchange described by Peter Michael Blau. Blau (1964) argues that, to gain benefits in social exchange, low-status members are willing to submit to the high-status ones and give respect and compliance in return, which indicates that they recognize dominant powers of the highs.

Though customs and conventions followed by gangs in the nexus are not deliberately turned into regulations, gangsters are under an invisible pressure to observe them. This pressure produces "rules", which are equivalent to "regulations" in formal organizations. There is no clear definition for having established a relationship in the nexus. But when two members keep to a certain rule, they develop a relationship; and their relationship is severed when they do not keep to the rule. For instance, if a subordinate refuses to engage in a fight for his big boss, the boss would no longer bestow favors on him, or even fire him. And it would be clear to everyone that the guy has been excluded from the nexus for having broken the rules for interpersonal relationships. In other words, members of the gangs are in casual relationships that follow unstated rules. The rules are unstated, but are by no means unclear. In other words, when rural gangsters form into a hierarchical nexus of connections, they are regrouped into something similar to a social organization.

Two types of relationships exist within the nexus: cooperative and dependent. In a cooperative relationship, gangsters are independent individuals who make independent decisions about what they do and where they do it. Even when they are faced with difficulties and "collective effort" is needed, cooperation is carried out without compromising their individual independence. That is not the case with a dependent relationship in which gangsters seek protection from one or several "big bosses". They take "favors" from big bosses, and their activities are carried out within the big boss's "sphere of influence" (territory). As the big boss often relies on constant intimidation and coercion to safeguard their territory, they need many gangsters working as their dependent subordinates. Though both the two relations are reciprocal in nature, there is no such personal control strictly exercised by means of rules and regulations as in organized crime groups.

In urban gangs, many big bosses have subordinates maintaining a dependent relationship with them. That's also true to a few at-town gangsters who have a relatively large territory. But most of the other at-town gangsters and at-village gangsters are independent individuals that maintain cooperative relationship with one another. Rural gangsters and big bosses in the city often work together when in need, though most of the time they would mind their own business in their own territory. Relationships between gangsters have something to do with the resources they hold onto. For instance, dependent relationships are established in *Linhu* City where it is profitable to occupy the lake area; at *Chujiang* City, *Linjiang* County or other places without such profitable resources, as the cost for maintaining dependent relationships may be higher than what gangsters may get from doing so, they would choose to stay in cooperative relationships for carrying out their shady business. But when they establish organizations similar to organized crime groups for pursuing more unlawful interests, they would make enough earnings for sustaining dependent relationships. In recent years such an opportunity was provided to those lottery vendors (*mazhuang*) when prevalent underground "Mark Six" (lottery game) allow them to support a group of dependent gangsters. In a word, as the lives of rural gangsters become closely intertwined in their cooperative and dependent relationships during the new era (1990s), they have gradually built up a loose alliance.

4.3.2 Social Mobility of Rural Gangs

Contrary to common understanding, the nexus of connections is not a well-structured organization with permanent membership. Rural gangsters are constantly moving up or down the stratum. Their social mobility is highly dependent on the cycle of their personal "growth". The beginners are often naught boys who are introduced into the nexus by their connections. When some of them become "known" for what they have done, they get picked by the superiors as those with high "potentials" to go through further "training". Some others who have accidentally jeopardized the gangster's interests would be "taught a lesson".

In the rural areas of Double-lake Plains, there are channels for making the transition from juvenile delinquency to joining the lower stratum of the nexus. This is probably true to the broad rural areas in central and west China, and even more so now because an increasing number of children are left behind as more rural residents migrate to work in cities. It may be safely concluded that juvenile delinquents, left-behind children and unemployed youth constitute a fertile source for the lower stratum of the rural gangs. In junior highs, "the big brothers" in senior years often ask freshmen to be their "little brothers". The "little brothers" pay protection fees so that they may seek protection from their "big brothers" when being bullied. The "little brothers" who are admitted to the "cliques" would be protected by the names of their "big brothers". Members of the same "cliques" call one another brothers and would often party together. "Big brothers" often hang out with gangsters from outside school and blackmail or even rob students. After graduation, "big brothers" would establish their presence in areas around the school. Some of them would also make an entry into the gang's "circles" and become a gang member. In the beginning they would idle their life away in pleasure-seeking. Occasionally they get picked for fights or guarding a casino. There are 3 types of future for them.

First, "retire when coming of age". Those who cannot "make it" as a big boss will eventually go back to their old business—farming, industry or commerce. Most of the gangsters would exit the gangs when coming of age, especially when they get married. They are disciplined by their wives, and they face increasing pressure to support their families. When they quit the gangs and go back to live in the villages, they are still no ordinary rural residents. They may establish their presence in the villages and become at-village gangsters. Even if they choose not to do so, they may still be "a tough guy" compared with ordinary residents. So it is easier for them to become "an important person" in the village (Xuefeng 2003a: 14). Ordinary residents would have to give way to them; even rural officials would pay them due respect.

Since the 1990s, it has become more common for gangsters to keep being involved in gang activity.[14]

Second, be hit hard by public security agencies. Those brutal and ruthless fighters are easy targets for public security agencies, because they cannot evade punishment if they have caused serious injury or fatality. Drug addicts and gamblers are also easy targets, because they need a lot of money to feed such habits, they would often choose to steal or rob. With the increase in the number of such cases, they would eventually be caught by public security agencies. But it is also common for many to go back to their old business once they get released from prison, because old habits die hard. However, things may have changed dramatically if they are in prison for a long time. So, it would not be easy for them to reenter the nexus of connections. That's probably why many would choose to start working alone.

Third, "make it" as a big boss. Those who have no long-term plans cannot make it because they would squander whatever they have earned. To "make it" means to have what it takes to establish a reputation and become independent. But that does not mean severing their relations with the hierarchical nexus of connections. Instead, this means people are willing to follow them. They have what it takes to work for themselves by running a casino or undertaking a project. They can protect their subordinates.

There are also a few different "futures" for those big bosses. They may choose to "to quit while one is ahead" and "make an exit". Or they may be hit hard by public security agencies. They may also "embark on the right track", which means to eventually stop being a gangster and become an entrepreneur or a businessman. Once primitive accumulation of wealth is complete through fighting and gambling, they may start a business. Being a good fighter is itself "an asset", for with it, no one would dare to provoke him, and one may harness such name for coercing others into giving him projects or establishing a monopoly in a certain trade. In this process, as they make more connections, they may get a loan more easily; as they have more chances to share benefits with others, they may win over a group of brothers who are willing to do anything for them; as they become more experienced, they become a better fighter, a better con man, and even get away with that. When they become such seasoned veterans, they may give up being a gangster and become an entrepreneur. That's what people call "embarking on the right path". Once on the right path, they no longer rely on their gangs or violence for running their business, and they may encounter less trouble compared with other business owners. Considering their past, no one dares to cause them any trouble. Naturally some of them still choose to make use of their gangs or violence for keeping their business afloat. If that is the case, with a reliable and legitimate source of income, they have what it takes to win over more followers or protect their followers. This will lower the turnover rate of their groups and send their groups on a slippery slope towards organized crime.

It would be difficult for public security agencies to crack down on those gangsters who have "embarked on the right track". During fieldwork at police stations, I was

[14] This topic will be discussed in detail in Chap. 9.

told by several police officers, that they are often at their wit's end in their investigation, though being well aware that an entrepreneur or businessman who used to be a gang member is involved. There are mainly two reasons for that. First, these entrepreneurs and businessmen do not even have to be personally involved. Sometimes they do not even take part in the planning. All they have to do is just give a little hint and his followers will take the hint. Therefore, it is almost impossible for the police to find any evidence in their investigation on criminal liability. Second, the entrepreneurs and businessmen have the financial capacity to buy off their followers so that their followers will take full responsibility. In a township of *Linjiang* County, an entrepreneur's follower assaulted one of their competitors and ended up in prison for that, but the entrepreneur kept giving him the same pay. Therefore, for those people who have "embarked on the right path", it is very unlikely for them to be caught. For big bosses who honor camaraderie as the one in this example, their people would be more willing to serve them unreservedly.

In their nexus of connections, gangsters are constantly competing with one another. No one is the permanent "boss". The best performer will naturally be the boss. The boundary of a boss' territory often matches the natural boundary of a town. Within this area, the boss may call up all his followers when in need, and know all the occurrences. Therefore, even those non-affiliated gangsters should give their shares of gift money (*fenzi*) to pay respect to the boss. Otherwise, the big boss will ask his men to report the guy to the police and have him arrested. That's also the reason why police stations would need assistance from inside the gangs for cracking a case. However, a big boss's absolute power is limited only to the town area and not to the county area. The big boss's influence in a non-local town (township) is acquired only based on their connections with the local gangsters through social networking. Their connections with government officials have been earned on a one-to-one basis. That is, they could not possibly have established a good connection with the whole officialdom of different towns (townships) at various levels. Therefore, they would become easy targets once they enter the area of other towns (townships).

The "boss" in the nexus of connections is a rotational position. No one can beat time. As the boss grows older and more beginners join the club, the old will naturally be superseded by the young. The position is not chosen or appointed to anyone, but is earned as one builds up a career in the gang. All the resourceful beginners would try their best to secure the position. As the boss gradually gets tired of his gang life, he may choose to resign the position or be forced to leave after being defeated in the competition. But he does not officially leave and will not entirely severe his relationship with the nexus, because he may lose the ability to protect his own interests once he does so. However, even when he does make a clean break with the nexus, no one would dare to make trouble for him, for people may still be deterred by his past connections in the nexus. Some of the gangsters fail to make it as the boss even when they've become middle-aged. As more young men grow up and claim their turf, those middle-aged men would have a diminishing influence and a shrinking territory. But as long as they stay in the nexus, the nexus would still have something to offer them.

As "no one can beat time" also applies to ordinary gangsters, the balance in the nexus would be disturbed by their mobility up or down the stratum. Take *Wan Jin* and "*Liu Ye*" (literally Lord surnamed *Liu*), who both used to be gangsters active at *Shaqiao* Village of *Chujiang* City. In recent years when the underground Mark Six lottery became prevalent, *Liu* took the chance and opened a lottery outlet. He hired a dozen gangsters as his help and made two or three million Yuan. In contrast, *Wan* was imprisoned for a few years and just got released this year. When *Wan* came back, he looked at *Liu* in the same old manner. In a casual conversation with others, he dismissed *Liu* as "incompetent" and said, "I could easily break his ankle!" *Liu* was furious when hearing about his comments and sent eight gangsters to beat *Wan* up. He forced *Wan* to treat him to a nice meal and buy him nice cigarettes before he would forgive *Wan*. He coerced *Wan* into kneeling down while proposing a toast to everybody present and said to *Wan*, "You have no territory in the town! You can only do your old business. Go back to being a thief!" Like it or not, one would have to accept inferiority to former "comrades", accomplices or peers who have moved up the stratum.

As the gangsters move up or down the stratum in the nexus of connections, members belonging to the same stratum are also constantly competing with one another instead of just keeping up their old relationships. Such overt or covert competition would lead to both minor disagreements and major conflicts. The nexus is essentially just a temporary alliance based on common interests. Without shared interests, there is no need for sustaining the nexus. They would treat members having no common grounds very cruelly in the way they treat their enemies. In a word, gangsters would never work together if it were not for the shared interests; there is no camaraderie without the exchange of favors.

4.4 Social Stratification and Unlawful Interests

As mentioned previously, the second half of the 1990s offered plenty of opportunities for the gangs, and those who have seized them made illegal gain in doing so. No matter what method they use to make illegal gain, they would generally maintain a cordial relationship with town and village officials. Gangsters of different strata have different methods of making such gain, which may be categorized into 3 types: (1) legitimate methods by the upper class; (2) unethical methods by the middle class; (3) criminal methods by the underclass. As their methods differ in terms of legality, gangsters of different strata face very different risks.

4.4.1 Legitimate Methods by the Upper Class

Most of the upper class have "connections" in government agencies, either because they have relatives or friends who work as officials at the town or village level, or the officials need assistance from them. "Connections" are established following three rules: first, gangsters do not cause disturbances within the precincts or to oppose, attack or retaliate against village officials; second, gangsters would assist village officials in overcoming some difficulties or challenges, such as collecting taxes and public dues, land acquisition and house demolition, etc. with their aggressive and scrupulous temperaments; third, rural officials may benefit from the official-gangster alliance.[15] Because they have connections in the grassroots government, gangsters would be capable of "earning a living" by undertaking government projects, and so they also need to make such connections work. Naturally there are also a limited few of them who would rather work for themselves than to depend on officials, but they would not have what it takes to join the upper stratum. If they have jeopardized the interests of government officials, or are classified as "those who have crossed the line", they would easily become targets of serious "crackdown" or "special campaigns". And as they are not protected by the upper class, nor have they gained approval from rural residents, the government may easily remove them before they could develop into anything important.

At *Shaqiao* Village of *Chujiang* City, at-village gangsters would maintain a cordial relationship with *Wan*, the village Party secretary. Secretary *Wan* is a big bully in the village, who not only tyrannizes over rural residents but uses his forces to suppress all dissent whenever possible. He has held the secretary's position for long because he has "connections": he comes from a powerful family with many brothers and cousins, and he has a relative who used to be the mayor of the town. He is very popular among town officials, because with his tyrannical rule, he could always accomplish various tasks, including collecting taxes and public dues before the tax-for-fee reform (TFR).[16] After the reform, he is also good at keeping the village under control and making sure that no residents would appeal to the higher authorities for help and that no incident that undermines social stability or solidarity would take place. In 2005, when the chairman of the village committee was to be re-elected, *Wan* had to win the re-election and remain chairman if he intended to stay in the secretary's position, because the stipulation is that both the chairman and the secretary positions should be held by the same person. Contrary to expectations, he won the re-election with a big majority. This is because he has put some of his "own folks" (some gangsters) on the election board, who could pressurize the rural residents to vote for him. In addition, they even got to directly fill out extra ballots. Besides, he always let the gangsters know when they are about to be arrested. For instance, town officials have made several attempts to arrest *Wan Min*. However, because town officials would often let main village officials know when about to pay the village a visit, *Wan Min* has always managed to hide out because every time he got the message from the

[15] The Chap. 8 will discuss in detail why town and village officials would maintain a cordial relationship with rural gangs.

[16] Translator's note: In response to rural discontent and unrest over tax burdens, the central government enacted the tax-for-fee reform in 2002 that abolishes local fees levied on individuals and rural households in favor of a single agricultural tax. The government also plans to eliminate the agricultural tax after 2006.

village secretary. When the town officials came to know that, they decided not to notify village officials if they want to arrest gangsters.

Many gangsters who are active in the town area around *Shaqiao* Village would maintain a cordial relationship with the former chief of the town police station, who recently got suspended and is currently under investigation for his relationship with gangs. In the above-mentioned incident where *Liu Ye* gave *Wan Jin* a lesson, *Liu* told the police chief about his intentions, and all that the chief said was, "you can do what you want as long as you do not go too far". When *Wan* treated *Liu* to a nice meal, the police chief even drove there to propose a toast to *Liu. Liu* felt very flattered and bragged to *Wan Jin*, "do you know who this is? He is the police chief! My brother! It's only a piece of cake for me to get rid of you!" The police chief also tried to handle gang disputes privately. When one of *Shunzi*'s men cheated in gambling and got exposed by *Liu Ye*, the police chief was present at the hotel and tried to use his connection with *Liu* for mediation. As they did not see eye to eye, *Shunzi* broke a plate and stabbed *Liu Ye*, while *Liu* pushed a beer bottle into *Shunzi*'s lung and seriously injured him. The police chief was very embarrassed for being caught in the middle. It may be inferred from these incidents that local police and rural gangs are very close. There is probably some truth in the general public's presumption that the police and gangs are family. Things may have not been as bad in other places as the area around *Shaqiao* Village, but cordial relationships between gangs and officials are indeed a common phenomenon.

The upper class in the gangs would also try their best to maintain a cordial relationship with businessmen and make illegal gain from doing so. Being on good terms with businessmen will guarantee that someone would intercede with public security agencies if things go wrong for them. Businessmen also count on gangsters for help. The latter may act as the former's "security guard", whereas the former may count on the latter for handling some disputes. For instance, when businessmen go on a trip and encounter problems that may not be properly handled by the police, gangsters may take advantage of their connections and easily solve them with a single phone call. In this way, businessmen are incorporated into the gangster's nexus of connections. These businessmen are often of different types. Those who run a legitimate business maintain a cordial relationship with the gangs to stop gangsters from causing trouble for their businesses or in their business scope; some others try to use their connections in the gangs to make special profits through establishing a unique position in the market. Either way, businessmen are willing to transfer a share of their profits to the gangsters, which may be referred to as "sharing of profits". This is true with the above-mentioned gangster *Xiao Qing* at *Jingmen* City. At first, he intended to blackmail the coalmine owner. Later it occurred to him that coal truck drivers were competing fiercely for more trips, which often led to disputes and in turn caused traffic jams. The interests of both the truck drivers and coalmine owners have been affected. *Xiao Qing* decided to deal with this problem. On one hand, he proposed to the coal owner to pay based on the weight of transported coal and promised to the owner that he would keep the coalmine in order; on the other hand, he asked the truck drivers to pay "protection fees" in exchange for their safety in the coalmine area. The coalmine owner found him to be resourceful and immediately made an agreement with him.

4.4.2 Unethical Methods by the Middle Class

Those who may draw illegally from national wealth are mostly influential and resourceful gangsters with long term plans. They may eventually establish their own business entities, such as companies and restaurants. Some of them may establish a monopoly in their scope of business with the help of their influence in the gangs. The less wealthy ones, who intend to find a decent employment and quit the gang, may open business entities such as shops and retailers, or undertake some profitable projects. *Wan Qing* managed to monopolize the selling of grain alcohol and beer in *Shaqiao* Village. Gangsters in the area around *Shaqiao* Village once collaborated in an attempt to monopolize the selling of beer in the *Chujiang*-to-*Jingmen* section along the provincial highway. Though they failed when other gangs did not allow them, they still managed to do so in the town section along the highway. At *Longyou* Town of *Jingmen* City, investigators were shocked by the unreasonably high price of pork there, which was higher by 2 Yuan than in other town markets just seven or eight miles away. The extra price was caused by an "all street pork alliance" monopoly established by a gangster named *Huang Laoliu* (literally the sixth child in *Huang*'s family). The bus line connecting *Longyou* and the city is also monopolized by local gangsters, but the number of buses controlled by them fails to meet the villager's actual needs for buses and for that reason, villagers going to the city must transfer in a neighboring town (Leiming 2007). We were also told that rural gangs have controlled the selling of beer produced by all the manufacturers at *Jingmen*.

Another unethical method for them to draw illegally from national wealth is to coerce others into selecting them as contractors. These are mostly better-off gangsters, but not wealthy enough to start their own businesses. Having established a cordial relationship with government officials, they would be selected as the contractors, or more often than not, they would simply coerce government officials into selecting them as contractors. Project sponsor (the government) would often give in because they do not want to get embroiled in disputes with gangs. If a contractor is selected following the lawful procedures and the gangster's attempts fail, the gangsters would still cause trouble or get back at the project sponsors and lawful contractors. The gangsters are often not equipped with the required technologies for carrying out a project, or they do not intend to follow pertinent technical standards. Therefore, the result is often shoddy work, or they would subcontract the project to others and profit without having to do anything. That's the way many gangsters of *Shaqiao* Village have made a fortune. With the gang presence, the government could not possibly follow the normal tendering and bidding process for many projects in rural China.

Gangsters may also make illegal gain through coercive monopoly or selling. This is the road they must take if they wish to build up a fortune in rural areas with better developed markets. They detect the scent of "profits" in the rural markets, and with some simple "dressing up", transform themselves into market sellers, trying to get a piece of the pie without paying any effort. They set prices as they wish, regardless of the original market prices—a serious violation of the basic "legal transaction and fair competition" principle. Gangsters in *Shaqiao* Village bought over all the eel on the town market during eel season, ruling any other competitor out of the market. They even beat up law-abiding buyers and uncooperative farmers. They've now become the "eel monopolist" in the area around *Shaqiao* Village.

Some gangsters would get involved in others' debt disputes and earn commission from collecting debt using violence or threats of violence. Debt disputes arise almost every day in production management, commerce, trade and other economic activities. The government has formulated clear laws and policies for handling such disputes, and both parties to the disputes and law enforcement agencies should follow them. Debtors often delay payment or even refuse to pay to benefit from delayed payment or even make illegal profits from doing so. In addition, enforcement of law in such disputes often entails, among other things, high costs. Therefore, creditors would seek help from a third party, the government excluded, in their debt collection efforts. That may be the reasonable choice in certain places (Xi 2005). However, being known as a group that would "stop at nothing", gangs would often use physical assault on debtors and earn what is known as "fist fees" (payment for hired thugs). Since the 1990s, the involvement of rural gangs in debt disputes has constantly led to criminal offenses. As a matter of fact, commissioning rural gangsters to get things done is very common in the whole country. Illegal debt collection is just one of the many tasks that rural gangsters are hired to do. Many of them have been put in labor camps (*laogai*) several times. With such *laogai* experiences (also mischievously referred by them as "entering the palace") that they take pride in, they would command payment from debtors within certain time limits. Some of them also take pride in having a large group of their own peers, so they would keep pestering debtors, e.g., by putting debtors under house arrest. Worse still, they would damage debtors' properties, imprison and hold hostage innocent people, use verbal abuse and physical assault. For that they have become a major factor destabilizing the rural society and causing psychological terror on the public.

There are many other ways for them to draw legally from national wealth. When they or their relatives and friends get "provoked" by someone's careless behavior, for example, being scraped by a motor bike rider, they would "negotiate" on behalf of their folks. Such negotiation often involves extortion. When a major project is about to be carried out in their village, they would extort money from the lawful contractors on the pretext of "having shared interests". Before the tax-for-fee reform (TFR), gangsters are hired by some government agencies at town or township level to collect taxes and public dues. In addition to being paid for "doing the hard work", they may also ask small private business owners to be their "sponsors", who are coerced into satisfying their demands, and charge "turf fee" and "protection fee" in their own "territories". In this manner, they are functioning as "profiting brokers" (Duara 1988).

In recent years, as "underground Mark Six"[17] lottery spread widely in the Chinese mainland, rural gangs with their "keen sense of smell" detect a good opportunity in becoming lottery vendors (*mazhuang*). They sell Mark Six lottery at home and receive commission for lottery sales. Mark Six cost people large amounts of money, whereas gangsters get to "build up their family fortune and become better-off". In the Double-lake Plains, many gangsters managed to make several hundred thousand or even a million Yuan in just two or three years. Though those involved in the lottery business are not necessarily all gangsters, most of the lottery vendors are. In addition to receiving commission, gangster vendors also make revenue by "keeping some amounts of bets" (*tundan*), or not turning over all the amounts of bets, commonly referred to as "*chi hei ma*" (literally pocketing "tickets"). If the kept bets happen to win, lottery vendors have to pay the prizes themselves. Therefore, "*chi hei ma*" is both profitable and risky. They often prepare some money for the winning bets they have kept, but they often deal with big winnings by simply evading payment. They are always ready to run away whenever winnings exceed proceeds. When they come back a few months later, lottery buyers would not claim the unpaid prizes considering their gang background. To guard against police investigation, the turning over of bets is completed over a simple phone call between buyers and vendors and between lower-level and upper-level vendors. There are no real lottery tickets or written agreement. Everything is done because buyers and bottom-level vendors have established a trust-based relationship (Yanhua 2006). The higher-level vendors do not have to worry about noncompliance from their lower-level counterparts because of their identity as gangsters and past experience of using violence and intimidation as such.

4.4.3 Criminal Methods by the Underclass

The underclass of the gangland has to use criminal methods for making illegal gain. They are subordinate to the big bosses, but such subordination only keeps them from starving. The more perceptive ones would keep an eye on chances that allow them

[17] The underground Mark Six in the Chinese mainland is an illegal gambling game that draws the winning numbers in the same way Mark Six run by Hongkong Jockey Club does. Vendors ask buyers to select a number from 1 to 49. If the number matches the extra number for a Mark Six drawing, the buyer wins 40 times of the amount of bet. Underground Mark Six functions in secret underground channels with secret transactions conducted between lottery vendors and gamblers, and a unique foolproof system of multi-level transactions—a pyramid selling scheme. There are many levels of lottery vendors. People become vendors through recommendation from someone they know that is also selling lottery, and they only get into contact with people they know. Lower-level vendors do not know those at the higher level. Vendors at the bottom level are responsible for collecting buyer's payment for bets, which are turned over by them to vendors at a higher level. Payment for bets is turned over from the lower to the higher level. The number of vendors becomes fewer as the level grows higher. Vendors subtract 10% of the paid money as commission. After each drawing, prizes are paid from vendors at the higher level to the lower level. Winning buyers claim prizes from bottom-level vendors, who is also the person turning over their payment for bets. Many unidentified high-level vendors would deliberately keep their distance from buyers.

to make a "leap" from the underclass to the upper class. But most of the underclass lacks such perception. Without a steady income and a good plan for spending money, they often run low on cash. So they often try to make money through such criminal acts as theft, blackmail, robbery, robbery by sudden snatching (*qiang duo*), etc. They are often gangsters with bad habits, e.g., drug users. They are often newbies who often visit ballrooms and karaoke boxes. When they have money, they would pay for sexual services; and when they run low on cash, they would kill time by surfing on the Internet. Hanging out in such entertainment venues, they start taking ketamine, ecstasy, among other things, and would engage in prostitution afterwards. This is a costly vicious cycle. For those newbies who have no income, the fastest way to make money is theft or robbery. In American cities, members to street corner societies are often involved in drug dealing, prostitution and violent crimes (Decker and Van Winkle 1996: 134; Hagedorn and Macon 1998: 141). As the gap between urban and rural areas becomes narrower with China's urban–rural integration plan, rural gangs seem to be moving toward the same path taken by their counterparts in America's street corner society. As theft or robbery often involves serious violence, evidence for such crimes may be collected more easily, and thieves and robbers are more likely to be caught. However, even with the state's relentless efforts in "crackdown" and other "special" campaigns, these crimes can never be stopped altogether, because juvenile delinquents, left-behind children and unemployed youth constitute a fertile source for the lower stratum of the rural gangs.

In fact, theft, robbery and blackmail seem to be inevitable for all beginners in the rural gangs. But the targets of such crimes have become quite different if we compare the second generation with the third. With a small "appetite", the second generation would idle away and be easily satisfied with taking some things or money from residents who are not at home. They would plan during day and steal chickens, cattle, and sheep during night. When they get caught doing so, they would try to keep the stolen goods using force, which often intimidates the owners and keeps the owners from letting others know what they have done. With economic development and the prevalence of hedonism, by the time they become gangsters, the appetite of the third generation has grown larger—they are no longer satisfied with petty theft, or simply stealing a chicken or dog. Such little tricks can no longer meet their expenditure, so they go to places that have more things to offer, or try more effective means to tap into social resources. So it seems that migration of the gangsters into cities leaves the rural society with more peace and tranquility.

Gangsters at any stratum could use gambling to make illegal profits. Gambling is also one of the fastest ways for making gain and one of the main forms for gangster gathering. Therefore, their nexus of connections is almost replicated in a casino. Those most successful gangsters are responsible for running a casino, which is referred to in "gang jargon" as "offering a course". Those "offering a course" are referred to as "the principal". The principal cannot possibly lose money, for he would take cuts of each game, referred to in gang jargon as "*chou shui*" (literally pumping water). The share taken by the principal is referred to as "water", and the "water fee" is covered by the winner and it equals to one tenth of the winnings. Gamblers could not possibly win at each game, but the principal always makes some earnings.

The principal is responsible for ensuring safety at the casino, by covering the fine levied on the gamblers when they get caught by occasional police raids. Therefore, one has to meet up to two requirements to be the principal. First, one has to be financially capable of shouldering the possible losses. Second, one must have established a cordial relationship with government officials, and would "have a say in matters" when dealing with the police. When gamblers get arrested, the principal should negotiate with the police and propose fines as an alternative to detention and pay the fine on behalf of the gamblers. Those who cannot meet these two requirements may not win over others and could not possibly become the principal. Even when a disqualified gangster somehow ends up as the principal, he would only ruin his own reputation and make his life even harder in the gangs.

Only those popular and "respected" members of the gangs may act as the principal and he often have a group of beginners "working their fingers to the bone". In the casino, the beginners are responsible mainly for "keeping an eye on the venue" or guarding the place. This often includes, first, staying on the lookout and tipping off the principal about police raids; and second, keeping order and acting as physical security force in times of disputes. The principal gives them a handsome pay and also nice meals and treats. The principal may "build a cage" (*zuolongzi*) or "use a cheat" (*chulaoqian*) on the outsiders (often non-local businessmen), and when his scam gets exposed, the beginners are the physical force and the order keepers that the principal may rely on. Casino staff would collude in a scam in which the outsiders are allowed to win at first, but when they look to some more gain and are reluctant to quit, they would lose a fortune. Sometimes people are lured into a game, and the winners are coerced into staying in the game until they lose all their money.

The bankers in the casinos are also those more successful gangsters. The position of the banker may be held by the principal and sometimes also by some others. Bankers are often better-off gangsters who would personally be financially capable of handling the wins and losses of the games. Bankers would often cheat (*chulaoqian*). Otherwise simply taking their chances may not be profitable, because the chances of winning are the same to anyone. The method of cheating is generally to load dice with lead and place a magnet beneath the roulette. Loaded dice and hidden devices in the roulette allow the banker to control wins and losses. Once the cheating gets exposed, the banker would have to refund the money to the losers. A refund often requires the presence of a principal, who would "rule the case" like a law enforcement official. If the cheater happens to be the principal and things get out of hand, the principal would run away with the money he has won, but that may damage his reputation in the gangs.

Another way of making profits in gambling is loan sharking. Loan shark at the casino is often offered by the principal, bankers or any other well-off gangster. Losers who are coming close to losing their mind always fantasize about winning their money back, so when they have lost everything, they would borrow from loan sharks in the casino. Most of the loan sharks have many followers working for them, because high profits on loan sharking incur high risks. When the borrowers refuse to pay off the debt, the loan sharks would ask their followers to coerce the borrowers into paying. In fact, many borrowers have paid their debts because they feel intimidated by loan sharks. Once the parents of a borrower who happens to be also a gangster consulted me about whether they had to pay back the high-interest debts borrowed by their son. I told them that theoretically they did not have to, for illegal gambling debts are not legally enforceable. However, they still paid back the debts eventually for fear that the mobsters would get back at them. In a word, rural gangsters have indeed made plenty of profits from gambling, which are not equally distributed by people of different stratum. Their share of profits generally matches their status in the organizational structure.

4.5 The Rural Underworld Disembedded from the Acquaintance Society

Since the 1990s, economic growth and social development sustained by economic transformation have brought great opportunities for rural gangs. When they try to seize such opportunities, they have also made the transformation from "being embedded in the acquaintance society" to "being disembedded" from it. Economic reform enacted by the government has caused the market to prosper around the country, which in turn has strengthened market forces in the allocation of national and communal resources. However, because social order is yet to be established, rural gangs are able to profit from misappropriation of public resources and monopolizing markets. Besides, they are no longer happy with the meager benefits offered by the rural society and gradually move away from it. For that reason, direct harm inflicted upon the rural residents may have become less common.[18]

As mentioned previously, the renaissance and later development of rural gangs largely rely on profits from markets. For that reason, they have gradually broken away from restraints that used to be imposed on them by the acquaintance society. They have gradually established an invisible hierarchical nexus of steady connections that spans urban and rural areas and that keeps them in a loose alliance to effectively evade punishment. In other words, the rural gangs have gained independence from the villages, and the state may find it difficult to impose restraint on the rural gangs through managing the villages. The rural residents may find it even harder to deal with the gangs. Village officials also seem to be at their wit's end when it comes to dealing with the gangs. In a word, the rural gangs have grown increasingly disembedded from the acquaintance society.

The well-known terms "embeddedness" or "being disembedded" have been proposed by Polanyi. "The term 'embeddedness' expresses the idea that the economy

[18] This is probably an illusion. The same belief is also held by many rural officials, that what rural gangsters are after is financial gain in the rural underworld. Therefore, farmer interests are not at stake. So the gang impact on farmer life must be limited. However, it must be pointed out that whatever gain that may be offered by the rural underworld eventually should be traced back to the society. And because rural gangsters are often involved in village governance and dispute settlement, we must reexamine the gangster issue in village order. (Xingzuo 2007).

is not autonomous, as it must be in economic theory, but subordinate to politics, religion, and social relations. Before the nineteenth century, the human economy was always embedded in society. … Polanyi (2001) does say that the classical economists wanted to create a society in which the economy had been effectively disembedded and came to dominate the society. Yet he also insists that they did not and could not achieve this goal. Here an analogy is drawn between the economy-society relationship in Polanyi's terms and the gang-society relationship. During the 1980s, rural gangs are still embedded in the rural society; but since the1990s, with the development of the gangs and the expansion of their organizational structure, rural gangs are increasingly disembedded from the rural society.

Chapter 5
Changes and Predicament in the Public Security System

The 1980s is the most important period for social transformation when the most radical changes have occurred, and when the state power has turned out to be a paradox. On one hand, compared with the times of feudalism, continuous expansion of state power, both vertical and horizontal, has led to its deeper penetration into the rural areas. On the other hand, compared with the commune period (1950s-1970s) when state control over rural areas climaxed, such control has somewhat diminished. After the founding of the People's Republic of China, the Communist Party introduces a land reform to realign the power structure in rural China and finally manages to "embed" state power (establish control over) through the people's commune system. Having established across-the-board control over Party activities, politics, economy, army, people and people's study, the commune system is omnipotent (Dang 1994:3), being both a form of political organization and social structure. Political power has almost unlimited control over any part of the rural society at all times.

The paradoxical nature of state power during the 1980s has resulted in two contradictory requirements for the rural public security system. On one hand, the state power must back off from what used to be an omnipotent system and allow some liberty for the rural society. On the other hand, the state must also fulfill the people's needs and expectations on public security, for such needs and expectations are closely linked with people's opinion on the Party and the socialist ideology. As China makes its transition from a traditional country dominated by socialist doctrines to a modern market economy, how should the state regulate its public security efforts? Under the new circumstances during the 1990s, especially when rural gangs have organized into an alliance, how will the officialdom and law enforcement conduct themselves in public security efforts? This chapter will discuss changes in grassroots public security efforts and the impact of such efforts on managing rural gangs.

B. Chen, *The Underworld of Rural China*,
https://doi.org/10.1007/978-981-19-8710-6_5

5.1 "The Mass Line" in Public Security Efforts at the Dawn of the Reform

The "Mass Line" (*qunzhong luxian*) is adopted by the Communist Party of China based on its motto of "serving the people wholeheartedly". As the Party's basic leadership method, it includes the essential elements of "for the masses, by the masses, from the masses, to the masses". "For the masses" means the Party must take responsibility on behalf of the masses and serve the needs of the masses. "By the masses" means that the Party must believe that the masses are capable of liberating themselves and it must be respectful and supportive of the masses' creations; that it must respond to the masses' call and listen to the masses' opinions; and that it must call upon the masses in the Party's work. "From the masses, to the masses" means, to quote from *Mao Zedong*, "take the ideas of the masses (scattered and unsystematic ideas) and concentrate them (through study turn them into concentrated and systematic ideas), then go to the masses and propagate and explain these ideas until the masses embrace them as their own, hold fast to them and translate them into action, and test the correctness of these ideas in such action. Then once again concentrate ideas from the masses and once again go to the masses so that the ideas are persevered in and carried through. And so on, over and over again in an endless spiral, the ideas become more correct, more vital and richer each time" (Mao 1965b: 120).

The mass line in public security efforts was originally an ordinance made by Chairman *Mao*. Right after the founding of the nation, to consolidate the new-born people's political power, the Central Committee of the Party decided to launch a nationwide campaign to suppress counterrevolutionaries. Chairman *Mao* formulated a clear line in the work of eliminating counter-revolutionaries: "leadership by Party committees, mobilization of the entire Party membership, mobilization of the masses, participation by the democratic parties and by personages from all circles, unified planning, unified action, strict examination of the lists of persons to be arrested or executed, attention to tactics in different phases of the struggle, widespread propaganda and education (holding various kinds of conferences, cadre meetings, forums and mass rallies, at all of which victims can bring their accusations and evidence of crimes can be displayed, and making propaganda through films, lantern-slides, stage performances, newspapers, pamphlets and leaflets, in order to make the movement known to every household and individual), a break with the practice of working behind closed doors and being secretive, and determined opposition to the deviation of rashness." (Mao 1977) In other words, Chairman *Mao* provided important theoretical guidance for political and law enforcement workers to implement the mass line.

With the promulgation of the first *Regulations of the People's Republic of China on Penalties for Administration of Public Security* (hereinafter "the Regulations") in 1957, the Central Committee of the Communist Part intended to "contain acts jeopardizing public interests or disrupting public order by a minority group of people, through thorough mass work, relying on willingness and support from the masses of rural residents or from the majority of the people". The Central Committee demanded that all levels of public security agencies should "follow the mass line. In enforcing 'the Regulations', they must rely on the masses. When requiring the masses to comply with 'the Regulations', they must stick with persuasion and education. For that purpose, they must widely and repeatedly publicize the significance of 'the Regulations'. They must call upon the masses to comply and to urge others to do the same. They must call upon the masses to keep an eye on the bad guys, keeping them from disrupting the public order. All levels of public security agencies must constantly listen to the opinions of the masses (Ruiqing 1957). The 1957 Regulations did not get invalidated until the promulgation of its new version in 1987, and the heavy emphasis on the "mass line" has extended to the whole 1980s.

Though it has been constantly pointed out that "the mass line must be enacted together with special campaigns", public security efforts during the 1980s still largely depend on the mass line, because special campaigns do not have the required professionals and technologies to be fully effective. The mass line in public security efforts is mainly reflected by the following 3 measures. First, respond directly to the demands of the masses. This measure has resulted in the establishment of moral standards on penalties against criminal offenses by rural gangs. Second, rely on the masses to maintain public order, which turns public security efforts into a nationwide campaign. Third, give the masses a sense of security by being identified with them and keep up public security agency's intimidation on criminals, which helps public security agencies to gain approval from the masses. These interconnected measures are often taken at the same time as parts of a whole package. Due to public security agency's heavy emphasis on the mass line, public security efforts are essentially "justice by the masses" in nature. The whole society has been mobilized in public security efforts. Having mutual influence on one another, people would eventually come to have a universal attitude towards punishments for transgressions.

Though legal elites have already begun their debate over rule of law and rule of man in the 1980s, people at the grassroots level still bear the same old thinking pattern as they did in the revolutionary period, and do not really distinguish between a legal offense and a moral transgression. With just a vague and ambiguous understanding of crimes, people would often put those they find to be unbearable in everyday life into the group of criminals (Shigong 2001: 389). As commune-era appellations like "four categories (landlords, rich farmers, anti-revolutionaries, bad individuals)" and "five categories (the above four plus rightists)" become less commonly used in people's daily life, new words have taken their places, such as "speculators", "thugs", "swindlers", "thieves" and "robbers". Also actively used are phrases that are passed down from the revolutionary period, such as "anti-revolutionaries"; and everyday phrases such as "slackers" and "easy people who sleep around". No matter what appellations people use and how different the underlying evils are, they all represent various transgressions that people find hard to bear with. Such aversion turns transgressors into marginal groups of the society.

Transgressors become marginalized for a variety of reasons. Some have violated stipulations in laws or regulations; others have directly harmed public interests. Some have gone against traditional ethics of the masses. Still for some others, the reason why they are marginalized is that people have become more politically conscious of their behaviors after going through "persuasion and education" and thus have developed "class hatred" to them. In a word, people during the 1980s generally have the same attitude towards the due punishments for deviations. There are two reasons for this. On one hand, all sorts of emotional bonds towards the commune system, which people have developed during the commune era and the longer period when revolutionary ideals have a continuing influence, have lingered to the 1980s. On the other hand, under the new circumstances of the 1980s, new emotional bonds shared by the people are still yet to be established by the state and the Party. Since people are still emotionally bonded to the commune system, they tend to rely on their ethical standards for forming an opinion on criminal offenses and have eventually come to have a universal attitude towards punishments for transgressions.

The "crackdown" campaign has created an even stronger climate of ethical judgment. The paradox about the "crackdown" is that its aim to establish the rule of law is achieved through mass movements—a typical rule-of-man instrument. It is an instrument used by the state during a transitional period when social order may not be kept by legitimate political systems; it depends on a considerable level of political legitimacy that the ruling Party has earned during the revolutionary times; it uses effective propaganda for instilling political ideologies; it uses infiltration tactics, calls upon the masses of the people to act—tapping into any available resources for maintaining public order. It turns out to be quite effective, in light of the social environment back then (Huangfeng 2007). The key to mobilizing the masses is to incorporate their ethical concerns into the rule of law, which entails the use of "extended case method". In addition to transgressions, this method considers the transgressor's regular behaviors, and also officials and other people's views on them.[1] One of the rural residents told me, "With a strong revolutionary atmosphere, everything is associated with politics. Sometimes the law does not apply. People do not really tell legal crimes apart from ethical evils." Severe punishments have been imposed on cases that should have been ruled as ethical evils because of considerable "indignation by the people".

Though the people's Commune system has been dismantled by the 1980s, interpersonal relationships in the rural areas still remain the same as during the Commune era. Therefore, rural governance does not go into a serious crisis until late 1990s. Before the 1990s, the village as a community still exerts strong control over rural residents; village officials feel morally obligated to do their jobs properly; rural residents seem to share and cherish the same moral emotions; the Party and the government seem to have effective channels for instilling ideology. Therefore, rule of virtue turns out to be effective for maintaining order before the 1990s. Directed by action plans of governments at higher levels, the local government carry out public security efforts in five orderly steps. First, keep officials informed. Multi-level meetings, especially village official meetings at the town and township level, are held to keep officials informed on government policies and local conditions, and to identify important groups or individuals as targets for investigation. Second, call upon the people to

---

[1] Discussed in detail in Sect. 3.5 of Chap. 3.

assist with thorough investigation. This also requires holding meetings, including public arrest meetings where the commune secretaries deliver messages directly to the masses, mass meetings in production teams and mass representative meetings in production brigades. The people are encouraged to assist with "catching the boss", "arresting them all", and "breaking them up" to solve cases as soon as possible. They are also encouraged to disclose information to the police so that the police may have clear targets. Third, organize special teams for key targets and special cases. Further investigation is conducted on suspicious persons and goods. Fourth, hold legal education classes. Once recent criminal acts by important criminals and suspects have been ascertained, they are encouraged to study the laws. Keeping them informed on policies to influence their attitudes will eventually divide them up and break down the gangs. Fifth, the responsibility system and "help and education" system are improved through the formation of organizations or the signing of contracts, which clearly describing the obligations and responsibilities for those who help out criminals.

Village officials play a very important role in public security efforts. They are guardians of public ethics and effective government agents. At a time when police stations are understaffed and when "special campaigns" inadequate, public security is mainly maintained by officials at the town and village levels. Village officials are the cornerstone of public security efforts. Legal education classes are held at the village level to help and educate juvenile delinquents. During the "crackdown", village officials are almost acting as the subordinates of public security agencies, as they have the power to practically arrest transgressors. At *Puxing* Village of *Jingzhou* City, a dozen youngsters got arrested in the 1983 crackdown. Most of them were seized and turned over by village officials. The same happened at *Xinwang* Village of *Chujiang* City.[2]

Based on the previous examples, it may be inferred that in the early 1980s, when administering public security punishment the government must respond to the needs of the masses and meet their wishes; sometimes the penalties are even directly introduced by the masses. The execution of penalties also relies on the masses and keeps the whole nation actively involved. What's more, penalties are executed in a way to be seen by the masses. A typical example would be frequent public trials held across the county and before and after-trial parading of criminals. To rectify negative influence in the social climate, some villages and towns would also parade minor offenders or moral transgressors. The same method is also used by some village officials. The following is an account of a public trial:

> On the tenth of November, under the general guidance of the municipal Party committee, under the correct leadership of the county Party committee and the county government, we have held a public trial attended by a group of 10,000 people including both commune officials and the masses. For its broad participation, solemn ceremony and deterrent effect, it has gained approval from the general public as the unprecedented public trial in the history

---

[2] Unexpectedly all of these young people were given a heavy sentence. When they got out of jail, they tried to get back at the village officials, which displaced the former village secretary of *Xinwang* Village's whole family.

> of our county. Having greatly deterred criminals and inspired the people, it has taken the crackdown campaign to the next level, marking an initial success in advancing rule of law.
>
> Before the trial, we held a grand and solemn parade with two exuberant fully armed police officers riding motorcycles to lead the way, followed by a publicity vehicle with a loud-speaker. The procession also included 8 cars carrying leaders at all levels acting as instructors and supervisors at the trial, 4 prisoner transport vehicles flanked by 5 trucks carrying fully armed soldiers and militants. All that gave the procession its grandeur and solemnness.[3]

The masses are mobilized to execute penalties on criminals. At this public trial, the county Party secretary gave a speech on behalf of the county Party committee and the county government. He called upon the county people to act as follows:

> The people must have an enhanced awareness of this ongoing crackdown and take part in it by bravely reporting criminal offenses to the police and taking a firm stance against the enemies;
>
> The people must work closely with law enforcement to conduct the campaign and give it their strong support;
>
> We must promote legal awareness among government officials and the masses, especially the youth, so that they would voluntarily reject the exploiter's world views and any other negative influence, and so that they would refrain from involvement in criminal activities.
>
> As an important instrument of people's democratic dictatorship, law enforcement must give full play to the functions of the dictatorship as "the sharp blade". They must have shared anger and hatred against the common enemies and show them no mercy. They must take a firm stance in enforcing the law as the guardians of people's interests.
>
> In the end, he gave all criminals a serious warning: all criminals must make clear judgments of the realities and refrain from committing any more crimes. The only way out is to turn in to the police as soon as possible, give a factual account of crimes and those of accomplices, and turn over a new leaf. And that is the precondition to lenient treatment. Family members of criminals must make a clear distinction between right and wrong; make timely cooperation with the government in educating and transforming criminals and encouraging them to confess for lenient treatment. If criminals would take chances by staging on a stubborn fight or keep covering up their crimes or even become persistent offenders, they would surely be subject to more severe punishment.

This speech by the county Party secretary bears distinctive features of the times—similar versions may be easily found in publicity materials. It reveals the perspectives of the Party, the government and the general public on criminal offenses. The state regards any act disrupting the socialist order as criminal offenses and anyong disrupting the socialist order as "enemies" or "bad element" (*huaifenzi*). The serious offenders are the people's enemy, and thus should certainly be severely punished. In the campaign against "enemies" or "bad elements", law enforcement will act as the "blade" that leads the charge and safeguards people's interests. The campaign against enemies must also gain support from the masses and mobilize the masses. Enemies will not stay enemies forever. They may be transformed by penal labor into useful people for the society. Therefore, lenient treatment may be given to those who have made clear judgments of the realities, turned over a new leaf, turned themselves in, or given a factual account of their and their accomplice's crimes. When they

[3] Public Security Bureau of *Shanxiang* County: *Our County's Grand Public Trial Fired the First Shots in the War Against Crime*, the 8th issue of "Public Security Briefs", Sept 12, 1983.

have received due punishments and undergone reform through labor, they may be re-incorporated into the people's category, whereas more severe punishment should be given to those unrepentant persistent offenders. The masses are not perfect. Though they belong to the "good type", many of them lack legal awareness, hence the need for legal education. Activists in the Communist Party and Youth League must encourage the people to fight against criminals. Some members of the masses, especially the young people, are more likely to be drawn over to the enemy's or the bad element's side, and are thus more susceptible to the exploiter's worldviews and evils. Therefore, they must be taught the right worldviews and outlook on life to assist with the development of proper ethics and character.

The pattern of thinking embodied in the above speech is an inheritance from the Commune era. With the promulgation of *he Regulations* in 1957, the central Party committee indicated that: "among all those who should be punished for violating *the Regulations*, some have always been the bad elements. ... They are the target of our dictatorial control. When they break the laws, we must use power against them and punish them. ...This indicates that *the Regulations* are a weapon for the people to exercise dictatorship control on the bad element." "Among all the punishable offense, many are minor offenses by the people. ...They have committed these offenses due to errors in ethical judgment, loose conduct, or mistakes in their work or life" (Ruiqing 1957). Such an attitude towards the bad element has passed onto the general public. When the above-mentioned public trial ended, the masses expressed their opinions:

> When Chen Yinqiu, the accountant to Chen Jiafang Production Brigade, commented on the trial today, he said it was very "inspirational". Before the trial, I thought there would not be many people present, but it was quite a different story when it actually happened. We told all the households that at least one of each should be present, but there were actually over 1,600 people. The whole families from one third of the households were present—they had to shut the door before they came. Zhu Xianglin's whole family with 7 people of 4 generations was present, with the eldest being 78 and the youngest 3 months. Chen Zixiang had to finish the paperwork required for his leave and bought a ticket for a trip to Guizhou for a family visit, but when he heard about the upcoming trial, he returned the ticket and managed to be present. The big success of the trial has greatly boosted good people's morals and dismayed the bad people. It was a great relief! How we hope to see more of this. If only the main thieves got executed too. That will surely put people at ease.
>
> Zhang Guilin, a man in his seventies from Anjia Production Brigade, told us about his 3 experiences of being stolen from when he was sitting in for the township-level imperial examination. Two out of his three experiences he had got eye contact with the thieves but there was nothing he could do because was too scared to tell anyone or catch them. When he saw the notorious thief Tang Yunxiang being caught at the trial, he was very pleased that the government did it for him. If the government keeps up its efforts to catch them, the people would not have to worry about thieves any more.

Through mass movements, the country has effectively restrained the enemies and the bad element. Mass movements have greatly encouraged the good people and deterred the bad guys. It provided guarantee for order and security. In addition, mass movements also keep the masses informed so that they may refrain from making mistakes in both life and work. The mass line in public security efforts has also kept the people involved in the processes of law enforcement and has enhanced the legitimacy of the Party and the state during a transitional period. The success in the mass line may be attributed to the functions of the mass line, the ways in which state power works and the nature of rural society of China. And I will discuss this topic in detail in future chapters.

5.2 Progress on "Special Campaigns"

5.2.1 Social Environment for "Special Campaigns"

With the dismantling of the People's Commune system, interpersonal relationships in rural areas have changed fundamentally. The era when the village and the grassroots government are practically "tied up together" in pursuit of identical interests is gone forever. "The village" that used to be synonymous with the lowest level of government has gradually changed into a self-governing body. Now that it has grown independent from its higher level of administrative divisions (township), "the village" has their own interests to pursue. Village officials that used to be brokers of government agencies have changed into leaders of self-governing bodies, and are inclined to defend their personal interests. As a result, public security agencies find village officials less cooperative; it is much harder for them to maintain rural order by engaging village officials. Against this backdrop, the grassroots government requests the village to establish a public security committee as the "cooperative unit" (*duikou danwei*) for public security agencies. The committee is responsible for keeping public order and security and providing assistance to public security agencies. In the early 1980s, public security agencies often complain about a lack of serious effort from village officials, which may be exemplified by the following text.

> Grassroots organizations for public security are still to be established. It requires serious effort on the part of the village. Among the 28 production brigades of the commune (equivalent to the village), only 2 have established public security committees. The rest of them share a public security director, who has to take on everything all by himself. Public security organizations at XXX village have their own leaders but are seriously understaffed. In times of emergencies they have to make things work with a temporary team. Due to a lack of required staff, awareness and effort on the part of the village, we could not possibly carry out "help and education" for criminals.[4]

A lack of serious effort at the village level is nothing but a fact. With the dismantling of the people's Commune system, people have gradually forsaken collectivist values and revolutionary ideals. No longer fully committed to the commune or the government like they used to, village officials have grown "smarter" and more conscious of their prospects for life and start to go after their personal interests. When their job requirements have not been clearly laid down, they certainly couldn't have done their job properly. To deal with this problem, the government requests the

---

[4] "An Effective Way of Good Administration is to Enhance People's Awareness", Document No. 2 of 1983, People's Procuratorate at *Linjiang* County.

village to establish public security organizations (PSOs) on top of existing committees. In over 20 years since then, other village organizations have been disbanded, for example, militant organizations, or downgraded, i.e. Communist Youth League organizations and women's organizations. But PSOs have gained importance and have never been streamlined no matter what's happened. Initially, PSOs are established in a way that attempts to sustain the old government-village symbiosis. The government hoped that PSOs would act as "cooperative units" of public security agencies and keep rural areas in order. However, things have not gone as expected. The town-village relationship has undergone fundamental changes and may not remain the kind of simple superior-subordinate relationship.

In the early 1980s, as tendencies in the political system are inclined to linger, generally the town-village relationship remains a superior-subordinate relationship in nature. This includes the relationship between town (township) police stations with village PSOs and officials. Though occasionally the town-village relationship also deviates from this pattern, and the grassroots government would accuse village officials of being uncooperative, such deviations are not very common. In mid and late 80s, such deviations have become more common. Village organizations seem to have become less potent in ensuring order, and the grassroots government is constantly complaining. That's also when the rural areas are in a worse order. The grassroots government would generally attribute the worsening order to impotent village organizations. Here is an example:

> In recent years, public security organizations at the village level have become less potent. Failing to do their jobs properly, a fairly large number of village public security directors and mediation directors exist in names only. For some directors that do not have any assistants, even their name has ceased to exist. For that reason, disputes in families, between neighbors or village groups are not well handled. Some disputes that have not been handled timely or thoroughly have escalated into major disputes or even occasionally criminal cases.[5]

During the 1990s village organizations have a diminishing role in maintaining order for multiple reasons. First, with the development of the market economy, forces disrupting rural order have grown continuously and village organizations find it increasingly difficult to keep them in check. The gang alliance, even still in its infancy, has been more than what the village or the local government can handle. In addition, as people move around more often, local gangsters would become disruptive forces in some other places and vice versa—both are a hot potato for the village. Second, when farmers are faced with a heavier burden, the village goes into a governance crisis. The prevailing antagonism between farmers and officials has made it impossible for the local government to gain farmer's wholehearted support in its effort to keep rural order. The same could be said about village organizations. Besides, those upright officials who care about others' opinion on the village or the village's reputation have, one after another, decided to or been forced to resign. Their successors are interested in nothing but their personal gain. They are often the culprits that jeopardize public order or at least are not enthusiastic about keeping order. Therefore, village

---

[5] Public Security Bureau at *Shanxiang* County: "Investigation into Unnatural Deaths at XXX Town", 12th issue of *Brief Reports on Public Security*, May 7 1985.

organizations and officials of the 1990s find it very hard to accomplish anything or have basically accomplished nothing when it comes to maintaining order.

Considering how things are, town and township police stations that are responsible for maintaining order have gradually given up any hopes they had for village organizations and officials. Since the village level could not and have not done anything, police stations would rather act by themselves than through the intermediary of the village. By the mid and late 90s, most of the village PSOs are not regarded as the "branch offices" of police stations. So instead of assisting the police stations in keeping order, PSOs are mainly in charge of mediation in civil disputes. Thus they seem to have become the subordinate to the Township People's Mediation Committee. That's probably why, during the mid and late 90s, in their performance evaluation, public security agencies tend to leave out PSOs. They do not pin hopes on or criticize PSOs. Looking back at the 1980s, the public security agencies used to direct constant criticism on PSOs because they still hope PSOs would assist them, or PSOs have accomplished something back then.

As mentioned earlier, because village organizations and officials are basically dysfunctional in order maintenance, and the local government have no effective means of calling upon the masses to keep order, the mass line that used to be effective in the Commune era and the early 1980s has gradually become unfeasible in mid and late 90s. On one hand, as the nation goes through a transitional period (Shigong 2003), it has adopted a new mode of rural governance and has given up the outdated mode of governance through mass movements. On the other hand, governance crisis has resulted in antagonism between rural residents and the local government, making it impossible for the government to mobilize the masses. Any gatherings of rural residents, no matter what the reason might be, would somehow direct criticism towards the government and cause mass incidents[6] (*quntixing shijian*). Most of the "peasant movements" or "peasant revolts" for which foreign media showed unwavering enthusiasm in their reports during the 1990s may be summed up as mass incidents. Under such circumstances, it would have been very unwise for the local government to stick with the old "mass line".

When the masses may not be effectively mobilized, public trials are not as effective as it was during the 1980s. One of my interviewees from *Linjiang* County told me that in 1995 when he was studying in the county's secondary school, the county had a public trial, but people were not interested, for most of those present were secondary-school students who were told to be there. Besides, as the society embraces more diverse thinking, public trials may not arouse shared emotions among the public. Especially when corruption has reached an unbearable point, people often comment by saying that those being held on the scaffold are supposedly criminals who "have no connections" and that, with a corrupted judicial system, those who have are still at large. Thus for criminals on public display, people feel a shared inexplicable compassion. Foucault (1995) argues that as the penalties of the modern age changes from cruel to civilized, public execution have been abolished because the volatile

[6] Translator: Mass incidents are defined broadly as "planned or impromptu gathering that forms because of internal contradictions". https://en.wikipedia.org/wiki/Mass_incidents_in_China.

roles played by the people in the rituals of public executions often lead to a result where "rules were inverted, authority mocked and criminals transformed into heroes, the shame turned round". Perhaps the gradual withdrawal of public trials and parades from China's penal system has something in common with the decline of the public spectacle of execution in the West.

5.2.2 Police-People Public Security Alliance

When the old "mass line" and the village-level organizations fail to deal with public order and security problems, grassroots public security agencies have to step up their efforts with "special campaigns". The old framework is superseded by new tools, which mainly include the police-people alliance and the "informant" system.

The police-people public security alliance (The "Alliance") has been widely built in the late 1980s and the early 1990s. In a meeting held in June 1981, the Central Committee of the Party came up with "a comprehensive policy for social security" for the first time, laying equal emphasis on prevention and resolution—a shift in the old thinking pattern that lays more emphasis on crackdown over prevention or on treating symptoms over treating root causes. In 1986, the Department of Public Order Administration of Ministry of Public Security issued "Opinions on the Establishment of a Public Security Patrol Network in Cities". In 1988, approved by the state council, MPS issued "Instructions on Building a Stronger Mass Alliance on Public Security". In 1991, the Standing Committee of the National People's Congress issued the "Decision Concerning Strengthening Comprehensive Management of Social Order"; the CPC Central Committee and the State Council issued the "Decision Concerning Strengthening Comprehensive Management of Social Order" in the same year. To "strengthen comprehensive management of social order", both Decisions requested governments of all levels to mobilize urban and rural residents, students and employees of government agencies, public service units (*shiye danwei*) and businesses, in establishing "public security organizations of self-defense and self-autonomy by the masses", in engaging in all sort of crime prevention attempts and in forming a police-people alliance. The promulgation and implementation of these new documents have effectively set the Alliance in motion.

In late 1980s, when village organizations and the political system at both the town and village level have failed to ensure public security, and when the police department is generally short-staffed, the Alliance has been established as a remedy. Mr. Li, a member to the Alliance at *Linjiang* County, described the relationship between Alliance members and police officers as one of "mutual benefit" because "we are their eyes and ears, or extensions of their hands and feet. They are mostly non-natives. They cannot work properly without our assistance." That is nothing but the truth. For instance, when a non-local police officer has to track down a suspect in the village, he needs help from the local folks. He has to seek help from village officials if it were not for the Alliance. But village officials may inform the suspect because they have a good relationship. So the suspect would have run away before the police could arrive. Even if they do not have a good relationship, the officials would be reluctant to cooperate for fear of displeasing their own folks and putting their own lives in danger. So having a local person as the Alliance member would make it possible for the police to "double the results with only half the efforts". Mr. *Li* also said, "If bus-riders get scammed by fake pull-tab prizes, I could track the scammers down to their own homes, for I know this town well; but the officer has to 'catch them doing it' in the bus." To some extent, the Alliance members are taking the place of village officials.

However, Alliance members are not supposed to enforce the law on their own. They cannot take any action until being told to do so by a police officer. When they come across a criminal activity while patrolling alone, they may do nothing more than protecting the crime scene and reporting it to the police as soon as possible. Even when they have kept the suspects under control, instead of handling the situation by themselves, they must take the suspects to the public security authority. However, the rural residents may not be able to notice much difference between the daily routines for the police and the Alliance. And that is probably why the rural residents would confuse the police with the Alliance. Still, when it comes to the right of law enforcement, Alliance members are very careful to ensure compliance with legal procedures, which may often entail the presence of police officers. Some of them may also have the experience of or have heard about other's experience of being questioned by a seasoned suspect on the right to enforce the law. Some even got sued.

We should not underestimate the role of the Alliance as an auxiliary for the police force, but it is also not short of problems. To start with, one of the main ones is its budget. The earliest form of such alliance exists in certain metropolises in the 1960s. In the 1970s, some teams are formed by "professionals" in urban areas of some cities and towns or townships, mostly staff members of businesses and public service units. In the early 1980s, instead of directly "staffing" it, businesses and public service units support the alliance by "sponsoring" it, which marks the beginning of "fee collecting". Since then, the teams have been supported by the fees they have collected and they take care of their own "revenues and expenses". When teams are formed in the rural areas, they have inherited this practice. However, the fee collecting practice is intrinsically expansionary, which may easily cause unreasonable fee collection. And that is also indeed what has happened in many places. Now fees for "public security teams" or for "temporary resident administration" have been canceled for being unreasonable by regulations promulgated by the central government. After canceling public security fee, these regulations also stipulate that public security funds should be properly handled by governments of all levels through normal channels for fund allocation." Presumably government of all levels must stick to its budget for finance. However, this is very hard for the town (township) government with a growing financial pressure.

A more important problem would be illegal action and rights infringement by Alliance members. As non-governmental officials, these members are often under-qualified. And since they are also not effectively bound by regulations, their misconduct is thus very commonplace and often violates legal rights of rural residents. The negative impact incurred by such misconduct has attracted close attention from the government, which requires public security authority to better manage the teams. In February 1993, being involved in a dispute with deputy chief of his village, a rural resident named *Ding Zuoming* from *Lixin* County of *Anhui* Province was locked up in a room by county police officers, and then brutally beaten up for two hours by 3 Alliance members using electric and wooden batons. Being seriously injured, he died the second day. In May that year, to deal with misconduct by Alliance members, the Ministry of Public Security issued a notice that requires local authorities to tighten control on the Alliance by putting the Alliance teams under police leadership. In March 2003, a suspect was beaten to death by an Alliance member at *Shenzhen*; an Alliance member at *Zhengzhou* committed a series of robbery and murder cases. Heads of the government including President *Hu Jintao, Luo Gan* and *Zhou Yongkang* successively gave instructions on these incidents. In September 2004, the Ministry of Public Security issued a notice calling for special efforts from all local authorities to dismiss existing members, following the principles of "stop rehiring and gradually reducing members to achieve ultimate abolition". By January 1 of 2008, public security authorities have dismissed all public security members and there will be no justifiable grounds for rehiring any of them.

The central government hopes to use such reform to abolish the system once and for all, but things are not that simple. In the 2005 reform to restructure the government, all Alliance members at *Wangzhen* Town of *Pingyu* County have been dismissed, leaving behind just 6 police officers at the township police station. Apart from taking care of the daily routines and making responses to reports, they have no more energy for other things. Police officers could not possibly cope with the rise of crime when the 2006 spring festival ended. The station chief complained to the bureau chief about under-staffing, but the bureau chief said they already had more people than other stations. The station chief then went to see the town's Party secretary, who happens to be a police academy graduate and has been very supportive of the police station. The secretary told him, "You may secretly get those competent alliance members back. The town authority will reimburse some of the costs. But don't let county leaders find out." The same method is also used in some neighboring towns and townships. If the station chief has a good relationship with the chief or Party secretary of the town (township), his claim to rehire former alliance members will be supported. The rehired members are given a miserably low monthly wage of only 200 Yuan. During my interview trips, they were busy with petitioning the government to give them one-time dismissal compensation or to give them a more generous paycheck if the government planned to keep them. In fact, firing all the Alliance members once and for all is unrealistic: on one hand, as an auxiliary for the police force, they are irreplaceable; on the other hand, serious under-staffing in rural police departments will not be alleviated any time soon. With rising rates of crime in rural areas, if the police force does not increase, major reductions in the number of alliance members will do no good to crime deterrence and order maintenance.

5.2.3 The "Informant" System and Rural Gangsters

If the Alliance system may be regarded as the "overt" alternative to the old village-level organizations under the "mass line", the "informant" system may be regarded as the "covert" one. Informants are very common in police investigation, but before the mid-1980s, when thorough implementation of the "mass line" allows the police to acquire active assistance from almost all masses and absolute support from village officials, the police do not really need any informant. Things have changed in the mid and late 80s, and that's when the informants become increasingly important. In 1984, the Ministry of Public Security compiled "the Guidelines on Criminal Intelligence", which is still a confidential document till this day. The information the author has consulted in police stations reveals that the first informant appeared in 1985:

> … Luo, a repeat offender and an ex-convict, knows something about counter-intelligence tactics. To keep our intentions from being revealed, our station sent two undercover informants to blend in and get close to Luo. After continuous efforts, the informants finally got to earn Luo's trust and …[7]

In this record, "informants" are a group of special officers disguised to lure criminal suspects into "taking the bait", similar to undercover officers. They are different compared with "informants" in its general sense. In general an informant refers to a person employed by the public security authority to provide information, also known as special intelligence officer, or the eyes and ears. Based on the manner of working, they may be classified into red and gray informants. Red informants are helpful residents who are willing to provide information to the police. Some act as informants out of a sheer sense of justice: these people would correct the wrongs and take the criminals in disdain. Some act as informants to keep up their relationship with the police. Others do so for financial gain. That's what distinguishes rural informants from urban informants, for the latter work with the police mostly for financial gain (Jianfeng and Xi 2007a; 2007b).

A gray informant is a novice gangster under arrest on petty offenses and may be legitimately treated with leniency. So the police would try to talk him into working for the police as an informant. If he is willing to do so, the police would immediately release him and drop charges. Sometimes a gray informant may also be a convicted offender, but the police would develop him into an informant so that he may "make amends for his crimes by doing good deeds" (*daizui ligong*), though it may be risky to release a convicted offender. A gray informant has his own means of acquiring information and they would frequently report back to the police. For instance, they may come back to the police and say "XX asked me to help him in a stealing job, but I did not do it. He may be responsible for the recent thefts." Informants are also planted by the police in *Falun Gong*[8] organizations. If they have any plans for going to *Beijing*, these informants would call the police and disclose details about

---

[7] Public Security Bureau of *Linjiang* County: "Good Services by Informants at *Tanqiao* Village", published on the 18th issue of "Reports of Current Events", April 20 1985.

[8] Translator's note: *Falun Gong*, literally "Law Wheel Practice", is a spiritual discipline which China banned in 1999 and calls an "evil cult". It was founded in north-east China in 1992 by *Li Hongzhi*.

the trip. Every month the police would have regular talks with these gray informants, mostly over dinner, greet them and keep up a cordial relationship with them. The police would also make an effort to help them with things like applying for a needed certificate or an ID card, saving them the trouble of queuing and having the things delivered to them. Deeply touched by the police's thoughtfulness, they would actively work with or for the police.

Currently the police stations in rural areas have planted in almost every village their informants, who would report latest occurrences to the police when necessary. For instance, if there is a murder case, they may inform the police of the dead person's enemies. They would also report to the police on a newcomer or an abnormality, e.g. a secret plan to start a gang fight. In solving both trans-provincial "kidnapping over debt" cases, the chief of police at *Wangzhen* Town of *Pingyu* County got informed early on by a snitch that "a stranger got locked up". After solving a case, officers may apply for a prize for informants from the special intelligence fund, for example 500 Yuan for assistance with arresting a wanted fugitive listed on the Internet. All officers have their own contacts and do not exchange informants. Even with redeployment, the officer would keep contact with his informant. All this is done to keep them safe.

Since the mid and late 80s, the police have always stressed the importance of combining the mass line with the informant system, as shown by the following two paragraphs.

> Third, we may call upon the masses to provide leads to track down criminal gangs. A successful example would be XX Police Station, which went to Jinglan and Sanhe Villages to collect leads from the local folks, knowing that three machinery theft cases have taken place there. Forth, we may track down fugitives with the help of informants or snitches. A successful example would be XX Police Station. This month, with the help of leads provided by their snitches, they have successively arrested 16 criminals in a temporary theft gang led by the notorious fugitive Zheng Yinnian and 3 criminals in a roaming theft gang led by the notorious fugitive Su Renzhong.[9]
>
> More efforts should be devoted to intelligence gathering. As shown by years of practice, no matter what happens, successful intelligence gathering will give us an edge and allow us to properly handle a case. We must use all sorts of means of intelligence gathering and deploy all sorts of law enforcement officers to gather information from all sorts of channels. We must combine overt and covert means of intelligence gathering to keep us informed and ensure fast response.[10]

Xinhua News Agency, the official news organization of the Communist Party, declared that *Falun Gong* is "opposed to the Communist Party of China and the central government, preaches idealism, theism and feudal superstition".

[9] Public Security Bureau of *Linjiang* County: *Eleven Gangs Under Arrest Within Half a Month in the County's Spring Campaign Against Crime*, Published on the 13th Issue of "Reports of Current Events", March 25 1985.

[10] Speech given by Director of Public Security Bureau of *Linjiang* County at a meeting for across-the-bureau section heads, station chiefs and team leaders, Oct 8 1992.

Since the beginning of the 1990s, public security authority has taken the informant system even more seriously, and has given more publicity to cases successfully solved by informants. The following is one of them.

> Strict management and bold deployment of special intelligence informants by XX Police Station have produced instant results in this campaign against crime. Within ten days, informants have provided leads for solving a total of 35 criminal cases, and have arrested a group of fugitives who have been committing crimes after escaping from labor camps, including the notorious Ding Mingsheng.[11]

Currently the informant system has become one of the basic systems for town and township police stations, and their higher authority has also made specific requirements for it. In 2002, performance evaluation for police stations at *Linjiang* County requires that each officer should have at least 2 informants and should build personal files for them. "Failure to meet requirements will lead to loss of points for an officer: 0.5 for each informant short of the required number and 1 for each non-functioning informant." Some information needed in public security efforts is only accessible through informants. Officers explained to the author that "when we are present, a lot of things seem different. This has something to do with our identity". Thus the police would do their best to keep up their relationships with some gangsters, hoping to secure some key leads from them. Officers also told us that "village officials dare not to help us, so we have no hopes for them. They are often really timid and too cautious. They are often afraid of gangster's revenge." In fact, village officials would even help gangsters because they are "their own people". Officers continued, "We all want to be 'the good cop', not 'the bad cop'. This is how things are nowadays. People will not help unless they get paid for it. We should all be grateful that they are not causing us any trouble." Some village officials would snitch to the police that a suspect has come home and secretly tell the suspect to run away. Most officials choose to keep their nose clean, refuse to take sides, and turn a blind eye to gangs. Those who are willing to work with the police and be the police snitch are only a very limited number in village officials.

As village officials are often inactive in public security efforts, police are faced with the pressing need to build up a public security alliance and the informant system, which seems to be the only practical alternative to inactive officials. However, they are also faced with the budget problem. In my field research, local police officers often complained to me about having limited funding, which was inadequate for effective performance, whereas their superiors often criticized them for not "thinking out of the box", and for not having enough informants for crime control. Other problems for the informant system, as scholars have pointed out, include severe shortage, under qualification, lack of deep infiltration, etc. (Yufan 1996).

[11] Public Security Bureau of *Linjiang* County: *Bold Deployment of Informants*, published in the 13th issue of "Brief Reports by *Linjiang* Public Security Authority", Jan 8 1991.

5.3 "The Mass Line" in Public Security of the New Era

As described previously, during the 1990s, the second and third generations of gangsters have successfully "seized" opportunities offered by the times, resulting in deteriorating rural order. Developments in the public security authority's "special programs" has paradoxically coexisted with the rising of rural gangs Radical changes in the rural areas have confounded the rural residents, whereas rural gangsters have already found their positions and paths for "personal development". The rural China is undergoing an unprecedented governance crisis.

In the context of governance crisis, the public security authority becomes embroiled in disputes between rural residents and the local government. This has in turn resulted in much tension between the residents and the public security authority. In dealing with agricultural affairs, it would often side with the local government, and is regarded by rural residents as the "hired thug", hence its level of legitimacy continuously diminishes. Therefore, since the 1990s, the higher public security authority has constantly forbidden the local authority to interfere with agricultural affairs, as proven by the following paragraph.

> Recently, in its interference with agricultural affairs, improper action on behalf of the local public security authority at Jianli County has caused the involved party to commit suicide by hanging. The terrible negative impact has attracted attention from the central government and the provincial Party committee. Also in recent days, "Suizhou virgin prostitution case"[12] reported by Hubei Daily and other news media has caused sensation among the general public. We've learned from these two incidents that public security authority must be extra cautious in handling agricultural affairs, family planning affairs, mass incidents, conflicts between the people and town (township) officials; and in treating vulnerable groups such as laid-off workers and employed people living in extreme poverty. The slightest carelessness could harm the fundamental interests of the people, and in turn tarnish the images of public security authority, party committees and the government. ... As an important instrument for people's democratic autocracy, the public security authority must work as the guardian of fundamental interests of the general public. It is not the hatchet man for a few bureaucrats at the local level....[13]

The letter has also specified procedures and rules for requesting police deployment in handling agricultural affairs. The higher authority generally requires that the local public security authority should make an effort for "reconciliation", and never "add fuel to fire".

With the deteriorating rural order and exacerbated conflicts between farmers and public security authority, coupled with ineffective "special campaigns", public security authority is now faced with "internal disturbances and external threats" (*neiyou waihuan*). Thus the public security authority gradually moves away from what once

---

[12] Translator's note: A police officer made false accusations of prostitution against a girl and several men, probably to look good on performance evaluation.

[13] "A Letter from Comrade *Chen Xunqiu* to Directors of Public Security Authorities at the Prefecture, City, and County (District) levels", August 8 2001. Chen is Standing Committee Member of the Hubei Provincial Party Committee, Secretary of the CPC Provincial Political and Legal Affairs Commission, Director of the Public Security Bureau).

used to be an excessive emphasis on "special campaigns" and revisits the idea of the "mass line". In 1993, the Minister of Public Security published an article in *The People's Daily*, proposing to "review Chairman *Mao's* teachings to ensure public security in the new era" (Siju 1993). Herein "Chairman *Mao*'s teachings" refers to what *Mao* laid down right after the founding of the nation as "the ground rule in ensuring public security", that is, "to educate the masses and mobilize the masses". The article also proposed a combination of special campaigns with the mass line as the right working method for the new era. Guided by this method, public security authority at all localities start to make new attempts at carrying out the new mass line. They re-establish their close contact with the people, and call upon them to participate in public security administration. The police and the people work together in crime prevention and reduction and in restoring social order. In the rural areas, these attempts mainly include home visiting and community policing.

Under the home visiting policy, police officers pay regular visits to people living in an urban street community[14] (*jiedao*) or in a rural village group. This allows them to understand the local environment for public security, identify patterns and features in criminal activities, and design specific measures to deal with main local concerns. This policy aims at preventing, controlling and fighting against all sorts of crimes. First established by the *Ningling* County Public Security Sub-bureau of *Henan* Province, it has been regarded as a new policing approach based on some new tentative efforts. It brings reform to the traditional administrative model of rural police stations, and points out a new path for combining special campaigns with the mass line under the system of market economy. Basically the policy requires officers to: follow up clues, find possible hiding places for fugitives fleeing from arrest, accusation or custody (the three types), inspect key targets, collect information for public security, instruct local security committees (*zhibaohui*),[15] publicize the socialist law, and listen to people's opinions and suggestions. The basic method is to combine home visits with the following four aspects (the four combines), namely case investigation, infrastructure building, team building and mobilizing the masses. Specific rules include: chief of town-level police stations (*suozhang*), party instructor (*zhidaoyuan*), senior police officer (*jinzhang*) and police officer (*minjing*) are required to make several home visits every month and record their visits within their respective jurisdiction. When making those visits, they are required to walk or ride a bike to be more approachable. Also a system of weekly presentation, monthly report, seasonal contest and yearly assessment has been established to encourage police officers to actively engage in home visits (Can 1994; Shouxin 1995).

---

[14] Translator's note: A subdistrict is typically part of a larger urban area. In general, urban areas are divided into subdistricts and a subdistrict is sub-divided into several residential communities.

[15] Local security organizations were established in both urban and rural China in 1950. By 1951, over 170 towns have established their local security committees to assist public security authority.

The home visit system has produced positive results: it has produced more extensive impact of the "crackdown" campaign and improved the police-people relationship; it has helped to build a more vigorous police force with a stronger moral rectitude (*zuofeng*: otherwise translated as *style of work*) and greater professional competence (Can 1994). Its advantage lies in its two-pronged approach—combining "proactive" and "defensive" approaches. Instead of "responding" to criminal activities, it "proactively prevents and combats" crime. The model for public security administration is characterized by more flexibility instead of invariability. Frequent visits by police officers in their jurisdictions allow them to timely discover clues and solve cases, thus enhancing the impact of crime control and also making the crackdown effort a daily routine for the police. The public security authority also has a better control of the whole society. As the home visit system has produced remarkable results, in 1995, the *Henan* Provincial Public Security Bureau ordered public security agencies across the whole province to replicate the *Ningling* practice and establish the home visit system. The home visit system is installed as a way for adopting the mass line in the new era. Under the new mass line, the local police officers are required to blend in with the masses, keep close contact with them, provide them with loving care and convenience, and keep up a close police-people relationship; they are also required to discover and collect case leads through home visits, and use home visits for crime prevention and control and criminal investigation.[16]

Apart from *Ningling*, many other places around the country have made similar attempts while considering their local conditions, which have turned out to be quite successful as well. For example, in 1997, *Zhushan* Police Station under *Hongjiang* Municipal Public Security Bureau of *Hunan* Province extended the home visit system to the village level. Police officers are required to make visits on the following 6 occasions, including making inquiries, responding to alarm calls, delivering permits, licenses and certificates, legal education and help and educate program, mediation for security enhancement purpose, and providing assistance to underprivileged people. Home visits by police officers who work and live in the village and stay close to the rural residents would allow them to accomplish administration purpose and to control and prevent crime (Chengshu 2000). In 2003, *Anxian* County Police Station of *Sichuan* Province assigned areas of responsibility (AOR) to its officers and evaluated all officers' performance based on the effectiveness of patrols and visits. Their main jobs have been summarized as "one hundred visits, two officers on one case, three unifications, four tasks and a 'five-five' system[17]" (Daigen et al. 2003). In

[16] See also the written summaries for *Pingyu, Huaxian* and *Anyang* to study the *Ningling* experience. The author collected these summaries on field trips to *Pingyu* County of *Henan* Province.

[17] One: Each AOR officer is required to make one hundred visits. Two: two officers are supposed to work together on one case. Three unifications: uniform looks on bicycles, working bags and working procedures. Four tasks: four tasks (suppression, prevention, control, management) are assigned to AOR officers, who would be held accountable for completion of the tasks. Five-five system includes 5 delivers (officers are required to deliver contact information, responsibility letters, legal knowledge, safety tips and the Party's "farmer enrichment policies"), 5 inquiries (make inquiries about financial status, behaviors of key targets, factors undermining security, social security status, advice and suggestions), 5 checks (check *hukou* and other identifications, check rented homes,

Linjiang County of Hubei Province, the pioneering policy by one of the police stations to "make a hundred visits per month" gained approval from the county PSB and was introduced to many other places. This policy is deemed to have unlocked the people's potential to the maximum, encouraged the people to participate in crack-down campaigns and brought the police force closer to the people. "Through the 'a hundred visits per month' policy, the general public may get into close contact with police officers, feel their constant presence, and somehow become more courageous and confident in the fight against criminals." All of these are tentative efforts to implement the "mass line" in public security efforts of the new era.

Community policing adopted at the village level is a new mechanism for implementing the mass line in the new era. "Community policing" refers to collaboration in ensuring security and crime prevention and control through establishing close partnerships between community members and local agencies. It is guided by local agencies and taps into community forces and resources. It is problem-oriented policing that aims to prevent and reduce crime, build a safer community and eventually boost the overall well-being of the people. It is generally believed to a concept that has been passed into China from Britain during the 1980s. However, this is untrue because it was actually brought to Britain from China at the beginning of the 1960s, when British policing experts visited our country, being amazed by the low crime rate in a country with the largest population, they went back and proposed "community policing" (Li 2001). The government and law enforcement agencies did not pay special attention to it until it was again "imported" into China. This probably has something to do with the worsening social order after reform and opening-up.

In the 1990s, many places in China have piloted community policing. For instance, in 1996, during the community policing reform at *Meixian* County of *Guangdong* Province, local agencies shifted their focus to administration and crime prevention. "Law enforcers" are turned into "community workers". Specific methods include: first, divide each precinct into several sections and assign each section with several officers, fully responsible for routine work in the section; second, and establish substations and assign police in remote areas or areas disturbed by social disorder. Establishment of substations gives people the convenience to get things done without having to travel afar. By keeping the police close to the people, it ensures instant handling of cases and disputes, and keeps local gangs from unchecked growth. The division of precincts into smaller sections and establishment of substations have ensured across-the-county coverage of the public security control network (Enshun 1996).

---

check explosives and dangerous items, investigate public security cases and criminal cases), 5 helps (help the people with permits, licenses and certificates, help key targets reform themselves, help with dispute mediation, help rectify safety hazards and help people become better-off) and 5 builds (build up human resources, infrastructures and technologies, build up security organizations, build up security and patrol alliances, build up people-based crime prevention networks, and build safe neighborhoods).

In a meeting on the work of police stations held in *Hangzhou* in March 2002, the Ministry of Public Security proposed a plan to adopt community policing in rural areas across the country by the end of 2005. In September 2006, MPS promulgated *the Decision to Adopt Community Policing and Village Policing*, requiring its rural agencies to adopt community policing at the village level to build a new socialist new countryside. The Decision laid out the main tasks for community and village policing: engaging the masses (mass work), understanding local conditions and people's opinions, managing population, adopting security precautions, and maintaining order. The Decision requires community and village police to: make visits and investigations, publicize community policing and engage the people, conduct patrols and provide protection, make site inspections, and report on the work of the police, etc. The Decision requires officers to focus on mass work and establishing a harmonious police-people relationship and effectively prevent and reduce crime in both urban communities and rural villages. Village policing then was promoted to the whole country. By August 2007, in rural areas around the country altogether over 80,000 police stations with over 90,000 officers have been established, with one or several administrative villages designated as a precinct. The establishment of this new policing mechanism allows officers to blend in with the people, handle disputes first time, and prevent and curb crime from a close distance. Their constant presence has effectively enhanced the sense of security for rural residents.

My fieldwork reveals positive results from village policing. First, it unites the village and non-governmental organizations. Having the police around, people feel safer and may count on the police and upright villagers to stand up against criminals. Second, it provides more sources of information. Before village policing, main sources of information include victims and informants. Ordinary residents are reluctant to provide information to the police, because the police have very little contact with the people unless they are arresting someone or giving someone a fine. They are strangers to the people, to whom the people are also reluctant to provide information. Having more contact with the people allows the police to have positive interaction with the people. The people would consult the police, ask for help from them, report cases to them. Naturally they become more willing to furnish information to the police. This even allows the police to pick up some leads by accident. Third, it allows the people to see a change in the police's attitude towards work. Before village policing, officers do not visit the villages unless they have to. They do not stay in the village unless they have something on their hands. After the adoption of village policing, being around the people has rid the police forces of bureaucratic arrogance.

In a word, village policing has been used as one of the best ways to adopt the mass line in the new era of public security. However, we should not overestimate its role either and I will further discuss this in later sections.

5.4 Changes in the Exercising of Political Power at the Grassroots Level

5.4.1 *"The Mass Line" and "Governance Through Supererogation"*

In the early days after reform and opening-up, with worsening order in both urban and rural areas, the national leaders made a resolute decision to adopt the "mass line" in public security efforts. The mass line has turned out to be quite successful in restoring order, and this probably has something to do with the revolutionary ideals that the people still believe in. People still remember vividly "the Cultural Revolution that claims to rely on the masses for strength", and would look at the mass line as a continuation of the governance paradigm that had existed before reform and opening-up. To use an expression with political overtones, the government officials and the masses have been "well-taught" with revolutionary ideals. Therefore, the success in restoring order should not be attributed to the mass line itself but rather to the fact that it has been the right method for exercising political power at the grassroots level. Let's look at one example for crime control for this time period.

> ...When we heard about the case, we established that it was a serious case and decided to start investigation without any delay. Our leaders personally took part in the planning and led the investigation team. Within ten minutes, 18 officers led by our deputy director Deng Yunchu drove to the crime scene. We also made an emergency phone call to report the case to the commune. The commune later deployed over 160 people to intercept the criminals. Both the local party committee and the commune attached great importance to this case. The district Party committee secretary personally participated in the investigation and all relevant authorities were involved. One of the drivers who had already set off from Xiangtan City came back as soon as he heard about the need for vehicles. Local government agencies also contributed their vehicles. A special tribute should be given to business owners Liu Dingsheng and Li Shifan. Both in their fifties, they rode bikes to help track down the criminals—at the forefront of the search regardless of any danger or fatigue. At 11:35, they came to a bridge and came across a fidgeting young man in a white helmet who matched the description of the suspect. They got off their bikes, followed him and when they could, they charged towards him and caught him in the hands. He desperately tried to fight back, and broke his right hand loose, took out a dagger and attempted to attack them. ... With help from the public, they eventually got him under control. ... After proper legal education, he snitched on two other criminals, both from Liu Shiba production brigade. We organized two teams and drove to 20 miles away from the crime scene and then had to walk to the scene because there was no road for vehicles. Our officers walked for miles to get there. At 2:40 in the next morning, with help from village officials and militants, our officers besieged the suspects' housing and arrested them. We have not only taken back lost items and tactically arrested the suspects and solved the case.
>
> To sum up, the rapid solution of this case may be attributed to: considerable emphasis and effective measures by the Party committee, a combination of special campaign and the mass line, great support from the public for besieging the suspects, and immediate action and hard work by police officers.[18]

[18] PSB of *Shanxiang* County: "Five Hours of Hard Work Paid off in Successful Arrest of Robbers", the 10th Issue of *Brief Reports on Public Security*, April 6 1980.

The successful solution of this case may be attributed to the mass line, or more importantly, the good work ethic of the law enforcement personnel. When drawing lessons from this case, law enforcement may have used some expressions with political overtones, but that is a true reflection of the reality. That is, with limited technological resources, law enforcement has to be on standby around the clock, and to go beyond the call of duty. I refer to the former as "governance through physical presence", and the latter as "governance through supererogation". In my previous study, based on requirements, duties and responsibilities for law enforcement, I've laid out three different methods of exercising state power: governance through physical presence, technology governance and supererogation. Governance through physical presence means that state power is exercised with law enforcement or civil servants physically present at the rural society; technology governance means rural governance is realized through the use of advanced technology; and supererogation means that law enforcement often has to go beyond the call of duty or do more than what is required on the professional, ethical and moral levels. These are the methods of exercising political power at the grassroots level and the effect of governance is also determined by the effect for these three methods (Baifeng 2009b). The following case may illustrate "supererogation".

> ...To follow up case leads and ascertain case facts, on May 30, Chang Zhongyuan and two other police officers arrived at Wushan County, which was still more than 90 miles away from Luoping District—a hilly area with no convenient transportation. At five o' clock the next morning, they set off without having breakfast. The road outside the county area instantly became rough and bumpy. They had to climb up a 15-mile-long slope that spiraled around the hill. They climbed hills over hills. They were exhausted and soaked in sweat. They did not have the first meal in the day until 4 p.m. They arrived at Luoping Commune at 5:30. Having solved the case and retrieved the items taken, they reported to the Party committee of the commune. The good result may be attributed to their bravery, selflessness, stamina and hard work.[19]

Under the mass line, law enforcement is required to: get close to the masses, put aside their personal interests, always show the readiness to do hard work and seek the moral high ground—that is to rule by being on standby the whole time or going beyond the call of duty. Otherwise the mass line could not have been thoroughly executed. If law enforcement were not always present or did not seek the moral high ground, the masses would have lost faith in state power or felt insecure, which may have made it impossible to carry out the mass line.

The effective implementation of the mass line may also be attributed to qualified officials and people. Since the mid and late 80s, grassroots public security agencies have rightfully accused village organizations of inaction. I've discussed the reasons behind their inaction previously. However, we should look at their inaction in light of the social context. Rural order in the Commune era has been well maintained even without any public security organizations. But after the collapse of the commune system, many social order problems have arisen, because when people no longer

[19] PSB of *Linjiang* County "Revolutionary Committee" (committees that took over the functions of government during the Cultural Revolution): "No Mountain can Hold Back Law Enforcement", the 16th Issue of *Reports of Current Events*, June 24 1975.

believe in collectivism or revolutionary ideals, organizations that require action from rural residents will naturally become less binding on them.

The effective implementation of the mass line in the early days of the reform also has something to do with the acquaintance society nature of rural China. On one hand, the feelings people have developed for communities during the commune era has not worn off; on the other hand, the Party and the state have not yet found the new way to nurture feelings for communities. In the acquaintance society, out of strong feelings for their communities, people have effective penalties for transgressions, which reinforced social unity.

5.4.2 "Special Campaigns" and "Technology-Based Governance"

Since the mid and late 80s, as rural order worsens, governments at all levels have gradually abandoned the mass line as a way to rally the public. For one thing, it seems to be an idea that goes against the rule of law; for another, it becomes increasingly ineffective when social changes are under way. Faced with serious public security problems, the government and law enforcement have to step up their effort in "special campaigns". This has been done mainly from the following two aspects: establishing a public security alliance, which may increase law enforcement personnel; and boldly building up a system of informants, which constitutes a significant breakthrough in investigation tactics. While it is less likely for the government and law enforcement to mobilize the public, these two systems have to some extent alleviated some problems, e.g. difficulty in collecting criminal information, though both systems have not been short of their own problems, e.g. having a low budget.

The development and transformation in special campaigns have been made possible by: first, improved public infrastructure—road building has greatly enhanced the accessibility of rural areas; second, the introduction of new technologies to rural areas—full coverage of television, phone and the Internet has made it easier for the government to publicize national policies and engage in direct communication with the people; and third, improved equipment and enhanced technological capacity for the grassroots government—better means of information collection and storage means better control over rural areas. Just as Giddens (1985) has described, with the development of the "nation-state", people of rural areas will continuously break away from regional constraints and become subject to national regulations, and affected and controlled by the nation's administrative surveillance, industrial management and ideology. This process is achieved in two ways. One is (the nation-state) to enhance its soft control through penetration of ideology. The key to soft control lies in making people accept the basic values spread by the national ideology and enhancing people's recognition of the state's power and its orthodox order mechanism. The growing popularity of broadcast, television and the Internet is conducive to the enhancement of soft control. By using these means of soft control, national ideology may successfully turn villagers into citizens. The other is to enhance its hard control through using modern technology. Instead of public use of violence for punishment, the nation-state relies on infiltration of administrative powers, for instance through surveillance by policing and identity systems in a wider space and in more social circles. Enhanced hard control largely depends on the advancement in methods of communications, transportation, information retrieval and storage. As a result, the surveillance system achieves an unprecedented level of development.

As far as the rural residents are concerned, the government's more effective technology-based governance or its stronger technological control means that "the cops will arrive with one single phone call". For them, the police represent the state or law "going to the countryside" (*xiaxiang*). They may report a case any time whereas the physical presence of the police is not required; and they may count on the legal solution of disputes after case reporting. This has been made possible by better communications, more convenient transport and increased media coverage. Having obtained information about national policies from the media and been educated about the laws, rural residents become more aware of their citizenship, and thus more willing to cooperate with the state. That makes it easier for the state to achieve governance goals. Therefore, we might conclude that enhanced technological capacity enables the government to stop over-reliance on governance through physical presence or supererogation. Grassroots officials and police officers do not have to be physically present unless they are up to a specific task, or they may physically stay away. At the same time, the government only have requirements for general work ethics and no longer require acts of supererogation. Supererogation gradually loses its charm. Therefore, since the mid and late 80s, special campaigns mainly rely on technological governance. However, when technological advances assist the government with handling traditional deviant behaviors that disrupt the rural order, as factors influencing social order become increasingly complicated, even the latest means of technological governance may still turn out to be incompetent.

As the government abandons "governance through physical presence" and supererogation, "technology-based governance" becomes more effective, and it has brought with it procedure-based rule of law. Such rule of law is based on procedures established for legal actions—especially those by government agencies—to protect rights of individuals and keep the government in check, and to "provide an applicable theoretical foundation for reforming power structures and for solving problems related to government legitimacy" (Weidong 1999: preface). After reform and opening-up, China has been moving non-stop towards the goal of procedure-based rule of law. But the legal procedures are not without flaws. As it protects rights of individuals and keeps the government in check—keeping it from "doing the bad things", it also rules out many options and possibilities for the government. The government's hands are tied when it tries to "do some good deeds". Besides, as far as the rural residents are concerned, the government must be doing either good things or bad things—there is no area in between.

These flaws in the procedure-based rule of law have made it more difficult for law enforcement to carry out special campaigns, or in other words, technology-based governance seems to be not as effective. Many rural residents are well aware of the fact that the police have to act by the law, which means they may not arrest anybody as they wish. The police also commented by saying, "they (rural residents) know the law. They know our hands are tied. We must have proof when we try to arrest someone. We used to arrest anyone that seems suspicious. But now that's not an option. We may get a lawsuit." Therefore, the police would choose to do nothing unless they have sufficient evidence for arresting a gangster. In one case, in a raid on illegal gambling at *Qiaotou* Village of *Linhu* City, the police were assaulted by a rural gang. Having difficulty in collecting evidence, the police did not press charges and had to secretly investigate the attackers to wait for another chance for punishing them. This has undoubtedly created a considerable negative impact. One of the rural residents commented on the police's behavior as having "turned public clashes into private grudges". When law enforcement may not "survive" unless it takes unusual steps through "selective enforcement of the law" (Xi 2005: 246), the problems involved deserve our due attention and consideration.

Staffing shortage and tight budget for law enforcement does not allow it to handle everything. Just as one chief of a police station in *Chujiang* City has described candidly, "we must bring in some revenue. It takes money to crack a case. So we must make sure that we have income. We have an unwritten rule to overlook cases that do not bring us any income. That may be seen as 'whoever has money gets things down'. But the same goes by courts, procuratorates or prisons, for example, one may cut a three-year sentence to one year by paying the fine; one may pay 10 or 20 grand in exchange for commutation; and one may be released on bail for medical grounds. However, all these practices have produced negative impact."

It seems like when faced with legal procedures and financial distress, law enforcement has developed a principle of their own—to avoid trouble whenever possible through case screening and selective compliance with legal procedures. Generally speaking, law enforcement does not deal with disputes that have not caused serious harm or disturbances. They also choose not to deal with cases with no obvious leads or solid evidence, or whatever case for which they may get by with doing nothing. These cases often deplete human and financial resources but produce very little result. Though well aware of the unlawful nature of such selective treatment, they have no other choice with limited staff and budget. More attention is given to cases that bring potential revenues. Therefore, the public is under the impression that law enforcement shows more enthusiasm for arresting gamblers or imposing fines than keeping order. They are in a legally gray area—they tend to impose fines if both custody and fines are applicable; with financial penalty paid off, they may turn a criminal case into an administrative one. In a word, selective enforcement of law in rural areas has obviously resulted in unlawful distribution of limited resources—in a way that means no good for escaping the predicament for law enforcement in its security efforts.

In its effort to cope with financial distress, law enforcement has left the residents a bad impression and created a very bad impact in rural areas. The rural residents told the author that "gangsters are large in number nowadays. Those thieves, blackmailers and troublemakers get caught and released, caught and released. Always the same lot." "They may easily get caught and easily get out." A police officer in *Linhu* City told the author that "we have no other choice but to catch them and release them after. We need evidence for conviction. Without evidence, we had to release them. We know it's always the same lot up to some bad things, but getting evidence is impossible." Bearing that in mind, law enforcement gradually develops the habit of "not provoking" them, unless they have enough evidence to put gangsters into jail once and for all. Otherwise they would end up inflating the gangster's arrogance—for they have to release them after all; and they would also impress rural residents as being too lenient with the gangsters as if police and gangs are like families. "That's so not worth it".

Compared with before reform and opening-up, law enforcement has followed an entirely different strategy and inclination in exercising state power. It no longer unconditionally keeps order, but would first consider its self-protection and survival, while being bound by legal procedures and having limited technological and financial resources. As gangs have grown stronger, law enforcement has to ensure its own safety and development—ensuring nothing bad happens—before it may keep order and deter crime. Officers may choose to look the other way unless some serious business is up. In this way, law enforcement would be always dealing with "the weak ones" in the gangs and seems to be powerless against "the strong ones".

5.5 The Public Security Predicament and Its Implications

As rural order deteriorates, law enforcement has almost got into a predicament, being restricted by required legal procedures, limited technological competence and financial distress at the same time. The new mass line, in the form of home visit and village policing, is the main method in the state's attempt to solve the predicament. Both mechanisms have been promoted to more places, after some positive results have been achieved in pilot areas and approval has been gained from the higher authorities. However, these mechanisms are subject to the same restrictions. Not long ago, village policing has been promoted nationwide, but due to under-staffing in law enforcement, village policing offices have not been established in all villages—only in those more economically developed and more populous administrative villages with a more complicated public security setup. My fieldwork reveals that law enforcement could not possibly establish police stations in all villages. Therefore, despite some promising results, we should not overestimate the effect of these two mechanisms, and we should not expect the mass line to work in the same way it did during the 1980s or the way it climaxed during the commune era. It is not the ultimate solution to all the problems in rural China today, and the reasons are as follows.

First of all, the government may not use the mass line to govern a rural China that has grown disembedded from the acquaintance society. Having established a nexus of steady connections, gangsters have gradually organized into a loose alliance. The upper class of this nexus has developed legitimate methods for making illegal gain, which allows them to avoid legal punishment. Even in certain special occasions when they are forced to resort to criminal methods, they have the underclass of the nexus doing it for them. The underclass is the only group who still rely on criminal methods, but the government is faced with the inconvenient reality of not being able to reduce the number of them, because juvenile delinquents, left-behind children and unemployed youth constitute a fertile source for it. Even law enforcement in rural areas finds it hard to contain the rural gangs in alliance, let alone the rural residents. To make things worse, because rural residents do not know the gangsters well like they used to, they could not possibly provide relevant information to the police. So law enforcement would rather rely on informants—the insiders of the gangs—for information.

Second, the mass line has changed a lot compared with what it was during the 1980s. The mass line of the 1980s includes 3 aspects: first, respond directly to the demands of the masses—as a result penalties against criminal offenses are based on moral standards; second, rely on the masses to maintain public order—as a result public security efforts develop into nationwide campaigns and the public are directly involved in the judicial process; third, give the masses a sense of security by being identified with them and keep up public security agency's intimidation on criminals. The mass line of the 1990s has been reduced to just one of the 3 aspect—being close to the masses. The government would still "respond to the demands of the masses" and "rely on the masses to maintain public order", but to do so must follow required legal procedures. Specific spontaneous measures that are allowed for the public are also restricted by required legal procedures. The general public may work with law enforcement to maintain order, e.g. by reporting an offense to the authority or providing information, but may not act on its behalf and may not be directly involved in the judicial process. In other words, the rule of law in its modern sense stands in opposition to mass movements, which do not follow legal procedures and may easily lead to human rights violations. Therefore, in the new era, law enforcement stands against mass movement or administration of justice by the people.

So the current mass line is to some extent a revival of "governance through physical presence", maintaining order through keeping officers physically present. But it does not require law enforcement to go beyond the call of duty. That may cause a mismatch between "physical presence" and "ethical requirements". Without proper ethical requirements, law enforcement will not be as effective even when physical presence is guaranteed, which makes the mass line a mere formality. At the Double-lake Plains many county and city police stations have established offices in villages, but while I was on fieldwork trips, I did not see any officers present.

Of course, theoretically, law enforcement can also apply "pressure" on subordinates (Jingben et al. 1998: 25) to ensure the implementation of the mass line. In fact, "pressure" and "motivation" are used alternately by law enforcement to maintain public order. To meet the targets set by the upper authority, law enforcement will break down and delegate assignments to its subordinate authorities or individuals, require completion by a designated date, devise parameters for performance evaluation, and grant promotions or financial incentives based on performance. "Pressure" is well illustrated by the "leave no murders unsolved" motto and the system of parameters for performance on criminal cases. This evaluation system by the upper authority urges its subordinates to live up to standards. *Wu Yi*'s research reveals that (2007: 41) law enforcement is also motivated by the old "motivational" system.

This may be illustrated by the considerable efforts of towns and townships to "get ready for inspections" after the rural tax reform,[20] and also by all sorts of crackdown campaigns and other special campaigns by law enforcement, for which a set of evaluation parameters are also given—an indication for the inseparable relation between pressure and motivation. However, based on their long-term effect, "pressure" and "motivation" may have provided solutions to certain temporary problems but have failed to solve the overall predicament faced by law enforcement.

The predicament in rural order may be traced to the overall transformation of the Chinese society. As our nation transitions from a traditional socialist country to a modern market economy, the exercising of political power relies on modern technologies rather than the traditional models of "governance through physical presence" and "governance through supererogation". The shift in the model of governance matches the government's effort to establish a rule-of-law nation, but national finance fails to keep up with the required budget. In this way, while having discarded the old governance model, the government also fails to carry out the new model as planned. Based on the finances of the local government in central and western areas, this problem caused by limited budget may carry on for quite a long time. As it transitions from the old model to the new one, whenever the government is placed in a public-security predicament, it would reuse the old model in some new forms as a contingency plan for emergency or crisis. As it does so, the government attempts to inherit the merits of the old model without going against the basic rule-of-law principle. Thus the efficacy of the new model may not be ensured, since it may bring positive results, but may also stay just on the superficial level. Therefore, in the process of modernizing the nation, what kind of state power is to be exercised at the grassroots level is a critical question.

Mann (1986) distinguishes between two types of state power: despotic and infrastructural power. The former refers to "the range of actions that the ruler and his staff are empowered to attempt to implement without routine, institutionalized negotiation with civil society groups". The latter refers to "the capacity to actually penetrate society and to implement logistically political decisions". Based on these two types of state power, Mann compared four types of historic empires in ancient Europe—first, feudal states, low in both despotic and infrastructural power; second, imperial states, high in despotic power but low in infrastructural power; third, bureaucratic states, low in despotic power but high in infrastructural power; and forth, authoritarian states, high in both despotic and infrastructural power. Many scholars use Mann's paradigm for analyzing state power, including Charles Tilly and Joel Migdal, etc. We may also use his paradigm for contrasting state power in modern China. To build a bureaucratic state with weak despotic power and strong infrastructural power is an ideal in

[20] Translator's note: After the rural tax reform which aims at reducing taxation burden on rural residents, relevant authorities will visit rural areas for inspection. This requires rural residents to present all sorts of papers as evidence for reduced taxation on them. Law enforcement has to help residents get ready for such inspections by preparing various papers. More information may be provided by *Wu Yi*'s article "Getting Ready for Inspections" (*ying jian de you xi*).

our modernizing process. However, the predicament in establishing rural order and in implementing the mass line indicates a gap between reality and ideality, because state power still fails to penetrate and control the rural areas, and the building of infrastructural power is not yet finished. In addition, the building of infrastructural power may not be completed easily if it were to follow the current course of action.

Chapter 6
Rural Gangs and Public Order

During their renaissance in the 90s, an alliance is formed by rural gangsters in their nexus of connections. They gradually break away from the restraint of the acquaintance society. The alliance has formed and grown outside of the acquaintance society and thus has inflicted less direct harm on rural residents. However, we can never underestimate the negative impact they have brought on the rural order. The alliance of rural gangs become the super power in rural China and has greatly influenced or even fundamentally altered rural order.

We may analyze their impact on rural order by looking at the following 3 aspects—influence of local gangs on rural order, influence of external gangs on rural order, and resistance from local gangs against external gangs. To better grasp the rural order mechanism in the Double-lake Plains, this chapter compares rural order in the Plains with that in other rural areas. With differences in terms of development history, plantation crops, man-land relationships and cultural traditions, villages in different rural areas have distinct structures and social nature, and thus have reacted very differently to external changes and pressure. Once bringing out regional disparity in rural China, the author will present gang influence on the Plains more vividly. Compared with gangs in other areas, gangs in the Plains have exerted the deepest and most pervasive influence, often with an aim to make financial gain. Deviation from the traditional behavior pattern of rural land (*xiangtu luoji*) seems to be much more noticeable in the Plains.

6.1 Influence of Local Gangsters on Rural Order

Adverse influence on rural order by locally bred and born gangsters has been common to many of my investigation destinations, but their influence has been most distinctively noticeable in the Double-lake Plains. When faced with radical changes, rural residents are still at a loss, whereas the bold and crafty gangsters have already found

B. Chen, *The Underworld of Rural China*,
https://doi.org/10.1007/978-981-19-8710-6_6

their paths and positions for "personal development". Being both arrogant and aggressive, very soon they stand out as a special group. But they are no strangers to the other residents, because they and anything about them have become an integral part of village life.

Rural gangsters mainly rely on unethical methods for making illegal gain. In the beginning, these include stealing chickens and dogs, blackmailing, robbery, taking advantage of other residents and the village community. Later they rely on coercive monopoly and selling, gambling, illegal debt collecting, entering into contracts using coercion, and running businesses.[1] They may cause harm to people both in and out of their own villages. They use violence or intimidation on anybody standing in their way, regardless of whether the person comes from their own village. The late 1990s offers the rural gangs all sorts of opportunities, and gangsters are able to seize them for making illegal gain. In this process, they gradually migrate from their hometown. But because they are locally bred and born, they also cause harm to the local order as they grasp opportunities elsewhere. Of course, as gangsters of different social strata have different methods of making gain, they would cause different levels of harm on the local order.

In the Double-lake Plains, to make illegal gain, gangsters also often establish a cooperative relationship with the "powerful figures". For example, at *Shaqiao* Village of *Chujiang* City, gangsters would maintain a cordial relationship with *Wan*, the village Party secretary. *Wan* has also become a "village tyrant" because of his connections with the gangs. He takes advantage of such connections to manipulate elections, deal with disobedient residents and administer the village. He also obtains tremendous personal gain in doing so. The rural residents estimated that during his rule he may have embezzled about 500,000 Yuan. His family has only 7 or 8 *mu* of land and has no other source of income, but one of his daughters was given a job in the town-level office of the Financial Department and later in the town government—this may have cost him at least 150,000 Yuan; he got the other daughter of his to study in a nursing school and to work at a hospital in the city after graduation—this may have cost him at least 200,000 Yuan.

The tyrannical secretary *Wan* and gangsters at *Shaqiao* Village are both a result of the gang culture; when they collude, they are actually "different aspects of the same thing". The former acts the "ruling party" whereas the latter the "opposition party". The latter relies on the former's rule to do as they wish whereas the former uses the latter as a "thug-for-hire". In 2001, Secretary *Wan* intended to appoint a gangster named *Wan Sanyuan* as the leader of a village group. When the members expressed dissatisfaction with this decision, he explained, "All households have taken turns to lead this small village group (with only 25 households). *Wan Sanyuan* is the last one in turn. To borrow from a famous poem, I 'send down talents of more kinds than one'!" *Wan Sanyuan* helps Secretary *Wan* to "ensure stability" in the village group through using coercion. When the government tries to catch the gangsters, Secretary *Wan* would choose not to report them to law enforcement and keep them

[1] Gangster's methods for making gain are described in greater detail in Sect. 4.4 of Chap. 4. Chapter 7 also briefly touches upon this topic.

informed when law enforcement takes action. This has caused much difficulty for the government's crackdown campaigns.

Gangs also interfere with rural governance through intervening in elections and disputes between village officials and rural residents. In the 2005 re-elections for village committees in *Hubei* Province, it was ordained that the Village Party Branch Secretary and the Village Head must be held by the same person. Therefore, if the former secretary does not win in the preliminary elections for the village committee, he will also stop being the head of the village. This to some extent has made the competition for village committee election fiercer. Take Village Secretary of *Liji* Village of *Chujiang* City. Having gained support from local gangsters, the Secretary has served three successive terms. But he is actually a big fan of gambling, and he once contracted a lake (public property to the village community) to a few gangsters at a low price. Even before the election, gangsters had started a campaign to get the people to vote for him by letting out some "tough words". During the election when some people expressed disapproval, gangsters went to argue with them and even beat them up, which interrupted the election. In the summer of 2004, residents from *Liji* Village intended to appeal to higher authorities for help because of disputes in drought relief efforts. Secretary of the CPC Town-level Political and Legal Affairs Commission and the Village Party Secretary both came to talk them out of it. Residents accused the Village Secretary of having caused water shortage in their drought relief efforts due to his dereliction of duty. The Village Secretary soon started to swear at the residents, and when a female in her sixties contradicted him, he was furious and gave her a push, which caused her to fall over and suffer from bone fractures. He was soon encircled by about 70 infuriated residents. To keep himself from being beaten up by the angry crowd, he telephoned the town officials responsible for their village group and asked for help. Town officials arrived very soon, followed by the Secretary's son and his nephew, each carrying a one-chi (about 33 cm)-long steel knife. Seeing this, none of the people dared to speak up any more.

In their pursuit of financial gain, rural gangs would not hesitate to harm their own folks. When they have conflicts of interest with village officials, they would even treat the officials lightly and interfere with decision-making at the village level by using force or threats. At *Qiaotou* Village of *Linhu* City, fish farming contractor for the ponds is often selected by lottery. In Feb 2006, when the village was about to choose a new contractor, *Li Liu* (one of the local gangsters) insisted to be the contractor without going into the usual lottery drawing procedure. The Village Party Secretary did not agree with this and the two got into a brawl. Then *Li Liu* went to fetch a long knife and went after the Secretary. The Secretary did not get away until he had gone out of the village perimeter. In 2005, at *Linsha* Village of *Chujiang* City, when the village intended to launch a competitive bid for contract farming of one of the reservoirs, a farmer from the neighboring village offered 10,000 Yuan per year in rent to take up fish farming in the reservoir for 10 years. Some local gangsters planned to blackmail the farmer, so they proposed a yearly rent of 12,000 Yuan. People knew the gangsters did not really intend to go into the contract, and that letting them take over the reservoir was not really an option. Therefore, the village officials had to intervene by convincing the gangsters to give up and asking the farmer to treat them to a nice meal before they were willing to back out and allow the deal to happen.

Binhu Village in *Hefei* City of *Anhui* Province is in the same situation. Tyrannical local gangsters would use threats on village officials, misappropriate public property, and both village-level and town-level officials could do nothing about them. In 2004, the village gave 200 *mu* of land abandoned by families whose members had migrated to work in cities to a farmer under a five-year contract. In 2005, a gangster intended to take over the land. He went to tell the Village Head and gave the Head 1000 Yuan. The Head told him that they could not breach a signed contract and he would report to "Fastest News"[2] (*diyi shijian*) if anyone intended to cause trouble. However, the gangster replied, "if you don't allow me to take over, I will make sure that your kid 'loses some blood'. It's your call to make." The head had to report this to the town-level leaders, who promised to deal with the gangster but did nothing. The head also called the police station but the police could do nothing about the gangster. So the head decided to back out and refused to have anything to do with the 200 *mu* of land. When the land is contracted to some outsiders, the gangster would occasionally pester the outsiders. In the past few years, these gangsters have undertaken all sorts of projects at *Binhu* Village, but no one dare to comment on their project quality. People knew they have connections in the gangland—their "boss" may be even more powerful than the town mayor (Yuanyuan 2007).

At *Runan* County of *Henan* Province, rural gangs have also become an active force, bringing greater harm on the rural order with every passing day. The local folks call the gangsters "*lai hai zi*" (literally rascals, bad boys). Residents from *Songzhuang* Village told the author, "While the society is falling apart, many people lost their jobs and became bad boys roaming the streets." They coerce others into handing over projects to them, or monopolize the trade of quarry rocks and trees. They would threaten village officials by saying, "We will not cause trouble for you, if you give us the project. But if you don't, someone will be knocking on your door." Concession is the convenient way for the officials, since "it's not my own interests at stake", and "to put my whole family in danger for someone else's business is not worth it", and "who I give the project to makes no difference". The result is that all of the projects in the vicinity of the village are taken over by gangsters without going into the normal bidding procedures. For example, all the trees are sold very cheaply to the gangs.

In rural areas of *Runan* County, whenever leaders of village groups fail to settle a dispute, they would ask the "bad boys" for help. This means, the more "bad boys" they know, the more power they have. The "bad boys" also intervene in disputes between officials, and that was what had happened with two village-group leaders in *Fulou* Village. *Liu Hua* is a native at *Guozhuang* Village Group and was put in charge of it when village officials were accommodated into the responsibility system.[3] During

[2] An Anhui Television news program, broadcasting news on ordinary people's lives and well received among Anhui rural residents.

[3] Translator's note: in the "complete responsibility" system for rural governance in China, upper level officials are put in charge of an administrative division of the lower level, e.g. county officials

the commune era, *Guozhuang* villagers planted a row of trees in the area bordering *Fudong* Village Group. However, when public roads were built in the village, the tree area was assigned to *Fudong* Village Group. That was when the two village groups came to disagree on the belonging of this piece of land. In 2001, before the land dispute was properly settled, *Han Hong*, the *Fudong* leader, personally decided to sell the trees to his brother-in-law, who happened to be a travelling gangster. Chopping down trees requires applying for a permit with the village committee. *Liu Hua* happens to be the person responsible for permit approval in the committee and without hesitation he refused to grant the permit. In 2002, without proper consultation with *Fudong* Village Group, *Liu Hua* sold the trees to another gangster, who has more power compared with *Han Hong*'s brother-in-law. Hence *Han Hong* and the other villagers in *Fudong* Village Group could not do anything about it. *Han Hong* felt disgraced and resigned as the village group leader.

In contrast, rural order in Northeast China Plain is generally harmed by a group of people called the *kuang kun* (literally single men). There are tyrannical *kuang kun* in almost every village of Northeast China Plain. They disregard face and have no intention of entering the face-based competition for social ranks. They are a reckless group: they would steal chickens and dogs, seize collective and personal properties, and even enter into forced marriage with people' wives or daughters. People would say the *kuang kun* have nothing to lose because "barefoot walking men have no fear of shod walking people". They can be very persistent and would stop at nothing to get what they want. Many residents have cut off any connection with them for fear of what they might do. In his fieldwork, the author has gathered information for several cases.

At *Zhaoba* Viallge of central *Hebei* Province, a *kuang kun* first beat up an impoverished villager and then even forced his wife to commit adultery with him. When the villager sought help from the township government, the government did not really know what to do, so it simply detained the guy for a couple of days. When the guy was released, he beat up the villager even more badly, and warned him against seeking help from the government. No one else in their village helped him. Another *kuang kun* has many times raped his own sister. What an outrageous act! But no one in the village was willing to intervene. When his mother could not bear it anymore, she had no other choice but to kill him with rat poison. At *Dongxi* Village of south *Shanxi* Province, a *kuang kun* took over *Zhaijidi*[4] that belongs to the village community, took advantage of their impoverished neighbors and farmers, and refused to pay utility bills, but the residents and the officials could do nothing about them. Some other *kuang kun* often get into brawls with their own folks. With the slightest disagreement, they could pull out a knife on others. They do not fear death because they have nothing to lose. They are nicknamed by the local folks "red bricks", signifying that they do not have a sound mind, because in the local brick burning business, bricks

in charge of a town, town officials in charge of a village, and village officials in charge of a village group.

[4] Translator's note: *Zhaijidi* is a Chinese term for "land on which farmer's houses are built". Like farmland, *Zhaijidi* is also owned by the village community but is used by individual households.

turn blue when ripe (due to higher firing temperature) and bricks are red when not ripe. At *Guohan* Village of north *Hebei* Province, one of the *kuang kun* there often took advantage of others by forcibly taking over others' property. When caught doing it, he said, "You ask me for things you've lost. Who should I ask when I lose things?" He also refused to pay for what he bought at market fairs, and when people criticized him, he would break in to steal from the same person the same night. The local folks detested him, but also could not do anything about him.

Rural residents often could not directly punish the *kuang kun* in their villages. However, as everyone competes fiercely to "give face", when someone's evil acts harm the village, he could not possibly gain respect from the others. For those who have constantly disregarded the "face-giving" custom, the mainstream group in the village would abandon them, avoid making contact with them and wait for an opportunity to punish them. At *Liyu* Village of north *Anhui* Province, the village official once punished an unreasonable *kuang kun* named *Wang Wu* by "borrowing the law enforcement's hand". Once, *Wang Wu* stole a watermelon. The owner *Li Zhi* caught him doing it and called the village official for help. *Wang Wu i*s a habitual thief, and has got used to disregarding the village official's opinion even when sufficient evidence is provided. Village officials and residents are both frustrated by his unreasonableness. Though similar theft was settled through negotiation between the two parties or mediation by the village officials, the official refused to confront him, but told *Li Zhi* to directly hand him over the police. Apparently, rural residents and officials have "worked together" and "borrowed the law enforcement's hand" in punishing *Wang Wu.* Many people feel greatly encouraged by the settlement of this case. Though the *kuang kun* have brought tremendous harm to the villages, they are still restrained by the village and more susceptible to punishment by the law enforcement. In contrast, gangsters of the Plains are no longer restrained by the village.

In the vicinity of southern counties like *Shanxiang* County of *Hunan* Province and *Anyuan* County of *Jiangxi* Province, local gangsters seem to have limited influence on the rural order. They are mainly active in towns and townships, and not as much in the villages that are acquaintance society in nature. Villages of these counties all belong to clan villages where residents share a strong sense of identity with their clan. So all residents, including the officials, could not possibly use violence on others and become a village bully. Local gangsters would sometimes even protect their own folks. When some of their own folks get robbed at a market, they may even help get the money back by harnessing their nexus of connections. In the eyes of the folks, local gangsters do not "mess around". Instead, they respect shared values and codes of conduct, respect the old and the young, and have good manners. Local gangsters would make an all-out effort in the village-level scramble for interests or resources (such as mineral, tree, and water, etc.). They are ready to fight in the interest of their own villages. So sometimes in the eyes of the local folks, they are "heroes". At *Shui Village* of *Hunan* Province, when a gangster that was jailed for using physical force in a fight for the clan's interest got released, his own folks gave him a hero's reception ceremony (Hua 2009).

However, this does not mean that gangsters in the southern clan village do not cause disruptions in rural order. At a village in *Ji'an* County of *Jiangxi* Province, a gangster's arm was cut off in an armed clan conflict. Having his arm cut off became an asset that he possesses for coercing the others to also do things "in the interest of the clan". Once he asked all households to play movies as a way to gain prestige and admiration. When all households have played for 13 days in a row and did not want to continue, he coerced them into playing for some more days. He has in "fact" assumed control of the whole village in the "name" of honoring the clan—the name of clan has become separated from its factual object (Xiaotong 1992). When things like this happen, rural order has started to change nature; violence has been introduced into the rural life.

To sum up, when it comes to the level of harm brought by local gangs on the acquaintance society, rural areas in Double-lake Plains come the first, followed by those in the Northeast plain and south China. To go after financial gain, rural gangs in the Double-lake Plains would often collide with village officials, bringing harm to the village community. The same is happening to the Northeast Plain, but in a much less common and less serious manner. As for villages of south China, the prevalence rate and frequency of such incidents are both very low.

6.2 Influence of Local Gangsters on Social Nature

Currently the harm on rural order in the Double-lake Plains mainly comes from locally bred and born gangsters. How do the local gangs go about recklessly harming their own folks? Why have the officials colluded with them? Why are the Double-lake Plains distinctly different from south China when it comes to the level of harm on rural order? The answer to these questions may be constructed with a look at the nature of the rural villages in different areas of China. This chapter will provide an understanding of how local gangs have harmed the rural order in the Double-lake Plains by contrasting it with rural areas of south China.

In the acquaintance society, rural gangs come under the influence of multiple factors, and the level of harm inflicted by rural gangs is determined by all the factors combined. These factors include the government's administrative capacity, the village's ability to harness public opinion, the market and its impact on people's value systems and desires, potential behavioral gains, etc. Some act as encouraging factors while others discouraging ones. But which of these factors are the key ones or even the determinant? Our field work reveals that the effect of rural governance probably has something to do with the nature of the rural society, as the villages of different natures may react differently to external pressure. But what has determined the social nature and caused social nature to differ? How has such difference caused the different ways for gangs to harm the rural society?

To analyze the causes for disparities in rural governance, *Xuefeng* and *Zhihui* (2002) proposed endogenous organizations as one of the influential factors and characterize them with "social connectedness". Villages with a higher level of social connectedness may more easily establish an internal order, internally achieve cooperation and externally put up resistance, whereas the same does not happen to villages with a lower level of connectedness. Based on further empirical study, *He* (2006b, 2007e, 2005b) proposed another term for characterizing endogenous organizations—"the unit of farmer identity and action"—and uses it to explain issues of rural governance, e.g. loan sharking. There are mainly two types of "unit of farmer identity and action": the family—the basic unit; and other units based on unions of family, such as clans, small kin groups (*xiao qin zu*), or household unions (*hu zu*). The effect of rural governance is determined by whether or not farmers act on the second type of unit.

At the Double-lake Plains, with a low level of social connectedness, there are no effective unit for strong action above the family level—the area is composed of "socially atomized" villages where self-interested residents act like disperse atoms and households fail to rally against gangs. In south China, such as *Shanxiang* of *Hunan* and *Anyuan* of *Jiangxi* Province, though clan identity has already been weakened by political movements and the market economy, it is still much more influential compared with the Double-lake Plains. The clan has become the basic unit of identity and action. This means a higher level of social connectedness. The clan as the unit of farmer identity and action has effectively kept the rural residents from acting as separate and disperse atoms. A clan is formed on the basis of both consanguinity and regionalism (ties to a specific place). It is often composed of many households—its size varies from several dozen to several hundred or even several thousand. As residents share a strong sense of identity with their clan, they act on the unit of the clan and therefore would more easily achieve large-scale cooperation internally and externally put up resistance. When gangsters enter such villages, they will encounter resistance from the whole clan.

When the state fails to effectively contain the gangs, the key to maintaining rural order boils down to whether the village level could rally against them. Ordinary residents detest them but no one dare to confront them single-handedly. Though rural gangsters tend to restrain themselves when facing village officials and upright elite members of the village, even these people would still find it hard to handle the gangster's sneaky and single-minded action. To better keep a grip on the gangs, they need collective action—having the whole village devoted to deterring them. This is only possible in villages with a higher level of social connectedness and a larger unit of collective identity and action. Collective action is the only solution to gang intimidation. Otherwise officials and elites would feel isolated and fatigued when facing up against gangs. When taking action for the village's common interests, officials and elites would end up with upsetting the gangsters, while the others are cold-hearted on-lookers "sitting on a hill watching two tigers fight" (*zuo shan guan hu dou*).

The level of resistance against rural gangs may also be explained by the people's understanding of public and private spheres, or more specifically, the boundary and the relationship between public and private spheres. Such an understanding constitutes the starting point for how people handle public and private life. If the common unit of identity and action goes above the family level, they would share the same public–private conception, which may be referred to as a kind of common sense shared by the locals of an area.

When they wrote about the human nature of the Chinese people, many scholars have touched upon the people's public–private conception, including *Xun* (1981: 567), *Yutang* (2006), *Shuming* (1999:157), and *Xiaotong* (1992), etc., with *Fei*'s *Chaxu geiju* (the differential mode of association) theory being the classic paradigm for research on public–private conception. After visiting *Xicun* Village of *Fujian* Province, *Mingming* (2004: 74) furnished a more vivid description of such conception. *Xuefeng* (2006c) found that the parties in *chaxu geju* are nonequivalent, so he proposed the "double units for identity and action"—this may be regarded as a development on the *chaxu geju* theory. He proposed that in different rural societies or during different time periods, the main unit for collective identity and action may be varied. The chief unit for collective identity and action will become a regional consensus and thus become the dominant factor determining the characteristics of rural governance and behavior patterns of the rural residents.

At the socially atomized villages of the Double-lake Plains, there is no unit of identity and action above the family level. The family belongs to the private sphere, which is highly atomized, whereas whatever goes beyond the family level belongs to the public sphere. Thus the family and the village do not have shared interests. The family or individuals are not incentivized to safeguard the village's interests at stake. And the fact that family and individuals are not obliged to do so has become a regional consensus. As a result, when public interests are at stake, individual residents have developed a kind of incompetence and that the boundary between public and private sphere is very distinct. This will further cause people—practically anybody—to take a share of public good without others intervening. "Mind your own business" becomes the legitimate excuse for whoever that has benefited from public good to get around other people. In the context of such regional consensus, it has become a common practice in irrigation cooperation for people to use personal connections to get water quota (this is called "*ren qing shui*") and such regional consensus provides a reasonable explanation for both the beneficiary of "*ren qing shui*" and other villagers whose interests are indirectly harmed (Xuefeng 2006d).

For the same reason, ordinary residents do not have a legitimate reason to stop the local gangs' encroachment on public interests. When external gangs are introduced into the village, ordinary residents are not motivated to safeguard public interests. Those honest and upright residents would be struck dumb with a simple question like "what's in it for you?" because it has become a regional consensus for them not to meddle with things that do not concern them. Any act of meddling aims not at safeguarding public interests but at provocation. Therefore, ordinary residents would not report corruption and unethical behaviors in the public sector or encroachment of public interests by the gangs, unless their own good is at stake. When that does happen, an individual—with his limited power—would have to give up before putting up a fight because he could not handle a corrupt official or a tyrannical gangster. In an socially atomized village, whenever state power fails to intervene, public interests would be devoured. Public order could not be autonomously maintained, unless some morally tough guys with high moral standards would be willing to safeguard public interests. But they may also line up their own pockets doing so. That's why occasionally some "good gangsters" would turn up or gangsters could do a good deed.[5]

In southern clan villages, people have a sense of identity with the clan; both the village and the clan belong to the private sphere. The village is their "own" village. That's what people all believe in. They all have a sense of belonging to the village. They regard village affairs as private affairs. Therefore, whatever happens in the village does not just happen to the parties directly involved but also happens to everyone. According to *Yangjing* (2008), the village is a private sphere composed of the clan, clan branches (*fang zhi*),[6] lineages, and residents. Individual households belong to the smaller private sphere (*xiao si*), whereas the village is a larger private sphere (*da si*). Everything that needs to be done for the village will be done by the larger private sphere itself. The same is true with the clan. It faces the outside world as one single unit—it is "privately" owned by its members; anything that concerns the clan is publicly shared by the members—in this sense it also has the "public" nature. So anyone is in a position to stop a local gangster harming public interests and to accuse an official of corruption or immorality, because harming one man's interests equals to harming everyone's interests, and harming public interests also harms every individual. In other words, instead of being purely private, an individual person also carries some "public" nature. Every individual is a part of the public sphere, which gives them both some rights and responsibilities. In a word, the clan is a private sphere carrying some public nature, so naturally anything concerning the clan is subject to supervision and accusation from all its members. As a result, local gangsters would have to think again before they attack the village, and the residents would have to do the same before "bringing a wolf from outside".

The degree of harm inflicted by the local gangsters may also be explained by the village's intrinsic level of productivity for both economic and ethical values (Xuefeng 2005c). A village has a higher level of such productivity if: local residents have long-term expectations for living in the village; they would not just focus on temporary financial gain or loss, but value interpersonal affection and obligations (*qingyi*); as they try to make financial gain—this is done through "convincing others with reason" (*li xing ren*), they would also like to gain on the ethical value level—this is to be rated as "having high moral standards" (*dao de ren*). Our field work reveals that the village's intrinsic productivity level is fundamentally influenced by the unit of identity and action. In a socially atomized village, people have reached a consensus that views the "family" as the unit of identity and action. For that reason, interpersonal relationship is very distant. When households work as individual atoms, they could not exert pressure of public opinion on transgressors to effectively restrain their behavior—in this sense, the village has a low intrinsic level of productivity. In contrast, people in clan villages have reached a consensus that views "the clan" as the unit of identity and action; when people feel close to each other, they have the

---

[5] Chapter 8 will discuss this in detail.

[6] Translator's note: *Fang* refers to a group relationship that mixes consanguinity with geography, and *zhi* is a sacrificial unit.

collective power to handle transgressions. Hence such villages have a higher intrinsic level of productivity.

The Double-lake Plains is going through a serious outflow of human resources, financial resources, goods and materials. Residents go out of their way to get away from this area. They'd be like "I'd rather pee somewhere else". They would like to sever any connection they have with their village. They run away whenever they can. The village is nothing more than a temporary habitat. Village life is not desirable or reasonable. They do not have long-term expectations on living in the village. They don't seem to care much about whether others speak positively of them or whether they are reasonable and respectable people, for such "word of mouth" does not generate economic value. Thus public opinion becomes useless. People do not care about their reputation in the village. They look forward to life away from the village and expect to find value and meaning of life from participation in the market economy. Therefore, they do not value the interpersonal affection developed through living together for generations. Nor do they make attempts to build up such affection for their future generations. The rules they would like to follow are the kind of reason that also applies to the market economy. In this way, when someone becomes a gangster, he does not feel the pressure of public opinion or may disregard it altogether. In the eyes of gangsters, the village has changed in nature—it has changed from a permanent living space where people live from birth to death in the same place, to a temporary dwelling from which they would rather get away from and may easily do so. To borrow the words of Émile Durkheim, such a village has no "collective emotions" (Durkheim 1997), for the people there do not care about their village any more. Rather they only care about their own interests.

Gang activity is much less common in southern clan villages where people share a sense of identity with the clan and the village has a higher intrinsic level of productivity for both economic and ethical values. People would love to be rated as "reasonable or respectable" or to gain good "word of mouth". And that's what makes public opinion powerful. Village officials would refrain from misconduct for fear of being condemned by public opinion. People care about their reputation. They look forward to long-term living in the village and seek agreement on the value and meaning of life. They cherish interpersonal affection developed through living together for generations and take care to further nurture such affection for the sake of their future generations. They respect the traditional behavior pattern of the acquaintance society. In this way, when someone is looking to become a gangster, he would have to face the pressure of public opinion. Just like everyone else, the gangster would also find the village to be a permanent living space where people live from birth to death in the same place. Even though they may be away from the village while partaking in gang activity, they may not easily get away with doing so. "Collective emotions" still exist in such a village. When gangsters make evil attempts, they may provoke the public and face severe punishment. When people share a strong sense of identity with the clan, local gangsters would have to face the powerful public opinion pressure and could not afford to disregard their reputation in the village. Sometimes to build up a good image for themselves, they would even help out their own folks, in a way similar to what "social bandits" have done to have built up their image as justice-fighters

(Hobsbawm 2010). When gangsters attempt to harm their own village, they would be psychologically discouraged because of the village's high level of economic and ethical productivity.

6.3 Influence of External Gangsters on Public Order

A comparison between the North China Plain, south China and the Double-lake Plains reveals that the North China Plain is most severely affected by external gangs. In clan villages of south China, people would ally against external pressure. In Double-lake Plains, despite their atomized relationships, people would still take an active part in uniting against external gangs.

Despite great harm brought by rural gangs to villages of the Double-lake Plains, the level of security threat faced by the residents is not very high. This is because people may avoid becoming victims of violence for having known something about the local gangsters or those brought in by local gangsters. In other words, as gangs assume control of the Double-lake Plains, "rural residents have got used to having them around"—to quote from a town official from *Chujiang* city. However, in the North China Plain, gang influence is less pervasive—the *kuang kun* are still bound by local rules—yet the level of security threat seems to be higher, mainly from bandit groups from outside the village. People have many appellations for bandits, sometimes also "*hunhun*" (gangsters).

Our field work in the North China Plain reveals that people there find it very hard to handle bandits. For more than a decade, bandits have been a serious problem for residents in *Wangzhen* Town of *Pingyu* County and *Lvcun* Town of *Anyang* County. In some places, "the talk of bandits causes the change of the color in the face". Some local officials told us about some gruesome villages along the border of *Hebei* and *Henan* Provinces where all the residents sleep during the day and become bandits during the night. The locals never care to surface the roads because they do not want police cars to enter the village. Almost all households have "entered the occupation of banditry", otherwise they could not afford to find a wife for their sons. When local girls consider marrying someone in their village or neighboring villages, they would first consider banditry families. In the 1990s, a so-called rectification (*zhengdun*) campaign was introduced to "make things right" for these villages. The local police bureau planted "informants" in the villages, and when time was ripe, dispatched thousands of armed police into the villages and arrested the bandit gangs in one single action. Since then, public order has greatly improved, except that there seems to be resurgence of bandit gangs in recent years.

Simply listening to local residents' accounts of bandit gangs sent a chill down my spine. Bandit gangs often bring knives and guns with them. When caught stealing, they would use extortion instead. Thus banditry often involves the use of violence. At *Wanghan* Village, a local resident named *Wang* told me about his experience of being a victim of burglary about ten years ago. He and some other residents were drinking at *Wang*'s home, when some bandits had been lurking outside his wall. When they were almost drunk, the whole bandit gang broke in; two of them brandished long knives, while others took all the valuables in his home. Before leaving, they gave *Wang* and his friends each a kick. A recent bandit case happened in *Guohan* Village, where some bandits snatched the water pump while the guard was present. A guard watches the pump for people often come to get water at night. The guard yelled "help!" seeing the bandits but he got shot in the foot and the pump was still taken away. No one dared to chase after the bandits even though they all heard the gunshot. We were told that at the end of the 1990s bandits would come to snatch quilts, TV sets and meat whenever there were wedding receptions and funerals. Once at a wedding reception, bandits stormed in and told everyone to squat down, and no one dared to say no; they had to look on as bandits took away everything.

In most villages of North China Plain, all households have high walls and big yards. At *Pingyu* Village and *Anyang* Village, people build their houses with walls almost three meters tall and on raised foundations of one meter high. Building higher walls is for showing off and also for preventing burglary.

Also to prevent burglary, many villages in North China Plain have followed instructions from their higher authority to organize alliance patrol teams, as in *Pingyu* Village and *Anyang* Village. A 5-man patrol team has been established at *Guohan* Village of *Pingyu* County since 1998. During nights of cold seasons, they sound out the hours to remind people of the time and patrol all over the village area. With long and cold winters in North China, people go to bed early at night, which makes it convenient for bandits to do their jobs. Later bandits are also active during other seasons, so the night men have to patrol all around the year. Bandits would steal or snatch any valuables, including livestock, motorbikes, three-wheeled motorbikes, well pumps, transformers, etc. During our ten-day stay at the *Guohan* Village, we heard about successive cases of transformer theft at *Guohan* Village and its neighboring *Maohan* Village.

The patrol team of *Guohan* Village was established based on requirements laid down by the township government. Patrolling is mandatory and offers no pay. Patrol shifts are arranged based on a name list, with one person from each household and five households in a subgroup. Those whose names get called in the broadcast must do the patrol job that day. However, some would be absent because they have "forgotten" about it or have not heard the names clearly. Some individuals even refuse to participate from the very beginning, but most people would not be bothered by their reluctance for they are "shameless". The patrol team has been active for several years, which proves the public's need for security assurance. In recent years, as more people from *Guohan* Village migrate to urban areas in search of work, there are fewer men available for patrol and they must take more frequent patrol shifts. The staying men feel unhappy about it, so the team has to break up. In 2006, when bandits became increasingly active, the police force sought support from the village and raised over 10,000 Yuan for introducing technology to prevent burglary by "installing security alarm in each household". Ten households are formed into an alliance and would help out any household whose alarm has been set off. The use of technology has indeed made up for the shortage of manpower for patrol. No sooner had they installed the alarms than a burglary got caught. One day at 3 a.m., *Guo Junsheng* set off his alarm when he heard someone stealing his motorbike in the yard. The burglar tried to rush for the door but neighbors arrived on hearing the alarm and successfully caught the burglar in the north fields of their village.

Since 2001, *Liyu* Village in the north of *Wan* (abbreviation for *Anhui* Province) has effectively used villager patrol teams. To encourage people to be actively involved, the village has decided to give each patrolman 2 Yuan per night as compensation. When the village committee fails to give such compensation after the lifting of agricultural tax, more people are reluctant to carry out patrols. To pass regular inspections by the township government, the village requires the poorer households to carry out patrols and in return they may get annual basic income from the state. However, because the basic income does not match their workload, they do not take the patrolling job very seriously. In the winter of 2006 when I was on a field trip to *Liyu* Village, the village had been chosen as a random destination for inspection by the county government, and for that reason the village officials repeatedly stressed that patrol members should not stop patrolling until 1 a.m. One night the village security committee member (*zhibao weiyuan)* and I were following the patrolmen. When we parted with them to run some errands, the committee director repeatedly urged them to patrol to at least half past 12. But when we went to check on them one hour later, they had already gone back home to sleep. We may imagine how things are when no one is checking. The village officials had no choice but to ask the patrolmen to man their posts through broadcasting, but it was actually "a bluff" or a message for the burglars, since by that time the patrolmen were already asleep.

The elder people in the village told us that before the founding of P.R.C., pubic security had been weakened by increasing burglary and ongoing wars. The village has to set up militia groups, purchase guns, build stockades (*yu zhai*) and moats, and form security alliance with neighboring villages. That is how basic order has been maintained at *Liyu* Village. Villages with no stockade have no choice but to give in to the outlaws. In areas around *Pingyu* and *Anyang* Villages, they have organized "*kan qing hui*" (literally "teams for guarding young crops") and have night watchman hitting a gong to fend off burglars. Unlike small villages in south China—some consisting just one or three households in seclusion, natural villages on North China Plain are generally larger, probably because fending off burglars requires cooperation from a larger crowd. As those who need protection are generally the richer people, security forces are organized in a way to safeguard the rich people's private interests (as bodyguards for the rich). The richer people hire the poor to be night watchmen or to guard young crops. Considering that the poor are the main source of burglars, Perry (1980) believed that the building of security forces "has devised a way to use potential outlaws for benefiting the society".

Perry (1980) pointed out that there are two survival strategies for people to deal with the fragile ecosystem in *Huaibei* (a prefecture-level city in northern *Anhui* Province): an aggressive one and a defensive one. As *Huaibei* has a fragile ecosystem (which is also true with many other areas in north China), when crop failure makes the competition for resources more fierce, the poor would set their mind on the rich and the rich would have to put up defense. Thus the fragile ecosystem has exacerbated the prevalence of burglary. The farmer's habitual tendency to resort to theft has also been passed on for generations. The young are forced to gamble, steal or rob others just to survive, as there are few opportunities for starting a business or making legitimate earnings. To handle such theft and robbery, the village must act; to ensure security for all, the whole village must act together. Methods for collective action include hiring people to guard young crops or as night watchmen (Uchiyama 2001).

However, burglars in present times are different from those before 1949. After the founding of the P.R.C., environmental problems in North China have largely been solved. For instance, the new China has solved once and for all the flooding problem in important rivers such as the Yellow River and the *Huai* River, which has also enhanced soil fertility in those areas. That keeps farmers of North China warm and fed and therefore farmers do not become burglars just to survive. What has caused some farmers to steal is perhaps explained by the their changing ethics and values. When some rural residents have been seduced by consumerism introduced by the market economy, there is always the temptation to "cut corners". Unlike the south and central China with abundant resources such as forests and lakes, the North China Plain has a more fragile ecosystem and its rural areas offer no other resources than crops. Such a difference has caused regional disparity in terms of rural gang development. Gangs in south and central China may make illegal gain through seizing forests, lakes or ponds, whereas those in North China Plain have no other choice than to form burglary groups.

A more important issue worth our attention is what has caused the villages of North China Plain to be so vulnerable to burglary groups. It's very hard to believe that a dozen burglars would dare to break into a village with nearly a thousand residents and might come and go freely. Before 1949, the rural residents are still able to put up a unified defense against burglars. How come such defense becomes impossible in the new era? As we have discussed earlier, the defense against burglary before 1949 has been organized in a way to protect the rich people's private interests. Public defense by the whole rural community is needed only in times of war. However, under the current circumstances, public defense is also needed for ensuring security on a routine basis. A comparison between North China Plain and south China or Double-lake Plains easily reveals that the former is most vulnerable to impact from external gangs.

At *Ludong* Village of *Jiangxi* Province (south China), the people are mostly simple and honest (*chunpu*) and seldom get involved in brawls, but they would take concerted action against intruders. Almost all households there have a dog, which in winter often becomes the target for intruders from the county. Once, a gangster tried to snatch a dog from a little girl. Seeing that, the girl refused to let go of her dog, the guy fastened the dog to his motor bike and dragged the girl with it. He did not stop until the girl was badly injured. This has greatly infuriated the local residents. Later, when a woman caught the same group of gangsters stealing dogs, she yelled for help and brought the whole village to pursue them. The local folks captured two of the four intruders and battered them as they screamed in terror. The local folks then threw them into the river, making them soaking wet and shivering, and even submerged their head in water to drown them. But that was not enough. The folks had to smash their motor bikes and hang them up in a tree next to a public road. The police arrived soon and accused the folks of illegal detaining and asked the folks to release them immediately. Hearing this, the folks were burning with rage. Based on past experiences of similar clashes with the local folks, the police had to give way and asked the gangster's family to agree to the folks' request for reimbursement. Even with police intervention, the two shivering speechless gangsters did not get released until their family reimbursed the folks as requested. (Xiaojun 2009:153).

At *Shuicun* Village of *Hunan* Province (also in south China), there are several underground Mark Six lottery outlets, which are the lower level for a larger outlet in town. The best selling outlet was the one owned by *Yang Shuxiao*, and a town gang asked *Yang* to sell lottery for them instead, but *Yang* refused to do so. One afternoon, the gang of five people rode motorbikes to *Yang*'s family to threaten *Yang*. Very soon this news was spread around and people were shocked to know that "some outsiders would dare to march onto our turf". So, all the male residents quickly arrived at *Yang*'s place. They besieged and bashed the gang with anything that came handy. Three of the gang rode motorbikes and got away. The other two were knocked down and couldn't stand back up. Their motorbikes got smashed too (Hua 2009).

As previously discussed, people's willingness to unite against outsiders in clan villages of south China may be explained by the level of social connectedness, unit of identity and action, people's understanding of public and private spheres and the village's intrinsic productivity of both economic and ethical values. In the Double-lake Plains, despite social atomization, people do not succumb to external pressure, regardless of what is at stake—their own, others' and public interests. Therefore, they would unite against outsiders and have demonstrated great competence to fend off outsiders.

Take *Shaqiao* Village of *Chujiang* City in Double-lake Plains. In 2004, *Chujiang* Petrochemicals Company intended to build an oil pipeline that would run through *Shaqiao* Village. In the three clashes between the contractor and the local people, despite efforts by the contractor to mobilize gangs, the local people have always won. The project manager is a man named *Liu Hong*, who happens to be the town-level Party secretary's cousin. With a powerful figure as his patron, *Liu* completely disregarded the rural residents' personal rights and interests.

First clash The building of a pipeline on an area that belonged to the Sixth Village Group of *Shaqiao* Village would destroy the village group's land facilities, which would affect 15 households. The village women's federation director, who happens to live in the construction area, demanded back-filling for what used to be three ditches, a road surface and a water pump. One of the construction workers said to her, "we will see to it that someone will come and fix this." But the director's husband did not believe what they said, so he asked people to stop the earth mover from working or leaving. Some of the members to the Sixth Village Group are tough guys who talk tough. After several rounds of "negotiation", the director of the town-level armed forces department had to step up and mediate between them. *Liu Hong* had to pay 3000 Yuan for the locals to finish the back-filling job.

Second clash The building of an elevated section of the pipeline required building abutments in the ground in the Fifth Village Group of *Shaqiao* Village, thus occupying a 4-square-meter land area that belonged to *Yang Suixing*. Based on the compensation standard of 2 Yuan per square meter, the contractor was willing to pay 8 Yuan. But *Yang* disagreed. He said, "This thing in the middle of my land makes everything inconvenient. And you try to buy me off with just 8 Yuan? Over my dead body!" The contractor was still mad about what happened with the Sixth Village Group. So this time he called on a few town gangsters to help him. One of them said, "We have what it takes to be a contractor. Do you have what it takes to mess with us?" But *Yang* was prepared for this. He had some cousins in *Chujiang* City as gangsters there. He had already informed them, and then with a single phone call, all of them came. They pointed fingers at one of the gangsters on the other side and said, "Who tried to buy my bro off with just 8 Yuan? I say 5000! Nothing less of that, or I will chop off one of your legs!" The contractor was intimidated by their manner of talking and was forced to seek help from the town officials, and eventually had to pay 800 Yuan as compensation.

Third clash When the building of the pipeline came to the Seventh Village Group's land area, it damaged a U-shaped cement pipe of about 20 m long. The locals demanded compensation, but the contractor said, "Without the pipe, water goes through all the same. So no need to pay them for this." The leader of the Seventh Village Group came forward to negotiate with the contractor. Since the Fifth and Sixth Village Groups both gained some benefits, he would like to do the same. But the contractor was already irritated by what had happened previously and thus refused to negotiate with him. The contractor drove the truck away, not to be stopped by anyone. The group leader immediately went back to his group and called on other households. The contractor had already gathered about 15 gangsters in three cars from *Chujiang* City. They stayed close, ready to fight any time. When more and more local people gathered around, the gangsters came out of the cars with long knives in hands, and said, "Who wants money? We are here to take his life!" One of the villagers said, "We are not afraid of knives. You must pay for the damage you've caused." And then he got stabbed in the thigh. So, someone called the police, and another villager called their local gangster *Wan Huaqing*. Two officers came trying to keep things under control. But the gangsters continued to talk tough. They said, "Calling the cop is no big deal! We will kill anyone asking for money!" When *Wan Huaqing* arrived, he called out to all his fellows, "Come on, fellows!" With that, about 40 or 50 villagers charged at the gangsters, who were forced to retreat toward their cars. The villagers smashed stones and sticks into their cars, breaking the windows and scratching the paint. When the gangsters got away, the villagers battered the contractor. The police could do nothing about it. The contractor ended up with a serious head injury and had his spleen removed. Eventually the county and township officials came forward to mediate between them, and the contractor still had to compensate the villagers for the damage he has caused.

In conclusion, rural residents in clan villages of south China and in socially atomized villages of the Double-lake Plains may unite against outsiders, whereas those in North China Plain may not. In both their daily life and the political sphere, rural people of North China Plain often form cliques that constitute their unit of identity and action (Baifeng 2008c). Other scholars propose small kinship-based groups as their unit of identity and action (Xuefeng 2009a; Duanfeng 2007b). But since these cliques or kinship groups are not small units, we may not simply explain the lack of unity against outsiders with having a smaller unit of identity and action.

6.4 Resistance from Local Gangs Against External Gangs

In North China, farmers are generally a group of obedient people, or there are no "badass people" in the village, making the village very vulnerable to both local unscrupulous *kaung kun* and to external gangs or bandit groups. With local *kuang kun*, people do not punish them immediately. Instead, people look for other opportunities for punishing them in the future.

At *Wanghan* Village of *Pingyu* County, *Henan* Province, a gang of over ten bandits got removed in 1999. In three years this gang committed more than 100 cases, including almost 100 cases of theft and robbery, and over 30 cases of rape or group rape. Considering such rampant criminal activities over such a long time, the author was astonished to have found no evidence suggesting collusion with government officials. The only plausible explanation is that local residents are really intimidated by their "badass" image. This image has been built up not by them as individuals, but rather as a group with skillful maneuvers.

At *Fugou* Village of *Henan* Province, rampant banditry greatly discourages the local residents from raising cattle or sheep, despite great profits from doing so. The residents detest bandits, but do not dare to offer a helping hand to a neighbor yelling for help. Once, two of a villager's sheep got stolen. He chased after them all the way to the road side only to discover that none of the cattle and sheep driven by the bandits were his. He decided not to "stick his nose into other's business" because his own sheep were not there (Chunxia 2007). The villagers do not dare to resist outsiders when group interests and other's interests are at stake, not even when their own interests are at stake. They fear all external influences, including gangsters, bandits and government officials. That's probably why external gangs and bandits could easily break into their village and do as they wish. Their perpetration of evilness has not been stopped until the state intervenes.

Another example would be *Songzhuang* Village of *Runan* County, *Henan* Province, where the villagers are also easily intimidated by tough guys. In November 2005, *Runan* County successfully solicited investment for a private experimental middle school to be built in *Sanlidian* Township. The planned area of this school was over 300 *mu* of land (about 200,000 square meters), which required expropriating land in *Linzhuang* and *Xuzhuang* village groups. A poll on methods of compensation for land expropriation showed that 80% of villagers agreed with being paid with an annual rent of 900 Yuan per *mu*, which allows the villagers to retain ownership of their land and increase rent with inflation. With one-off compensation, the farmers may have to starve, since their land per capita will decrease from 0.8 to 0.3 *mu* (533 to 200 m^2). But 20% of them still agreed with one-off compensation. Some needed money urgently, others wanted to build a house or start a business in the county. Such disagreement kept land expropriation from proceeding as planned, not even with county and village officials' reconciliation attempts. Right after the spring festival of 2006, the villagers were informed that a land expropriation "contract" has been signed, promising them a one-off compensation of 16,000 Yuan per *mu* of land. This also covered compensation for destroying young crops (*qing miao fei*). They were told by the village Party secretary that signing the contract was an instruction from the higher authority not to be disobeyed. People started to speculate that the village officials probably have gained some benefits from signing the contract. The construction crew had no sooner installed the bulldozers in place than the villagers stopped them. When the two sides reached a deadlock, a group of gangsters intervened. Their head said, "We will build up a wall here. We will get the job done! Tell your boss to leave the villagers with us!" So a new crew came in and drove the bulldozer to the land side, ready to start working, but the villagers stopped them again. Then the gangsters made a phone call and soon over a dozen people came each bearing a chopper. This sent the villagers running for their own lives. Some bold ones dared to argue with the gangsters but the gangsters lifted their choppers and came at them. One of the villagers got injured in the brawl.

Someone called the police. The police came in their truck, but caught no one. But the gangsters went to the hospital to pay for the medical bills of the injured villager. The villagers commented, "They came in broad daylight to pay for the bills, as if they were heroes doing a heroic act." On the second day of the incident, the bulldozer came again. Official construction started a few days later. People had to accept reality. Based on what had happened, they decided to drop the request of rent charging, for they knew they could not count on the county to pay the due rent every year even if their terms had been granted. Instead, they demanded one-off payment in real terms. Perhaps out of limited financial resources, the county refused to pay off the money in one-time payment. A respected villager from *Xuzhuang* village group and some ten others secretly set off on a trip to appeal to the provincial government and Ministry of Land and Resources at *Beijing*. The related authority of the county was pressurized into agreeing to their request. Knowing that *Xuzhuang* village group has succeeded, people of *Linzhuang* village group made the same request to the county level, but their request was rejected because no one appealed (Song 2007).

To sum up, the villagers are often too weak to fight for interests at stake. Their weakness is shown by their fear for both officials and gangsters. The village secretary was pressurized into signing the land expropriation contract, and contrary to common speculation, he did not gain profit from doing so. The villagers eventually accepted reality even though they did not consent to signing the contract. When the gangsters intervened, the villagers seemed to have "surrendered without putting up a fight", and may have given up the fight too easily considering that their major interests were at stake. When they had to appeal to the higher authority, they had to leave secretly at night. After the appeal, the village secretary was removed from his office because the township government rated him as incompetent. What's behind the villager's weakness, when they have to face such external influences as officials of higher authority and gangsters? Why not unite and "fight for their rights and interests"?

To answer these questions, we may compare north China with other areas. In clan villages of south China and those at Double-lake Plains of central China, people do not fear officials or gangsters, because among the simple and honest villagers, many are upright people who do not hesitate to stand up against others. With the intrusion of outsiders, gangsters who often misbehave would quickly become the main protectors of the clan or the village. At *Chujiang* City of *Hubei* Province, despite social atomization, people do not succumb to external influences, regardless of whose interests are at stake. Intruders would always have to deal with the local tough guys. These tough guys and gangsters are in a relationship of "regular relation" (Jicheng 2002)—a tough villager is likely to be a gangster, or the group of tough guys and the group of gangsters often overlap. Villagers of *Chujiang* do not fear outsiders or officials. In 1999, a new Party secretary was transferred to the town governing *Shaqiao* Village. The ambitious man had revolutionary visions for being "the new broom" or the "parental" official (informally known as the *Fumu Guan*). The first change was to reform local grain station (*liang zhan*). The cotton handed over by the farmers were often too wet to be used directly, so the grain station had to reprocess the cotton before turning it in to the higher authority. Reforming the grain station was intended as a good thing, but would jeopardize the farmers' vested interests. When the new secretary arrived at the grain station to supervise over cotton collection, a few farmers snuck upon him from behind, covered his head with a piece of clothing and bashed him.

Apparently, villagers of *Chujiang* do not fear outsiders. In previous chapters, we've come to understand that due to social atomization, people do not take care to safeguard the village's public interests, and the reason behind may be that public and individual interests do not overlap. Villagers are reluctant to safeguard public interests. It is also not part of their regional consensus to do so. And when they do step up to safeguard public interests, they would be told to "mind their own business". That explains why no one steps up when local gangsters harm public interests. And when external gangsters get introduced into the village by their own folks, the other villagers also feel reluctant to nose into others' business. But with total strangers in the village, people would be like "How dare you! This is our turf! Stay out of my business!" So when public interests are at stake, who are involved matters a lot—total strangers or local gangsters. With total strangers, anyone may criticize them and even beat them up.

At Double-lake Plains, whether people stand up against those that harm public interests depends on their legitimacy of access to the public interests of the village. If the intruder's method of accessing public interests seems to be locally legitimate, villagers have no choice but to be onlookers. If the intruders do not have such a legitimate access, they may be stopped by anyone. But to gain a legitimate access, simply being a bad ass is not enough. For instance, to contract (*cheng bao*) a pond at a low price, the local gangs may use coercion, but that may not work for external gangs. Instead, they have to find local gangsters as their intermediary. It looks like the pond gets contracted to a local gang, but in fact to outsiders. Finding a local intermediary gives outsiders a legitimate access. The same goes with confronting a person. If an outsider wants to confront someone, it will not do if he simply says that "I saw this man at the market yesterday and he rubbed me up the wrong way. So I have to confront him". The man's folks and local gangs would both help him out.

They would be like "how dare you to march onto our turf?" Instead, if the outsider says "you have rubbed up XXX (a local guy) the wrong way. He told me to teach you a lesson." With that, the outsider gains a legitimate access.

Therefore, external gangs must collude with local gangs before entering the village. Otherwise, they do not have a legitimate access for "doing as they wish" and the badass local gangsters would be the first ones to confront them. That's why people feel like occasionally local gangsters are also up to some good things.

The Double-lake Plains does not lack for such badass local gangsters. But for them, the purpose of standing up against outsiders is not to safeguard public interests, but to defend their own territory. If outsiders with no legitimate access are allowed to do as they wish on their territory, the local gangs become the indirect victims. If local gangs do not stand up against outsiders, people would be like "Being a badass to your own folks makes you no hero. To chicken out with outsiders only makes you a coward!" To be a hero or a coward: that's a choice local gangsters have to make in the presence of outsiders. There is a regional consensus among villagers that a person that is generally a tough guy in his daily life should live up to this image when facing outsiders. If they are allowed to exploit public resources for their own benefit, they are also obliged to protect their own folks. Villagers expect them to do so. Supposedly the gangsters should also bear this in mind. In other words, the village becomes the gangs' "sphere of influence". So when local gangsters stand up against outsiders, they would be like "How dare you fool around on my turf!" or "Mind whose dog it is before you strike. Messing things up on my turf is messing with me!"

Local gangsters are expected to treat their own people nicely, but to act aggressively towards intruders or other people at other places. For a gangster who does not seem as badass as his peers, dealing with outsiders is a great chance for him to establish his reputation. If he displays the kind of anticipated ferocity, people would willingly take orders from him. A good example would be the previous case with *Wan Huaqing*, who made an impact by yelling "Come on, fellows!" A gangster that treats locals nicely but turns into a tough guy for others would be like a hero to the villagers. They are similar to what Hobsbawm (2010) refers to as "social bandits". The term probably may be used to also refer to the "good gangsters"[7] the author had constantly heard about at Double-lake Plains. For instance, the Party secretary of *Fuwan* Village of *Linhu* City is basically also a gangster—colluding with gangsters in taking over the lake area and treating villagers from other villages very cruelly. But he protected his own village and people's interests. When he got arrested, all of the residents petitioned the local court and public security authority for releasing him.

Just like external gangs, state power is also considered to be an external influence, and hence it must come with a legitimate access. Therefore, for a national policy to be effectively implemented at the Double-lake Plains, it requires more than simply the power of the political system. The state must work with the "tough guys" and form a "clique of shared interest" (Xuefeng 2007a). The same goes with village officials,

[7] Chapter 8 will discuss "good gangsters" in detail.

being appointed as representatives of state power. When external influences buy over the local gangs, they would scratch each other's back. With that, they face no restrictions and soon public interests are divided between them. If the external influence fails to build up an alliance with the locals, they would become the target of blackmailing and extortion. At *Luoci* Village of *Jingmen* City, the owner of a harvester that goes into the village from other places is taxed by the local gangs at 20 Yuan per *mu* of land. During harvesting seasons, gangs may generate an income of more than 10,000 Yuan over half a month (Yiyun and Tao 2007: 17). Therefore, despite social atomization at Double-lake Plains, external influences may not get in the rural areas without collusion with the local gangs.

Different levels of resistance to external influences by rural areas of north, south and central China may be explained with the level of state penetration. North China is the cradle of the Chinese civilization and has always been the focus of central power. The open plain enhances its geographical accessibility, allowing state power to more easily penetrate to the rural areas. Farmers have to deal with the flooding of the Yellow River and the *Huai* River, but flood controls may not be accomplished by an individual or a clan. Farmers often feel incompetent in flood control and therefore look forward to state intervention. The result is that farmers have high expectations on the state and tend to recognize the state power and therefore, state power is able to deeply penetrate to the rural areas. In contrast, the hilly areas of south China is less accessible and also more distant from the central government. Farmers in south China mainly have to deal with droughts. As droughts often last for a long time, people have more time for coping with them. Individual households or clans may effectively avoid losses or relieve droughts through saving up water in ponds or managing water sources. This means that people in south China do not have high hopes for state power. In fact, the central authority did not enter the Double-lake Plains until the building of an imperial embankment on the north bank of the Yangtze River during the late *Ming* Dynasty. Since village community is still yet to be formed, farmers tend to provide public goods by themselves. In a word, in both historical and contemporary periods, state power has failed to deeply penetrate to the rural areas of Double-lake Plains.

The level of state penetration also has to do with the village's internal structure, people's demand for public goods and expectation for state power. On one hand, the internal structure of a village determines people's expectation on state power. In north China, people with different surnames live together, whereas in the south people live in clans or lineages. Therefore, northerners have to deal with two kinds of social relations: one is kinship ties between clan members, the other regional relations between clans. Inside a village, clan relations may be cooperative, competitive or hostile (Baifeng 2008c). It often requires a neutral third party to mediate between them in case of conflicts, and this third party role may be taken only by the state. Over time, people build up high expectations on imperial power and the government and look forward to the penetration of state. On the other hand, the internal structure of a village determines the government's manner of collecting taxes—whether to get in direct contact with individual villagers or not. Northern villages have a non-hierarchical structure comprising landlords, owner-operators and hired farm workers

(Myers 1970; Huang 2000a). The government must get in contact with a great number of small landlords and owner-operators. In contrast, southern villages have a hierarchical structure that allows the government to deal with only the landlords. The government does not have to get in direct contact with individual villagers. Most of the people there are used to getting in contact with the government through landlords and the landed gentry.

The level of state penetration has an impact on farmer's conception of the private and public spheres. The northern people's expectations on state power have made state penetration a reality, which has in turn kept the kind of influential community in south China from forming in the north. People do not have a sense of belonging to the village. Nor do they pay much attention to public affairs, because they assume public affairs should be handled by the government or the village officials and have nothing to do with individuals. The village is a "public" place instead of their own place. This matches what *Duanfeng* (2008) and *Jing* (2008) noticed in the south *Anhui* Province. This rule applies to almost everything, including weddings and funerals, dispute settlement, road building and dam construction, lack of filial piety (*buxiao*), piling of garbage, to name a few. People call the village group director "*Dang Guan De*" (literally a person being the government official), for in their eyes, village group directors are their officials, or a public figure in charge of public affairs. Therefore, people tend to be afraid of officials being transferred from other places; people expect local officials to deal with external gangs. If officials fail to do that, they could only pin their hopes on the higher authority, informally known as *Qing Tian Da Lao Ye* (literally an incorruptible official with his hands clean like the clear sky) or even President *Hu Jintao* (former president of P.R.C.). In our visits in south *Anhui* Province, people complained to us about the local gang problem, looked at us as "people from the higher authority" and hoped that we might report the problem to President *Hu* or the provincial leaders.

Resistance to external influences or the lacking of it determines whether gangsters may establish reputation. In north China, since people all pin hopes on the government, no "tough guy" would step up instead. In south China, many such people would stand up against external influences and would not mind other people taking advantage of their hard work. Therefore, in clans gangsters often get to establish their reputation. In Double-lake Plains, despite social atomization, gangs often resist external influences to keep such influences out of their territory, and also to establish reputation in the village.

6.5 Changes in Public Order

This chapter intends to present the influence of rural gangs on rural order at Double-lake Plains. In order to understand that, we may use south and north China as reference and the author has compared them with Double-lake Plains with great detail in previous sections. Here is a table summarizing the main points.

	Rural North China	Rural South China	Rural double-lake plains
Proximity to central power	Long been in close proximity to state power	Not in close proximity to state power	Somewhere in between
Villages are made up of	Families with different surnames living together, most owner-operators of farmland	Concentrated communities of powerful clan	Families with different surnames living on their own
Main public goods by provider of such goods	Land drains provided by the state	Drought relief by the clan (village)	Drought relief by rural households
Unit of identity and action	Kin group	Clan	Individual household
People's understanding of public and private spheres	The state is regarded as "the abstract public sphere" (the people still have hopes for state intervention); village officials represent "the specific public sphere" (they stand for the government) Clear distinction is made between the public and private spheres	Anything that concerns the village belongs to the public sphere but also concerns everyone (also private) Lack of distinction between the public and private spheres	Anybody may benefit from the public sphere whereas private sphere may not be hampered by external intervention Clear distinction is made between the public and private spheres
Intrinsic level of productivity for economic and ethical values	Fairly high	High	Low
Farmer mindset	People agree with, look forward to and rely on state power	People would rather rely on local forces than state power	People would rather rely on themselves
The form of rural gangs	Groups made up of *kuang kun* and bandits	Organized gangs	Organized gangs
Territory of local gangs	*Kuang kun*'s territory	Mainly outside the village	Both in and out of the village
Resistance to external gangs	Lacking resistance to external bandits	Strong	External gangs must collude with local gangs and will otherwise face strong resistance from local gangs
Sense of security	Lacking	Fairly strong	Average

Regional comparison reveals that rural gangs have a greater impact on rural order of the Double-lake Plains. Local gangs are no longer restricted by the acquaintance society, and in their pursuit of benefits, would often disregard other people's "face" or interpersonal feelings. People have a clear conception of the private and public spheres and gangs are allowed to take advantage of the latter and people are reluctant to intervene. To share benefits, gangs may collude with officials and the result is that public interests are divided up between them. External gangs form an alliance with their local counterparts and may do as they wish, and other people may do nothing about them. Local gangs will not stand up against external gangs unless they feel like their own territories have been breached or their vested interests are at stake. But when they do, they would take the lead in resisting intrusion, boycotting transferred government officials, and blackmailing non-local businessmen. All these indicate that rural order is increasingly controlled by rural gangs, that the behavior pattern of the acquaintance society has changed, and that the intimate community is falling apart.

As previously discussed, such increased gang control over rural order may be attributed to a lowered level of social connectedness, a lack of unified unit for farmer action, the village's lowered intrinsic productivity for economic and ethical values and lax grassroots governance. The alliance of rural gangs becomes the "super power" and has fundamentally changed the rural order. We may even boldly say that they are by stealth the really useful role in rural life. To pursue financial gain, they may use force or threats. Violence and financial status are mutually reinforcing for rural gangs, and as they take violence and financial status to the next level, they gain a higher status in the rural underworld and also dominance over public order. Generally speaking, officials find it difficult to restrain them. Those powerful and upright figures among the rural officials may barely manage to stay out of gang issues, but may never eliminate them. If the rural official does not have a strong influence, he may even have to suffer in silence. When threatened by the gang's use of force, many of the officials choose to give in or protect themselves. Compared with the "super power" of the gang alliance, villagers seem to have meager power; village officials and community are generally impotent; sometimes even the grassroots government is rather weak.

Chapter 7
Rural Gangs and Interpersonal Relationships

Having formed a loose alliance based on a hierarchical nexus of connections during the 1990s, rural gangs are no longer subject to the restrictions of the acquaintance society. Moreover, in turn they have significant control over the acquaintance society, which may not be overlooked. The notorious "super power" of rural gang alliance has changed people's thinking and behavior patterns, and therefore has fundamentally changed interpersonal relationships in rural areas.

The impact of rural gangs on interpersonal relationships may be summarized in three aspects. First, local gangsters have replaced traditional "bigwigs" of the acquaintance society and become the most "good looking" people. Second, in their interaction with their own folks, they do not observe interpersonal relationship principles that people have always stuck to in the acquaintance society. Third, they have also influenced the interpersonal relationships between ordinary villagers. This chapter will demonstrate the impact of gangs on interpersonal relationships at Double-lake Plains by comparing the region with other regions. The comparison has cleared revealed that interpersonal relationships and behavior pattern have both changed tremendously at Double-lake Plains.

7.1 A Group of People That Looks Good

On field trips to the Double-lake Plains, when I asked the farmers who was the most respected person, I often got the same answer, "nowadays we are all busy tilling our land; we all have meals in our own houses. That is, everyone is minding his own business, every man for himself. Who could possibly gain more respect from others?" But when I asked who has gained the most face (*mian zi*), they could name a few. It is generally believed that gaining face and respect are the same thing, however, farmers have offered very different answers to the two questions. So, I often asked again, "Why does a person gain face but does not gain respect?" They would again

B. Chen, *The Underworld of Rural China*,
https://doi.org/10.1007/978-981-19-8710-6_7

give me the same answer, "Respect comes from others, while face may be self-proclaimed. A person that gains face may feel good about themselves but may not be respected." Based on my observation, the kind of respect people show for people who gain face stays only on the face level. In private, farmers may still have a feeling of dissatisfaction, protest or even contempt towards such people.

The discrepancy between face and respect is essentially one between fame and fact, which often occurs during transitional periods of a society or when radical changes are happening. *Xiaotong* (1992) once pointed out that, during periods of slow change, social norms are not to be broken; but when social norms (the form) are ostensibly accepted, how the norms are being observed (the real matter) is open to interpretation and thus subject to change. Those un-defiable and non-applicable doctrines and orders will be deliberately misinterpreted and only the "face" is being saved for them. Save face for them means avoiding open defiance. Inevitably, one must outwardly agree but inwardly disagree. The result would be hypocrisy and dishonesty. As the society changes faster, the discrepancy between the form and the matter will grow larger. When the real matter is inevitably subject to distorted interpretation, a discrepancy would occur between position and power, fame and fact, statement and action, word and act, theory and reality. However, the discrepancy I talk about is not exactly the same to what *Fei* has mentioned in his works. My attention is not on whether the social norms are being observed, but rather on the separation of the form from the matter and the surface from the essence, when new things take on traditional forms.

The discrepancy between face and respect may be put down to the problematic separation of the surface of face from its essence or its form from matter. Such separation has resulted from the alienation (*yi hua*) of face competition from its essence. People may go after different things or compete in different forms, but "competition for face" always takes place over a certain object or thing ("the carrier"). Such competition often reveals, at a deeper level, a person's character, honour or main-stream values. If people aim solely for the carrier while disregarding the fundamentals or the real value of "face", people are going against the behavior pattern of rural land (*xiang tu luo ji)*. And that may lead to the alienation of such competition from its real value. Face is often associated with positive evaluation from other people, but such evaluation must also be associated with certain carriers of face. In the process of face competition, if positive evaluation has been overlooked and the carriers of face become the sole target of competition, face would be alienated from its essentials (Baifeng and Junxia 2007). For that reason, at Double-lake Plains, people who have gained much face are not respected by others.

To rural residents of the Double-lake Plains, there are three types of "good-looking people". First, the upper stratum of rural gangs, who have dominant control over a group of "brothers"; second, rural officials who maintain a good relationship with and have control over rural gangs; third, rural entrepreneurs and businessmen who maintain a good relationship with rural gangs and for whom many gangsters "throw their lives in" (*mai ming*). In both south and north China, there are still many traditional stereotypes of respected people; but at the Double-lake Plains, there are no such people any more. Almost all good looking people directly or indirectly have something to do with gangs. That's probably why the local people describe "having face" as "looking good" (*you mian zi*) or "cool" (*you wei*). These expressions imply that gangsters have control over others and that their behaviors are often open flaunts.

Rural gangs and the related groups look "good" or "cool" because they have control over ordinary people and more importantly over village elites. The elite groups include elites of the political system (political elites) and those outside of the political system (non-political elites). Political elites mainly refer to rural government officials. They are empowered by the self-governance body or the government's administrative system. Non-political elites are those villagers who have some influence both politically and socially. The difference between political and non-political elites lies in the source of power. The latter's power originates from shared cultural identity in the village community and relationships based on benefit or interest. Since they are not officially empowered, what kind of people may be regarded as belonging to this type is not clearly defined, but they mainly include: elites of clans, elites that have achieved financial success, intellectual elites (teachers, literate people), etc. (Zhihui and Xuefeng 2002). They would invariably form an alliance with gangs. Otherwise they would be harassed or controlled by gangs. In other words, rural gangsters "look good" in front of elite groups.

According to the law, the relationship between rural officials and residents is one of power relations within a self-governance body. Such a relationship is based on the coercive state power, the rules for self-governance and people's shared feeling toward the community. Rural officials should have maintained direct dominance over villagers. However, the same old approach to rural governance that has been used throughout the whole twentieth century has been passed onto the so-called rural politics and villager autonomy from the 1980s and on (Xuefeng 2007a). Therefore, such power relations have nothing left but a name to it. At the Double-lake Plains, officials fail to establish a dominant relationship over gangs. So the relationship between rural officials and residents is basically one of "administrative management". When rural gangs have formed an alliance, officials may not afford to neglect their influence, let alone have any control over them. One of the rural officials of *Shaqiao* Village said, "Gangsters are no pushovers. Officials have to suck up to them. That includes the village Party chief. A chief with the backing of an important person or with connections may get to keep the gangs at bay. Otherwise the gangsters would push him around." At *Linsha* Village, gangsters get to do as they wish, i.e. misappropriation of public assets, and officials can do nothing about them. In 2006, one of the local gangsters intended to contract a fish pond and spread out word to his folks by saying, "If the chief dares to say no, I would break his bone." Hearing that, the chief had no other choice than simply to live with it.

Even if officials maintain a relationship of mutual non-interference with rural gangs, officials would help them maintain their personal social networks (*guanxi*) by paying them casual visits and having a chat with them. When their families host weddings and funerals, officials would send gifts (*renqing*) to them to show them some face (*mianzi*). They are more powerful than officials, but they would also show the officials some face, since they still need help from the officials. Therefore, giving face is a reciprocal arrangement between gangs and officials. It may be concluded that the gang's bellicosity has made it impossible for officials to establish control over them and has forced officials to discreetly seek for self-protection.

Some officials choose to simply form an alliance with gangs. So they get to rely on gang violence or threats of violence to establish control over ordinary residents. The political system is partly responsible—in the process of modernizing the country, the political system applies pressure on officials and poses a great dilemma for them. Ostensibly the town/township level provides some kind of guidance to the village level, but in fact the village must do as told, i.e. collect taxes and execute the family planning policy. Village officials are overwhelmed by the high authority's ceaseless requests. In the context of the governance crisis during the 90 s, the town and the village form a community of common interest. Rural gangs are incorporated into this community, i.e. to help rural officials fulfill tasks such as tax collection. Tax collection is no longer a problem since the lifting of agricultural taxes, but rural officials still need gangs for solving problems they may counter in the provision of public goods.[1] The gang's sole reliance on violence or threats of violence for solving problems has distorted village power that should have been based on the delegated power and people's sense of belonging to the village community. In some worst-case scenarios, gangsters are directly appointed as rural officials, or officially incorporated into the political system. This has caused the official power to be superseded by undisguised violence. In the process of forming an alliance with officials or being incorporated into the political system, rural gangs become the most cool-looking group in the village.

The most influential non-political elites are economic elites that have become wealthy first, such as rural entrepreneurs and businessmen. They would either ally with rural gangs or "give face" to gangs (*gei mian zi*). Gangsters work for economic elites as their bodyguards or "thugs for hire". With the help of gangs, economic elites may gain an upper hand in market competition and may gain illegal financial benefits. However, since there are many gray areas, economic elites may also need help from the gangs to ensure the lawful operation of businesses. In a word, economic elites would maintain a good relationship with rural gangs, or even join the gangster's nexus of connections—these are ways for economic elites to give face to rural gangs.

Those most cool-looking ones of rural gangs may establish control over village elites and also officials. Everyone has to give face to them. We may refer to them as "forces of darkness and evilness" (*hei e shi li*). A gangster with the surname *Yang* of *Fucun* Village of *Linhu* City belongs to this type. Party Secretary *Hu* of *Qiaotou* Village intended to run for the deputy to the city-level people's congress and sought help from him. *Hu*'s main opponent was the Party secretary of his neighboring village, who happened to be an ex-inmate and a super powerful gangster in the area around *Qiaotou* Village. *Hu* knew he had no chance of winning, and he sought help from *Yang*, since *Yang* is one of the big bosses of *Linhu* City's gangs, and even the municipal officials have to give *Yang* face. *Yang* replied, "I'm willing to help. But the other side is also one of us. I should not displease him for your sake. So, how about this? You and he can each have half of the votes." Hearing this, *Hu* knew that he

[1] Chapter 8 will discuss this in detail.

would not win following normal procedures. *Liu Ye* (literally a lord surnamed *Liu*) of *Shaqiao* Village, *Chujiang* City also belongs to this type. The former police chief had to give face to him. When *Liu* was trying to teach his fellow gangster a lesson, he let the police chief know what he was going to do, and the chief simply told him not to go too far. At the dinner reception for *Liu Ye*, the police chief even drove there to propose a toast to him, giving him enough face.

At *Huangcun* Village of *Yueyang* City and its neighboring areas, those unruly casino-running gangsters often hold the police in contempt. They run a casino next to one of the main roads. The police know about that, but may do nothing about it, because they are understaffed. Six people in one police station have jurisdiction over 2 towns. In contrast, the casino is often visited by several hundred gamblers, guarded by several dozen "hired thugs" or "gang followers" (*mazai*). They have men on the lookout all the way from the doorstep of the police station. So the information about police force dispatch will be quickly spread out and cause the crowd to disperse. In case of conflicts, the police are also no rivals for the gangsters In 2005, in a police raid on the premises one of the officers got beaten and later the matter was unexpectedly dropped without being properly settled. This greatly discouraged the police. They would rather drink some tea or read some papers. All their attention is on "generating more revenues (*chuangshou*)". They often evade dealing with ordinary rural residents, or if they do, they are up to some money-making by scaring the people. They would not wear uniforms if working alone, to avoid being chased after by gangsters or being cursed by the folks. Once, one of our fellow researchers drove in a car with a police officer to a casino and triggered the siren just for fun, but the officer was very nervous and quickly fled the scene (Hai 2007). It seems like the police and gangs have reversed roles and the gangs have never been as proud and powerful as they are today.

Since the rural gangs have increasingly become the most cool-looking group, ordinary people cannot afford to treat them lightly. A middle-aged man from *Xinwang* Village of *Chujiang* City told me about his gangster nephew and displayed signs of pride. People do not feel ashamed over having someone as a gangster in the family. Instead, people regard them as admirable heroes who bring honor to the family. Therefore, rural gangs hold great charm to the young people. More and more young people are willing to get close to those "successful" gangsters. Many seek out the "famous" gangsters. At *Huchang* Village of *Linjiang* County, some parents even sent their children to the gangs. In the gang's network, these young people have discovered some distorted aspirations and a special way to climb the social ladder. Becoming a "big boss" (*da hunhun*) is the life goal for many beginners. On my field trips, many beginners told me that being invited in a fight gives them "a purpose in their gang life", and that the most honorable thing is to "rise above average". When that happens, money rolls in and they just sit comfortably at home, while others run the business, seize a dock and conquer a territory on their behalf.

7.2 Relationships Between Local Gangsters and Their Acquaintances

The acquaintance society is a place where people live from birth to death in the same place. They provide help and financial assistance to one another. They follow the reciprocity principle and the "not-going-too-far" principle, which has become the common sense in managing interpersonal relationships. The reciprocity principle is derived from people's familiarity and intimacy. It requires people to be sensitive, respectful, impartial, reasonable and tolerant (respecting others' feelings or *jiang ren qing*) in the way they conduct themselves in relation to others. The "not going too far" principle also requires "*jiang ren qing*". But in addition, it also requires people not to "stubbornly clinging to reason" when there is a conflict between reason (*li*) and feeling (*qing*). As members of the rural acquaintance society, do gangsters still follow these principles of interpersonal relationship?

The answer is apparently negative. At *Maocun* Village of *Changde* City one of the gangsters parked his car in the middle of the road. A villager passing by with a wooden hand cart got battered by over a dozen gangsters simply for saying "this car is getting in the way" (Junxia 2010). At *Wangcun* Village of *Tianmen* City, gangsters seized villager's land for house-building (*zhaijidi*) but villagers could only restrain their anger and abstain from saying anything (Xiongtao 2007). At *Huchang* Village of *Linjiang* County, Old *Zhu* (the person in charge of irrigation) has a son with many gangster friends in town. Taking on his son's influence, he often speaks rudely and acts recklessly. One night of August 2005, one of the villagers *Li Ming* heard the pumping noise coming from his farmland and assumed that the land has been irrigated. But when *Li* went to the land to check on irrigation, *Li* found that all farmland had been irrigated except his. So *Li* asked Old *Zhu* why, but Old *Zhu* simply replied that he would get around to it and then turned his back on *Li*. When *Li* was about to leave to talk to the director of their production team, Old *Zhu's* son yelled at *Li*, "What all the fuss is about? In this early morning?" Hearing his aggressive tone, *Li* asked back, "What's with the attitude?" The son immediately knocked *Li* over. *Li* snapped, "You son of a bitch!" Hearing that, Old *Zhu* came out and said, "You call some names, you get some kicks!".

Another reason for saying that gangsters do not respect principles of interpersonal relationship is that they may have a fall-out with others over trivialities and treat others in an overbearing manner. Taking on his gangster friends' influence, *Xue* from *Huchang* Village also has been used to acting rudely and recklessly and even starts to blackmail his fellow villagers. On one occasion, *Xue*'s wife came home from shopping and *Wang* gave her a ride on his bike. Knowing that, *Xue* was furious and went to *Wang*'s family and smashed *Wang*'s washing machine, TV set, cookers, among other things. The next day, *Wang* was afraid that *Xue* might strike again, so *Wang* treated *Xue* to a nice meal at *Huang*'s house. The incident did not come to an end until *Wang* gave *Xue* two packages of Red Golden Dragon (*hong jin long*) cigarettes. On another occasion, when *Xue* heard from his father that *Tu* asked them to pay off all due taxes and fees, *Xue* and his gangster friend *Tan* went to *Tu*'s family and on arrival started to beat up *Tu*'s family. They did not leave reluctantly until two of *Tu*'s brothers scolded them harshly. After that, *Tu* was afraid of being hit again, so he treated *Xue*, *Tan* and other gangsters to nice meals and asked his brothers to apologize for being rude. Later *Tu* gave two packages of Yellow Crane Tower cigarettes to *Xue*. The incident did not get properly settled until the final gift was given. On another occasion, *Xue*'s "farmland for flexible purposes" (*jidongtian*) was taken over by *Zhou Dajun*. *Zhou* planted wheat without removing cotton from the land first. When *Xue* came to know that, *Xue* went to *Zhou*'s and ask for reimbursement for the destroyed cotton—which made no sense since the land has been taken over. *Xue* broke *Zhou*'s cooking pot, and even gave *Zhou*'s mother a slap on the face. *Xue* did not stop until others tried to talk him out of making further damage. In a word, if his own interests are at stake, *Xue* always gets what he wants; he also blackmails his fellow villagers under all kinds of pretexts. For instance, when *Xue* came to know about a dispute between *Huang Ting*'s relative and *Lin* (*Xue*'s fellow villager) over duck raising, *Xue* and *Huang* ganged up in beating up *Lin* and demanding 1000 Yuan from *Lin*. *Lin* had to seek help from director of the village security committee (*zhibao*) and spend 500 Yuan on a nice meal at a restaurant before settling the issue.

However, not all gangsters would hang onto trivialities. At least it seems like some of them still stick to principles of interpersonal relationships. Some of them appear very polite and do not take advantage of others. But that is limited only to the surface level. When the others get involved in financial disputes with the gangsters or their close relatives, or when the others get in the way of the gangster's "career development", they would stop being polite.

Li Jian from *Qiaotou* Village of *Linhu* City belongs to this type. *Li* has many connections. Currently his family property is worth over 1000,000 Yuan. *Li* still lives in the village and treats others very politely. He does not bully the others over some trivialities. However, it is known to all that he is actually a badass. People say that it is better to keep a distance from him—what people have learned from a territory-splitting incident in 2002.

Not far away from *Qiaotou* Village lies a big freshwater lake. The ownership and usage of this lake has long been a controversial issue between *Qiaotou* Village and its neighboring villages, but gradually they have established a season-based routine procedure of harnessing the lake: during spring and autumn, people of *Qiaotou* Village may fish in the lake using *Mihunzhen*[2]; during winter, people from neighboring villages may enclose the lake for hunting wild ducks. However, by 2002, almost the whole lake has been enclosed by *Qiaotou* Village for raising crabs. *Qiaotou* Village has also reported to the city, claiming ownership of the lake surface. When *Qiaotou* Village seemed to have gained an upper hand in this, its villagers started to plan out the division of territory among them. However, *Li Jian* had his own plan. He intended to claim for sole ownership and threatened the village secretary to give him what he wanted. Without others' consent, *Li* called upon 5 households to enclose

[2] Translator's note: *Mihunzhen* is a banned traditional fishing tool that traps all sized fish with encircling gear. The name "*mi hun*" indicates that fish are lured into the tool as if their spirits are being controlled ("*mi*" means lure, "*hun*" means spirit).

the lake surface. This caused direct conflict of interest with over ten households who live on *Mihunzhen* fishing. So they tried to stop *Li*. *Li* then asked over 20 gangsters carrying long knives to chase after the fishermen. They injured 5 of the fishermen. The most seriously injured *Yin* got 12 stabs. After that, the village secretary was coerced by *Li* into mediating between them. People have also reported the case to the police but had to agree with private settlement. When *Li* paid the medical bills, fishermen backed out from the competition. The village secretary had no other choice but to agree with what seems to be an established fact. *Li* did not pay a dime for owning the lake surface, but anyone must pay him money for harvesting herbs there.

Gangsters who appear to follow interpersonal relationship principles have a bigger chance of achieving long-term development. They establish proper limits for both speech and action. They are both courageous and resourceful, and would not "unleash their ferocity" unless their essential interests are at stake. Since they are generally polite and courteous, others get nothing on them. They also keep a distance from others, which keeps them mysterious and intimidating. Therefore, they are both feared and respected by the others. This allows them to keep their dignity while pursuing important interests. In contrast, those reckless and foolhardy gangsters appear to be too narrow-minded for resorting to violence over trivialities. Sooner or later their faulty behaviors would cause them big harm, because they are more likely to break the law and thus become the target of crackdown campaigns. They may also face troubles from within the acquaintance society, in addition to all that trouble they must face in the rural underworld. In a word, discretionary use of violence or unethical methods allows gangsters to rapidly accumulate wealth, giving them a "respectable" life, and greatly lowers the risk of becoming the target of gang prevention efforts.

As a matter of fact, regardless of what appears on the surface level, truth is that all the gangsters have factually disregarded principles of interpersonal relationships in the acquaintance society, because those who appear to follow the rules would, when their essential interests are at stake, go against the face and not-going-too-far principles without any hesitation. What they do would often cause great harm to ordinary villagers. And the same goes for those gangsters who have migrated to other places (may be referred to as "emigrants"). They generally have found a legal means of making a living and live a "respectable" life. Though their lives and wealth accumulation are mainly done outside the village, they still constantly go back to take advantage of the village and interfere with village disputes. Among the 9 emigrants who are originally from *Shaqiao* Village of *Chujiang* City, *Wan Liang* is the only person who have never come back to treat their fellow villagers cruelly, probably because *Shenzhen* is too far away for that. I've been to almost ten villages of the Double-lake Plains, and incidents of emigrants harming hometown villagers have taken place at all of the villages, with no exception.

We've come to understand that in the traditional acquaintance society the attachment principle is followed by all those who have left the village. No matter where they are, people feel their roots in the land and have deep and enduring attachment to rural land (*xiang*). However, emigrants show a total disregard for such attachment. In the rural underworld, gangsters are used to honoring nothing but real benefits. Reputation and honor for friendship are both just means to gaining benefits. As previously discussed, when the form of face or reputation becomes separated from its matter, when face becomes alienated from its essentials, the kind of "face" or "style" pursued by gangsters is not the same to gaining positive evaluation, but something they may flaunt or control others with. They may look "good" or "cool" for having control over many followers, for having what it takes to host grand weddings and funerals, and for being so well-connected with both the government (that's the "white" side) and organized crime ("black" society, as it's called in Chinese) that they can "eat from both sides."

The result of how gangsters get along with others is that many villagers fear gangsters and their families. To ordinary villagers, rural gangsters may look admirably good but they are no good people after all. During my field trips to *Shaqiao* Village, in a semi-structured interview I asked 23 villagers this question: "Do you fear local gangsters?" The interview results reveal that: 16 of them clearly expressed fear for gangsters and would be careful not to provoke them; the rest 7 people said they do not fear gangsters or there is no need for that. According to an explanation provided by *Yang*—the local accountant and village official, 3 of the 7 fearless people have family members as gangsters, hence no need for fear; 4 of them lack a good understanding of what gangsters may accomplish.

Since most people fear gangsters, people would consciously keep a distance from them and avoid provoking them. A post-graduate candidate who is originally from *Shaqiao* told me that several friends he had as a child had all become gangsters, and that his parents specifically told him not to hang out with them. His parents said, "These people could have a fall-out with anyone any time. Who knows what they are capable of? People like us cannot afford to be fooled. You can never be too careful." Accountant *Yang*—a very smart person—expressed similar views. Many members of the upper stratum of the rural gangs are his classmates and playmates, and he used to have no fear for them. But when he came to know more about them, he becomes more fearful. When he recalled something he did—an innocent mind has no fear—it still sent cold chills down his spine. There was a time when he did not see eye to eye with *Wan Qing*. One day after a few drinks he took a bet on whether he dared to slap *Wan* on the face. All he did was to express his dissatisfaction while the wine kicked in, but *Wan* was furious and about to beat him up. Others stopped *Wan* by saying things like "it must the wine talking". After that, *Yang* and *Wan* went back to normal as if nothing had happened. But *Yang* became aware of what *Wan* might be up to if he offended *Wan*. Looking back, he said he did not know what to do if *Wan* was really up to something or asked some other gangsters to get back at him. He said, "Now I would be extra careful around gangsters, for they do not follow the rules. Though some villagers may also be quite bossy, gangsters can never be the same like the bossy ones of ordinary villagers."

7.3 Gang Influence on Acquaintance Relationships

With the formation of alliance in the rural underworld, rural gangs become unconstrained by the acquaintance society, but in turn also exert considerable control over the acquaintance society. Such gang influence firstly extends to gang-villager relationships; secondly to relationships between ordinary villagers.

As previously discussed, the way gangsters get along with ordinary villagers has apparently deviated from the interpersonal relationship principles followed by other members of the acquaintance society. They would either flaunt their powerful connections or directly bring external gangsters in on others. Surely they do not always need to bring outsiders in, because they could generally get what they want by directly use violence or threats of violence. This is well illustrated by *Xue* from *Huchang* Village of *Linjiang* County and *Li Jian* from *Qiaotou* Village, who have both brought in outsiders to beat up their fellow villagers. Similar incidents are so common at the Double-lake Plains that people would regard the coming and going of outsiders as a "regular" thing. Besides, bringing in outsiders seems to be very effective at problem-solving. At *Maocun* Village of *Changde* City, *Hunan* Province, a villager got knocked over by a motorbike. The motorbike rider knew the villager, but asked a gangster to intervene. The gangster also knew the villager himself. Right on arrival, the gangster said, "Since I'm here, I get to make the calls. I will do things my way!" Eventually, the motorbike rider got away with simply reimbursing the villager with 100 Yuan, far from enough to cover the 800 Yuan on medical bills (Junxia 2010).

It may seem as if the kind of disturbances caused by gangsters is not particularly frequent yet, but they have caused serious psychological intimidation on others and even have changed other people's behavior pattern. At *Hanshou* Village of *Hunan* Province, when people have an argument and one side apparently does not hold water, others would not intervene unless they may rule out the possibility of potential gang influence. We may safely conclude that rural security has been greatly compromised by the rural gangs and that the gang presence has impaired the political system and interpersonal relationships that underlie the security system.

Despite their fear for gangsters, ordinary villagers may seek help from external gangs as well. When their fundamental interests are at stake, they would not give in without putting up a fight. Ordinary villagers seem to have been "inspired" to seek help from outsiders by the gangster's disregard for affection between fellow countrymen and arbitrary resort to violence for gaining benefits. In the Chinese rural society where "the relationship between cousins may extend to three thousand miles away", a *lao shih* may be related to a gangster in the wider family. Besides, ordinary villagers may easily "buy" gang service as long as they are willing to pay the price. Gangsters are willing to provide their service as long as it is profitable. Therefore, gangsters may be brought in by another gangster and by the *lao shih* who have been compelled to do so. In 2005, *Wan Min* from *Shaqiao* Village of *Chujiang* City went into a dispute with his fellow villager *Wan Yi* over the right to use a piece of 2.5 *mu* (about 1667 square meters) of arable land. *Wan Yi* refused to give in, so *Wan*

Min sought help from Secretary *Wan*—the village bully. Secretary *Wan* threatened *Wan Yi* by saying, "If you refuse to give the land back to *Wan Min*, you may end up with even less land." When *Wan Yi* appealed to the town authority, Secretary *Wan* instantly changed his tactic by playing the good guy. He said to *Wan Yi*, "Don't mess with *Wan Min*. You stand to lose." Knowing that the Secretary has failed to convince *Wan Yi*, *Wan Min* went over to *Wan Yi*'s to threaten *Wan Yi* in person, "I'm a man of his word. I always keep my word." *Wan Yi* knew that he was no opponent for *Wan Min*, so he used his connections to seek help from gangsters in the city. They came to *Shaqiao* Village carrying shotguns and long knives, which greatly discouraged *Wan Min*. *Wan Min* soon dropped his request for the right to use the land. Besides, he had to treat *Wan Yi* to a nice meal and apologize for his bad attitude.

Apparently, whether the dispute may be properly handled largely depends on whether one can find a guy "tougher" than the opponent. In fact, rural residents not just bring in outsiders on gangsters; they may also bring in outsiders on other ordinary villagers, because doing so will help them gain an upper hand while evading punishment. With gangsters by their side, people on justified grounds may force others to apologize or make compensations; people with no justified grounds may seize the chance to blackmail others.

At *Shaqiao* Village, *Wang*'s cattle ruined *Zhao's* half *mu* of rice seedlings, either by eating them or rolling on them. *Zhao* kept *Wang*'s cattle—which is the regular thing to do—and did not release them until *Wang* compensated *Zhao* for his loss. Knowing this, *Wang*'s son was furious and went to the town to bring in gangsters on *Zhao*. They beat up *Zhao* and forced him to return the money, and on top of that, *Zhao* was blackmailed for an extra 500 Yuan. At *Huchang* Village of *Linjiang* City, *Luo* used to run an underground Mark Six lottery outlet. In May 2003, when *Luo*'s high-level vendor got arrested, *Luo* had to pay the prize himself. *Luo* was supposed to pay his fellow villager *Xia* 14,000 Yuan, but due to limited financial means, *Luo* was able to pay *Xia* only 7,000 Yuan. *Xia* was very unhappy, so the next day *Xia* brought his gangster nephew *Xia Tao* to *Luo's*. *Xia Tao* threatened *Luo* while holding a knife. *Luo* had no choice but to sign an IOU for the rest of the money. In January 2004, *Xia* came back again bringing another gangster to *Luo's* to demand payment. When *Luo* asked *Xia* to give him some more time, *Xia* was again very unhappy and picked up a knife on hand and cut *Luo* on the left hand. Several days later, *Xia* came back again bringing three gangsters to *Luo's*. Very arrogant and aggressive, they used harsh words to scare *Luo* and threatened to harm *Luo*'s children if *Luo* still refused to pay. *Luo* had no choice but to seek help from the police.

It is generally believed that when gangsters threaten people's life or property, the first thing that comes to mind is to call the police, for such threats may not otherwise be handled. However, at Double-lake Plains, such threats may not be removed once and for all, not even by the police. Just as a famous Chinese saying goes, "whoever hung the bell on the tiger's neck must untie it"—to remove gang threat, one must deal with the gangsters themselves. At *Huchang* Village of *Linjiang* County, in a dispute over injury compensation, one side brought in external gangsters to "negotiate" at the other side's house. The police was also called in and stopped the gangsters from making trouble. In the police's presence, the two sides agreed on 500 Yuan as injury compensation, but when the police left, the side backed by gangsters changed their mind and requested 1500 Yuan instead. Unexpectedly the other side did not bargain but obediently paid the money. When the police came to know about this, the police insisted being told the reason for paying the money and they were told that, "The extra money was for him not to come back for me again. The money was given to the gangsters. Since they have involved gangsters, I have to pay for that. You may protect me as long as I stay in the village. But I cannot stay here forever." Knowing that, the police were also convinced that paying the money was probably the only solution.

If both sides of a dispute are backed up by gangsters, the settlement of the dispute will depend entirely on negotiation between gangsters. If the gangsters already know each other, they may mutually give face and settle the dispute through friendly consultation. Such consultation is quite similar to the manner of dispute settlement in a traditional rural society, where the most influential persons of the clans are invited to settle a dispute between people from different clans. It seems like in contemporary times, gangsters—or the most cool-looking group of people—have become the most influential figures. However, the gangsters and the influential figures are in no way the same people. In traditional time periods, the virtuous people settle disputes through convincing people through the use of values and morals or moral persuasion. In contrast, gangsters are no virtuous people; though sometimes they appear to use values and morals as well, more importantly dispute settlement depends on their use of force or threat of force. Sometimes they disregard values and morals, and make no attempt to disguise the use of violence. In a word, the willingness to "give face" is based on the underlying power and influence, and when such power or influence is lacking or insufficient, others would refuse to "give face".

Sometimes what makes things worse is that gangsters have come from different cliques and they do not need to keep up a face-based relationship. They would directly resort to violence, which often leads to confrontations. At the Fourth Village Group of *Shaqiao* Village of *Chujiang* City, *Chen Xing* and *Zeng Fanhua* are neighbors with adjoining fields and use the same pond for irrigation. One day of June 2005, *Zeng* blamed *Chen* for pumping water from their shared pond to other places, but *Chen* did not think he was wrong. They first had a fall-out and then started to jostle each other in the fields. One of *Zeng's* ear and one of *Chen's* foot got cut by their shovels. When they went home and told their wives about what happened, the two families cursed at each other. Afterwards, *Zeng* was still brooding over what had happened, so *Zeng* called his son *Zeng Jian*, who works in town, and told him to come back. *Zeng Jian* brought home several gangsters and charged into *Chen*'s carrying knives. Hearing that gangsters were coming for him, *Chen* had already fled his house when *Zeng* and the gangsters arrived. *Zeng* left a note to *Chen*, threatening to "cut off *Chen*'s ears any time", and went back to the town. *Chen* was very scared, so he called the police. The town police station then asked the village *Zhitiao* director[3] to intervene.

[3] Translator's note: *Zhitiao* Director is in charge of both the security (*zhi*) committee and people's mediation (*tiao*) committee. The director's main responsibilities are ensuring security and mediating disputes.

In the presence of the director, *Chen* and *Zeng* agreed not to make any trouble before they were both fully recovered. Before long, *Zeng Jian* came back from town and went to *Chen*'s with a few relatives and aggressively demanded 6,000 Yuan in reimbursement. *Chen* was prepared for that. *Chen* told a gangster—who happened to be in his wider family—about what happened, and right after *Zeng's* arrival, *Chen*'s help also arrived. *Zeng* kept playing the "tough guy" card and threatened to cut off *Chen*'s ears. When the two sides were about to go into physical confrontation, the *Zhitiao* Director arrived in time and kept both sides under control. Soon the police arrived and carried out some routine questioning. The two sides again agreed to deal with it later when both have healed from previous injuries. The incident was left unsettled and no one intended to do anything about it anymore.

Based on previous discussion on the gang influence on acquaintance relationships, we've come to understand that villagers would often get embroiled in a dispute because they have gangster relatives or friends. During my field trips to *Shaqiao* Village of *Chujiang* City, I asked two questions to 23 villagers in a semi-structured interview, "In a dispute with your acquaintances, would you consider the influence of gangsters as the other side's family, relatives or close friends?" "Does this affect your attitude to dispute settlement?" Based on the interview statistics, all villagers have given an affirmative answer to the first question; 19 of them clearly expressed that their attitude will be affected by the likelihood of gang involvement. I was told by Accountant *Yang* that among the 4 villagers who think otherwise, 2 of them were quite bossy all along for they have gangster relatives; one of them was ignorantly fearless—*Yang* described him as "silly", "brainless" and "short-sighted"; the last one was described by *Yang* as "a typical stubborn mind whose deeds does not match his words". *Yang* told me that all of them, except the brainless one, would consider gangsters as an influential factor in their behavior. Even though they have gangsters in their wider family, they would compare the influence of their own folks and other gangsters.

In a word, rural gangs have already become an important factor that affects acquaintance relationships. That may be hardly perceptible when life is good for all. But when disputes occur, people would involuntarily consider the gang factor. In the autumn of 2007, a car accident took place at *Xinwang* Village—next to *Shaqiao* Village. *Li* was riding a motorbike carrying his wife; when they were going down a slope and trying to overtake *Wang* in his three-wheeled truck, *Yang* was going up the slope. When *Li* tried to avoid running into *Yang*, *Li*'s wife fell over onto *Wang*'s truck and was hospitalized for the fall. Before the traffic police's collision report came out, *Li* demanded that *Wang* reimburse him for his wife's injury, but did not demand the same from *Yang*. Both *Wang* and the on-lookers believed that *Li* did not dare to demand reimbursement from *Yang* because he was afraid of *Yang*'s gangster brother. Because *Li* had a gangster cousin but *Wang* had no gangsters to seek help from, *Wang* had to give *Li* 2000 Yuan as an initial payment. But the money was far from enough to cover *Li's* demand for 40,000 Yuan, so ever since *Wang* had been filled with fear that *Li* might bring his cousin into this. In fact *Li*'s cousin never showed up, and *Li* never threatened *Wang* by mentioning his cousin. The facts in this case aside, we may conclude that potential gang involvement—having gangsters in the social connections of the parties to a dispute—already poses great psychological intimidation for the villagers and will undoubtedly affect the result of dispute settlement.

7.4 Changes in Interpersonal Relationships

To ordinary villagers, local gangsters used to be their acquaintances but are turning into strangers. External gangsters have always been strangers to the acquaintance society, but they have never been discriminated against as outsiders. Instead, they have barged into the acquaintance society with their recklessness and arrogance, turning the acquaintance society into a half acquaintance society. We may safely conclude that locally born and bred gangsters and outsiders brought in by local gangsters are both "strangers at the doorsteps" to the *lao shih*. We say that they are strangers to villagers, not in the sense that the villagers do not know them, but that instead of following the face and not-going-too-far principles, gangsters would use the discrimination principle—the principle used by acquaintances on strangers—to handle relations that should have been treated as acquaintance relations. Therefore, the behavior pattern or code of conduct for the rural villagers has been altered; the behaviors of rural gangsters are entirely unpredictable.

Acquaintances are increasingly estranged. When people get into a conflict or dispute, they would constantly seek help from "the strangers on their doorsteps", that is, using violence—or what they use on strangers—on acquaintances. People that used to be treated as their brothers and seniors are more and more commonly being pushed away as strangers. Secretary *Hu* of *Qiaotou* Village said, "Running into road bandits is nothing uncommon to a man doing business in other places and thus psychologically acceptable, but frequent incidents of violence between acquaintances in the homeland is psychologically unbearable. With everyone living door to door, who bullies whom is as plain as day." As far as the villagers are concerned, using violence to treat strangers may also be unfair, but understandably so, because people have developed a regional consensus for treating natives and outsiders differently. However, now that acquaintances are becoming strangers, people do not follow the familiarity (or intimacy)-based code of conduct. Both "strangers at one's doorsteps" and "the estrangement of acquaintances" have clearly indicated the alterations in people's code of conduct, and are turning the village from a place with affection and care into a cruel battle field!

From the perspective of regional comparison, the above-mentioned phenomena are the most prominent in the Double-lake Plains, less so in the North China Plain, and even less common in south China. At Double-lake Plains, rural residents' code of conduct has been drastically altered; interpersonal relationships are increasingly dominated by gang violence. Compared with the Double-lake Plains, things are much better in south China where violence has not yet become the dominant factor in rural life. However, several signs also suggest increased susceptibility to violence for south China. First, external gangsters appear more frequently in rural people's life, which has caused people to be estranged in the grassroots market system; second, local gangsters are becoming active players in rural life under the pretext of protecting the village's name; third, the use of brutal violence is also seen in some particular villages of south China.

In traditional times, because rural order at the grassroots market system is mainly maintained by local elites, and such rural order is a natural continuation of the same order in the acquaintance society, local elites have inherited from the acquaintance society the face and not-going-too-far principles in handling interpersonal relationships.[4] However, in south China's clan villages of contemporary times, there are no respected local elites in the grassroots market system; the town/township government also has a diminishing ability to ensure order. Therefore, rural residents are increasingly vulnerable to violence or threats of violence from other villages of the same market system. Gangsters from *Shanxiang* County of *Hunan* Province constantly blackmail people from their neighboring villages. They also frequently forcefully intervene into disputes and conflicts between villagers. The villages of south China may still be the acquaintance society in nature, but at the grassroots market system people have become increasingly estranged. In other words, the rural society beyond the village border is no longer the acquaintance society in nature.

In certain clan-based villages of south China, though the reciprocity principle is still basically being followed, and that rural life is not yet dominated by gang violence, there are still signs suggesting increased susceptibility to violence whereas violence may have been used in disguise since its form has been separated from its matter. A case in point would be a gangster from *Ji'an* City of *Jiangxi* Province, who got injured in an armed clan conflict and then blackmailed his fellow villagers into doing things for him in the name of safeguarding the clan's interests.[5] It may be concluded that interpersonal relationships in the village are also partially changing in nature.

In our research on regional differences of rural areas inhabited by the *Han* Chinese, Student *Tan* (2007: 127) investigated on *Qiaocun* Village, which is also located in south China. Since 1999, there have been 3 cases of external gangsters brought in by the natives for settling disputes at *Qiaocun* Village. These cases carry important messages to us: gang violence is starting to affect interpersonal relationships in clan villages; clans have a diminishing role in curbing gang impact on rural order. That is to say, south China is following the same course as the Double-lake Plains, and is becoming increasingly similar to the Double-lake Plains in terms of social connectedness, unit of farmer identity and action, and the village's intrinsic ability to produce economic and ethical values. As time goes by, with the expansion of the market system, rural residents across the country would become more rational; the

---

[4] Section 2.4 of Chap. 2 has discussed this in detail.

[5] Section 6.1 of Chap. 6 has discussed this in detail.

acquaintance society's curbs on gangsters are diminishing and may soon disappear entirely. By then, the alteration of rural people's code of conduct would no long be a problem particular to the Double-lake Plains. As rural economy develops and social structure undergoes certain changes, it may be simply a matter of time before other rural areas encounter the same problem. But of course that remains a proposition to be verified.

Chapter 8
Rural Gangs and Village Governance

With the founding of the People's Republic of China, the Chinese nation has basically solved the survival issue, but to build China into a stronger nation remains a challenge. To make China stronger, rural governance is incorporated as a national strategy into the long-term plan for modernizing the nation. Rural governance must fulfill two main tasks, one is to provide resources for modernizing the country's industries; the other is to maintain order in rural areas to ensure stability and unity as the country moves forward towards modernization. These two tasks have been effectively fulfilled by the People's Commune in the three decades after the founding of the nation. However, when the commune system has been dismantled by reform and opening-up, the rural governance system has also changed. State power gradually backs out from rural areas, which causes a glaring contradiction in the state's need to tap rural resources and the need to ensure rural stability. This has in turn caused the agrarian crisis or the Three Rural Crises (*San Nong*)[1] during the 1990s. Rural governance is faced with great challenges. First of all, the government has increased difficulty in extracting rural resources, and due to escalating conflicts between farmers and the government, it has to pay a higher price for everything it takes from its rural areas. Second, maintaining rural order becomes increasingly difficult. Things are worsened by the shortage of public goods for villages, since the government fails to provide such goods and the village fails to be self-sufficient.

Under such circumstances, the rise of rural gangsters becomes particularly conspicuous. When faced with the Three Rural Crises and the predicament of rural governance, the grassroots government has no choice but to continuously incorporate rural gangsters into the political system. Those rural officials who are capable of fulfilling tax collecting tasks are often morally degenerate. In other words, more and more officials become "competent but unethical". But the ordinary villagers

[1] Translator's note: The Chinese government described the Three Rural Crises in 2003 as 'three rural problems' (*san nong wen ti*). These problems were: low productivity in agriculture, low income of peasants, and undeveloped infrastructure in the countryside–a crisis affecting the economy, people and space.

B. Chen, *The Underworld of Rural China*,
https://doi.org/10.1007/978-981-19-8710-6_8

seem to have encountered many "upright gangsters" in their effort to obtain public goods. People would even find the competent but immoral officials to be no match for "the good gangsters". Therefore, rural governance becomes inseparable from gang intervention. Even after the lifting of agricultural taxes—when gangsters are not needed for tax collection, they continue to be the de facto dominant force in rural governance. The abolition of agricultural taxes crosses out the task of "providing financial resources to the nation" from the rural governance agenda, but the other tasks such as "providing public goods" and "maintaining rural order" are moved up the agenda. Therefore, the abolition of agricultural taxes does not make rural governance any easier, because it has come together with reform on the political system, which brings new challenges for rural governance. The incorporation of rural gangsters into the village governance system reveals that: the acquaintance society has been seriously undermined in contemporary times; the behavior pattern of the rural society has drastically changed; and that gangster's incorporation is the result of all kinds of special efforts made by the village to simply maintain order.

8.1 "Competent but Unethical" Officials

ON field trips to the Double-lake Plains, I have constantly heard about people's opinion on rural officials. One of the most interesting choices of words would be "competent but unethical". People detest some officials and describe them as "corrupt and bossy" and "a swinger that knows no moral standards". But the same person is rated by other rural officials as "not the best ethical model but very qualified for the job", often suggesting approval. Before making a moral judgment on them, I intend to find out the reasons for such discrepancy in evaluation on the same person.

It occurs to me that most of the rural officials for whom people have conflicting opinions invariably have some kind of connection with gangsters. These officials may be classified into three types. The first type is gangster-turned officials, who are still active in the gangs. The second type cooperates or allies with gangsters. The third type has gangster relatives or friends and may take advantage of such connections to rule over the village. As discussed in previous chapters, rural gangsters would—in their pursuit of illegal gain—seek patronage from government agencies or maintain a cordial relationship with rural officials, whereas rural officials need gangsters to assist them in governing the village. Those officials that often rely on gangsters in their rule would naturally be detested by ordinary villagers. But they would gain approval from other officials if they can ensure good performance.

At *Shaqiao* Village of *Chuajiang* City, Secretary *Wan* has established an alliance with gangsters. Though *Wan* is a corrupt and unethical official who often misappropriates public assets, both town officials and officials in neighboring villages may not deny his work performance. During his "reign" over a decade, *Shaqiao* Village has been well administered with all tasks and responsibilities ranking among the top in the town. For that reason, Secretary *Wan* has gained the town's support in successive elections. The town's successive Party secretaries and mayors have all invariably supported *Wan* as the village Party secretary. The same is also true with the Party secretary of *Liji* Village, one of *Shaqiao*'s neighboring villages. The village secretary *Li Pinnan* has a gangster nephew, and the deputy secretary *Li Shan* has a gangster son. They both take advantage of their gangster connections to rule over the village.[2]

Officials of *Shangwu* Village, *Jinmen* City also have complicated relationships with gangsters (Leiming 2007; Caisong 2006; Lin 2006; Lina 2006). The director of the village's committee *Huang Lao Da* (the Eldest of the *Huang* Brothers) has a big family with six brothers and more than four cousins, all of whom are "tough guys". Director of the village security committee is taken by *Huang Lao Si* (the fourth child). No one dares to offend their family with so many brothers. Among the six brothers of *Huang*'s family, *Huang Lao Da*—a bold, aggressive and calculating person—is the first to have established himself, taking up the position as the director of the village's committee since the end of the 1980s and making "the first bucket of gold" being in this position. When a primary school was about to be built up at the village, he contracted the building of the school to his brother-in-law, but cracks appeared in the four-floor school building in just three years, causing two of the floors to be demolished. *Huang Lao Da* had had a three-room house; but once the school building was finished, his house was expanded to a stone and brick yard covering a 6-mu (4000 sqm) area. During his incumbency, he was responsible for misappropriating a large amount of public assets. Later he was miraculously transferred to many other lucrative positions, including director of the town's electrical substation, director of a neighborhood committee in the development zone, Party secretary to *Shangwu* Village, and director of the town's veterinary clinic. While *Huang Lao Da* worked as the director of the village's committee, he appointed *Huang Lao Si* as a chief of one of the village groups, and *Huang Lao Si* was later promoted as a village-level official. *Huang Lao Si* was rated as "one of the most honest men among his brothers". *Huang Lao Wu* (the fifth child) used to be director of the town's grain administration and is currently working for the Municipal Grain Administration Bureau. *Huang Lao Liu* (the sixth child) has been a butcher, a restaurant owner and now is in charge of collecting management fee at the local farmer's market. He also holds a monopoly over the town's pork producer association and uses gangsters to control the bus line between the town and the city.[3]

The *Huang* brothers are so well-connected with both the government and organized crime that they can "eat from both sides". So nothing is impossible for them. They were able to buy over the whole town-level farmers market with just 100,000 Yuan—the building price of the food market alone has exceeded that amount; and among the 50 stalls the first two at the front gate could have been sold at that price. The local people all fear the *Huang* brothers, for they always fight together, each of them reckless and ruthless. Even the town gangsters are awed by the brothers and would arrive at their service any time. Once, *Huang Lao Si* had a fall-out with a hot-headed young man. Out of everyone's expectation, the man scratched *Lao Si* on the face. The town police officer immediately arrived to arrest the man, but the

[2] Section 6.1 of Chap. 6 has discussed this in detail.

[3] Section 4 of Chap. 4 has discussed this in detail.

so-called "the most honest guy among his brothers" refused and threatened to "break an arm" of the man and did not intend to "bother the police". Hearing that, the man and his whole family knelt at *Huang*'s place and gave 300 Yuan for "medical bills". *Lao Si* demanded 1000 Yuan, and the man had to do as told.

Liao, former director of the Fourth Village Group, used to be a gangster. *Liao*'s gang career started right after he dropped out from middle school. He has been on the run for quite a few years, but the police failed to capture him. Back then, the village group director's job was a hot potato, for no one listened to the director's orders, and many directors got beaten up by villagers. From 1981 to 1898, the director of the Fourth Village Group has been changed four times. The last director quit his job after being beaten by his group members. Under such circumstances, *Liao was* appointed as the director, for only he may handle the gangsters in the Fourth Village Group. That was how *Liao* ended up as the director from 1990 to 2002 for 13 years. *Liao* was no moral model, and he racked his brains for ways to use his position for private gain. Every year, the taxes and fees collected from all households of the Fourth Village Group added up to 20,000 Yuan, and he would turn in half of the amount and keep the other half for production management and daily administration. Of course the money went into *Liao's* pocket. The shortfall was filled through requesting more money than real expenses for irrigation and drainage projects in budget proposals. He also coerced other villagers into doing his share of voluntary service. But *Liao's* tactics for handling gangsters have turned out to be very effective. Firstly, he gained prestige from keeping a firm grip on his "brothers in the gangs" and boldly punish the conspicuous troublemakers among them. Secondly, he showed enough respect for former "senior commune members" by treating them to nice meals at spring festivals, gaining him their support whenever he needed it.

It may be concluded from previous examples for gangsters at *Shangwu* Village that gangsters would strike up a rapport with party organizations through being directly appointed as officials or cooperating with them in rural governance. The purpose is to use government apparatus for private gain. However, truth be told, gangsters have done quite a good job in rural governance. At *Shangwu* Village, village groups take turns to clean out debris from irrigation channels, but because this is a voluntary job, slacking off or simply refusal to work are commonplace, but not in the Seventh Village Group, which is under the leadership of *Huang Lao Si*. *Lao Si* made it a rule in his village group that whoever does not want to work has to pay others for doing the job. And people do not dare to disobey his order. At the Fourth Village Group, *Liao* not only does a great job in collecting taxes—from 90% of the households, but also ensures villager's active voluntary service at public projects.

Needless to say, in return for good performance in one's office, officials must make certain trade-offs. Rural officials must concede some benefits to those powerful gangsters and "senior commune members", by reducing taxes due, allowing them to cut back on voluntary service, or contracting a public pond or forest to them at a low price, etc. Most of the delinquent taxpayers—except some extremely impoverished households—are gangsters and former senior commune members. On top of the delinquent list is a household that has owed over 7,000 Yuan, and other villagers must equally share the burden.

Since the beginning of the 1990s, law-abiding village officials often fail to handle rural gangsters and senior commune members. Their reluctance to cooperate with rural governance may be reflected by their refusal to pay taxes or offer voluntary service. Officials may do nothing about them using legitimate means, for they do not listen to reasoning, or bother to learn about laws and regulations. Officials are also no rivals for them in fist fighting. Therefore, officials have no choice but to let them be and often fail to fulfill tasks delegated by the higher authority. On the contrary, gangster-turned officials may efficiently fulfill tasks using their personal influence and connections. That's probably why rural governance becomes more efficient and effective once gangsters work with or directly become officials. They may resort to violence or threats of violence if people refuse to cooperate.

Once gangsters are incorporated into the political system, they would more blatantly misappropriate public assets and seek private gain. Previous chapters have shown the severity of the situation at *Shaqiao* Village of *Chujiang* City. At *Shangwu* Village of *Jingmen* City, by the end of the 1990s, an amount of 500,000 Yuan of public assets—which took several decades to build up—has been entirely depleted by the unethical officials. There are mainly three means of gaining private benefits for those "competent but unethical" officials. The first one is lining their own pockets with special funds, such as funds for relief efforts, poverty reduction and government projects. For one time, the higher authority allocated 150,000 Yuan for bridge-building at *Shangwu* Village. The village Party secretary contracted the project to one of his connections, and only got 3 little dams built—the real cost was probably no more than 70,000 or 80,000 Yuan. For another time, the county government intended to subsidize selective breeding with 24 Yuan per *mu* of land, but the village gave villagers only 10 Yuan per *mu* of land and the rest of money simply disappeared. The second one is selling public assets such as forests, ponds or fruit farms. Officials of *Shangwu* Village once sold a hilltop to one of the villagers. No one knew about it until the buyer stopped other people from cutting firewood there. The third one is free riding by fabricating fees when given the right to collect regular taxes and fees. In a word, a perfect official-gangster symbiosis has formed in their common relentless pursuit of private gain.

Despite good performance in terms of fulfilling required tasks, such a symbiotic relationship between gangsters and officials in rural governance would cause public assets to dwindle and also will greatly undermine the legitimacy of village community and government. A former leader of production team commented on this by saying, "Using a villain to handle a villain may keep the order for now, but is not the way for the long term. These people have an insatiable appetite for gain. Public assets will be used up in no time." The position of the "village group leader" has been canceled once and for all at *Shangwu* Village since 2004. *Liao* quit his job as leader of the Fourth Village Group two years before that, because at that time he already felt there was "no point" being the leader now that there was no prospective gain.

Among the eleven villages at Double-lake Plains I've been to, *Shaqiao* Village and *Liji* Village of *Chujiang* City, *Fucun* Village of *Linhu* City, and *Dengwan* Village of *Jingshan* County are the four villages where the Party Secretary's position has been taken by gangsters or gangster-related people for almost ten years; *Linsha*

Village of *Chujiang* City's gang-related secretary did not resign until three years ago; *Huchang* Village of *Linjiang* County's Party Secretary position has been taken by a "tough guy" who has many brothers and who treats others in an overbearing manner; *Puxing* Village, *Chencun* Village of *Xianning* City, *Xinwang* and *Wangcun* Villages of *Chujiang* City are the only villages where the secretary is taken by one of the *lao shih.*

At Double-lake Plains, despite widespread discontent among villagers at the corrupt and unethical officials, they are still able to assert their reigning authority. The reasons may be summarized as follows. First of all, the atomization of the villager relationships has kept the villagers from making concerted efforts. Second, when gangsters and the "tough guys" take up the leader's positions, they would bribe and formed a symbiosis with other gangsters, senior commune members, retired Party members or officials. But that's not the whole picture. To look at the broader picture, we must also investigate the town-level officials. Consider this question: if the town officials have nothing to gain from the village officials, would they stand side by side with these "unethical" and "gangster-turned" officials?

8.2 Town-Village Symbiosis and the Handling of "Holdouts[4]"

Gangsters maintain a cordial relationship with rural officials, firstly out of self-preservation instinct, secondly as a trial-and-error strategy for lowering the risk of joining the gangs. It is known to the gangsters that though the rural officials may have failed to cope with social problems, "pissing off" the government always brings bad consequences. "Nothing is impossible for the Party". They would say. When we look at the official's side, they would maintain a cordial relationship with gangsters, firstly to make private gain, and secondly, or more importantly, to ensure good governance. This is determined by the generally accepted approaches to rural governance by the town and village levels.

Although *Organic Laws of the Villager's Committee of the People's Republic of China* stipulates that the town/township level provides some kind of guidance to the village level (grassroots autonomous organization), any researcher who has been to rural China for field trips would know, the reality is a great deviation from the laws before the lifting of agricultural taxes. One of the most prominent evidence would be the "town-village symbiosis" (Xuefeng 2006e; 2007g). Actions by both the town and

[4] Translator's note: The essential meaning of "*ding zi hu*" in Chinese is resistance. In this book the term "*ding zi hu*" (generally translated as "holdouts") have two meanings. In Sect. 8.2 of this chapter, "holdouts" mainly refer to persistent tax evaders. In Sect. 8.5 of this chapter, "holdouts" mainly refer to a property owner who either refuses to sell or wants more than the developer would pay. The author uses the same term for both because both kinds of people resist doing something. "*Ding zi*" literally means "a nail that cannot be removed", and "*hu*" refers to a "household" or "person".

village levels are generally determined by this symbiosis, like a black hole "devouring" almost all of the officials. That includes even the rural officials who have been elected following legal procedures. Once they get "devoured by the black hole", they would not side with ordinary villagers. Therefore, villagers lose interest in village-level elections, and villager's autonomy becomes a political lip service. The forming of such symbiosis may be put down to the fact that the town/township government has held onto much more economic, political and organizational resources, compared with ordinary villagers. Anyway, they stand for the highly centralized administrative state power. Under the market economy system, small farmer households fail to unite with a bargaining power that matches that of the town and village level government agencies. So when village officials have to choose between the town/township government and villagers, since villagers are no match for the government, the rural officials would naturally side with the government (Xuefeng 2007g). Needless to say, village officials side with the government because it is in their interest to do so. When town/township government allows village officials to make private gain, village officials would side with the government, making it very difficult for villager's autonomy to be truly implemented.

The town/township government's main tasks include: collecting taxes and fees, implementing policies such as family planning and promoting cremation over burial, and reaching all kinds of performance standards. The county and town level government's appraisal on the town/township level is based on its performance, and whether it has fulfilled the tax collection task is one of the main indicators. If it fails to fulfill this task, the town/ township government does not have the needed financial resources to sustain itself. For example, the town government would not be able to pay salary to teachers. In addition, it will not be able to "free ride" in the tax collection or policy implementation processes. Especially after the 1994 fiscal reform,[5] the local government "has reduced authority and increased financial burden" and must "pay the bills for alleged treats by the central government"[6]. County and town level governments have been suffering from tight budgets, always trying to make ends meet. Things are particularly difficult for the central and west regions. Therefore, the county and town governments heavily depend on tax collection for revenue. Apparently, the county and town governments must ensure a reliable and steady source of revenue, or the higher authority could doubt their legitimacy.

The town government may not collect taxes or fees directly from so many households. Therefore, it may not go without village officials. However, theoretically village officials, who are elected by villagers, are not obliged or encouraged to assist the town government in collecting taxes and fees. So to fulfill this task in time and

[5] Translator's note: The 1994 fiscal reform was a Tax-Sharing Reform, a large-scale adjustment of the tax distribution system and tax structure between the central and local governments. After the reform, local governments mainly bore the expenditures for the operation of their own political organs and their economic and career developments.

[6] Translator's note: In the Chinese expression, "treats" refer to the central government's initiatives with good intentions, e.g. raising salary level for civil servants for local governments.

ensure tax payment from a fair number of households, the town/township government must encourage village officials to help it. Some regular administrative penalties include directly replacing or rebuking those who have failed to fulfill this task, but because the officials are not formal members of the *bianzhi* system (a system for determining the type and the number of posts needed), they care more about real gain than about being replaced or rebuked. Therefore, when village officials seek private gain while jeopardizing public interests or villagers' interest, i.e. selling public assets at a low price, or offer high-interest loans to the village community, the town/township government would often treat them with tacit approval or even secret encouragement. So eventually all of the officials—including those elected through legal procedures—would be willing to pursue private gain now that they have the town/township government's tacit approval or encouragement. Individual villagers may not unite against village officials. The result would be public assets and interests being divided up by village officials, and that villager's interests in both the short and long run would be harmed. That would cause the discontented villagers to appeal to the town/township government. But since what the village officials have done is nothing new to the town government, the town government would not investigate village officials and risk discouraging them in their work. This in turn would cause villagers to be reluctant in paying taxes and fees, and as collecting taxes becomes increasingly difficult, the town/township government must further allow village officials to make private gain and offer them some kind of protection. In a word, while the unbreakable town-village symbiosis exists, villager's autonomy would inevitably become nothing but lip service (Xuefeng 2007g).

As stipulated by *Organic Laws of the Villager's Committee of the People's Republic of China* that was promulgated in 1998, candidates to villager's committee must be pre-selected by villagers and villager's committee be elected by villagers, and villagers may remove committee members from office. These stipulations are made to protect villager's interests, but still fail to break off the town-village symbiosis. *Organic Laws of the Villager's Committee* has not fundamentally changed the de facto town government control over villager's committee. Nor have the *Laws* unite the highly self-interested villagers against the town-village symbiosis in any kind of effort to ensure villager's autonomy. It has occurred to the villagers that no matter how real the elections may have appeared, villager-elected officials only take orders from the town government. The failure to fully implement villager's autonomy may be attributed to the following factors: the continuous outflow of human, capital and physical resources, the lack of long-term stability, and villager's short-term anticipation for life in the village. As a result, both villagers and village officials tend to use things for their own advantage whenever possible. This means that the village has a lower intrinsic level of productivity for both economic and ethical values. Village elites work as village officials mainly for private gain, not to gain face or respect.

Though the town government supports village officials in many ways (e.g. allow them to "free ride" and allow for their corruption and misappropriation of public assets, and protect them when villagers appeal), such consent is not entirely unprincipled in that it is given on the condition that village officials are competent tax collectors. Village official's competency becomes even more important since the end of the mid-1990s, when the Three Rural Crises has worsened. Farmer's heavy burden becomes an unavoidable issue for the whole rural governance system; tax collection becomes the most difficult task; besides, village officials must effectively handle holdouts or persistent tax evaders (*ding zi hu*). All that requires village officials to "live up to expectations".

After the launch of the household responsibility system in the 1980s, the government still "treats all men equally" when it draws resources from rural areas, but since the people's commune has been dismantled, the unit of tax collection has changed from the production brigade and team to individual households. This means the government must deal with innumerable households, and the county and town governments would have to deal with persistent tax evaders. The problem with tax evaders does not lie in the shortage of a few tax payers or some tax revenues, but in the fact that the habit passes onto others in the acquaintance society (Dewen 2008). When one household refuses to pay taxes, other households are tempted to do the same—"why would I have to pay if he gets to go without"; tax evasion by anyone would invariably cause people to believe that "an honest man would suffer losses" and that "eager tax payers would suffer more losses".

There are two types of persistent tax evaders. One is "bargainers" who refuse to pay taxes unless officials solve their problems; the other is "overt non-payers" who refuse to pay out of no special reason or on the slightest pretext. Besides these two types, non-payers also include "impoverished households" (Dewen 2008), who could not afford to pay taxes. In the acquaintance society, it is easy for people to tell who belongs to what type, since people know each other well and keep no secret; but on the governance level, because there is no clear standard, the government finds it very hard to treat them differently. In theory, the government should give those impoverished households tax reduction or exemption treatment, in a way to "show the superiority of socialism"; and if the "bargainers" are looking to solve problems, the government should either solve their problems, if their requests are reasonable, or reject the unreasonable requests. As for the "overt non-payers", the government must treat them harshly and may be allowed to use compulsory enforcement. Therefore, in theory, what the government really has to deal with are the overt non-payers. However, in practice, treating the non-payers differently is much more difficult.

First of all, it is very difficult to determine "the impoverished households", or to make specific convincing standards for such. People may already know who are impoverished, but that is nothing more than a simple description, and there is no applicable standard for determining what kind of people is impoverished. When the "impoverished" households get tax reduction or exemption, those "poor" households would ask for the same treatment, and that may go on and on, and eventually cause people to evade paying. So the town government and village officials may not easily give "special treatment" to impoverished households. In practice, the town government and village officials would give special treatment to the fewest number of villagers; in fact, such special treatment is only granted when it does not get in the way of reaching the basic tax collection target. But however slim the chance of getting special treatment is, people would still delay paying and expect the government to grant them "special treatment". So whether to give "special treatment" or not becomes a game between defaulters and collectors. Sooner or later, granting special treatment to impoverished households would inevitably lead to the emergence of more "overt non-payers". Gradually those non-payers who claim to be too poor would become the so-called overt non-payers.

Second, it is also very difficult to determine the "bargainers". The government fails to determine whether the problems they look to solve are based on reasonable requests, because the government and the people may not reach a consensus on the reasonableness of requests. What villagers believe to be reasonable may be unreasonable for the government. For the government, paying taxes and fees is a legal obligation for any farmer that has contracted the collective land, and no "problem" may be used as a pretext for not fulfilling the obligation. However, villagers believe that the government must solve their problems before it may levy tax on them. The villager's belief is probably based on the notion of reciprocity: "I'll scratch your back, if you scratch mine". But the key question is that there is no clear boundary for what villagers ask from the government. They may come to the government for anything, but the government does not have what it takes to give them whatever they ask for. So where villagers and the government reach a consensus, the government may still fail to solve the problem, let alone the problems for which no consensus can be made. With limited financial resources, the government fails to grant people's reasonable requests—that's a very common thing in central and western China's rural areas. Besides, many are historical problems that are beyond the grassroots government. So the "bargainers" with unreasonable requests would easily be classified into the "overt non-payer" group, and thus become the target of punishment, though these requests may seem entirely acceptable for villagers. The government's failure to properly handle the bargainer's reasonable requests will cause great trouble for tax collection at both the town and village levels.

As a result, the government is placed in a tax collection predicament and the simple solution is to treat all villagers indiscriminately or use compulsory enforcement on all non-payers. And just as "the maximum force should be applied to the edge of the knife", the "overt non-payers" naturally become the key focus of the government's enforcement efforts. By handling the persistent "overt non-payers", the government would avoid the tax evasion practice to pass around and set an example for other households. Focusing on the wrong type of non-payers would, first, miss those tax evaders with a real impact, and second, by appearing to "bully the weak and fear the strong", undermine the government legitimacy.

One of the ways to handle the "overt non-payers" is to initiate legal proceedings against them. This is the simplest method that saves the grassroots government a lot of trouble. Once the non-payer is taken to court, things must be settled by the law. By the law, farmers have a legal obligation to pay taxes and have no grounds for tax evasion. The law does not consider poverty or outstanding problems in a judicial procedure, so to arrive at a court decision is simple. But that does not truly solve the problem. As enforcing judicial decisions still requires farmer payment, the court eventually ends up in a situation similar to that of the grassroots government. The failure to enforce a court decision will undermine the court's authority and judicial legitimacy. Compared with the tax collection task for the government, enforcing a decision for the court may be more difficult, because the court must depend on the local government for enforcement. The result would be that the court is incorporated into the "compulsory enforcement" task. When the court encounters resistance, the potential social impact would be worse. As the court tends to treat all non-payers indiscriminately, it is likely to encounter more resistance compared with the government. That was exactly what happened in the 1980s, when compulsory enforcement of court decisions led to resistance by large groups, making considerable negative social impact. What's more, if the court intervenes in tax collection, theoretically it should also intervene in other agrarian cases. In other words, the court must mediate between farmers and the local government. But that was an impossible mission for the court when governance crisis became serious in the 1990s (Ruiqing 2001). As a result, since 1993, the Supreme People's Court has repeatedly issued notices prohibiting courts at all levels from getting involved in tax collection, and from handling tax-related lawsuits filed by the executive agencies against farmers. But the courts are allowed to accept lawsuits filed by farmers against government agencies for increasing farmer's tax burden.[7] This means that when local government and farmers are in conflict, the court may "extinguish flames ignited by the conflict" but not "add fuel to the fire".

Another way to handle the "overt non-payers" is to ensure compulsory tax payment with administrative measures, such as: dispatch special tax collecting teams, offering compulsory legal education classes, which have turned out to be quite effective. But the result may be growing resentment to the government. This would in turn cause bad events in which either the farmers or the officials get beaten up. Either way, those events have a serious social impact. It is often reported that farmers get beaten or bullied by government officials. And during my investigation I've heard about officials being beaten. In one of the worst events, a female county magistrate got stripped naked and locked away in a cellar. In the process of "ensuring compulsory enforcement", the police would also join the officials. The joining of the police has also caused some bad events. For that reason, the department of public security also issued documents disallowing police involvement in tax collection.

Such diminishing assistance from other parties has made it even more difficult for the town/township government to collect taxes. The government becomes overwhelmed by the never ending tax collection work—what used to be a seasonal campaign in the 1980s has turned into a daily routine and one of the government's top priorities in the 1990s. When tax evasion becomes a more common problem, the government has to rack its brains for ways to raise average taxes and fees in an effort to fill the gap created by tax evasion. But this would cause more tax evasions—many are reluctant to pay or cannot afford to pay.

It is in this context that rural officials come to notice the rural gangsters and "tough guys" and start to court them. Rural officials incorporate rural gangsters into the political system and ask the gangsters to collect taxes and fees for them. Courting the gangsters will help rural officials effectively overcome resistance, because tax

---

[7] Examples for such notices are: *The Supreme Court's Notice on Immediate Acceptance of Cases Filed by Overburdened Farmers* in 1993 and *the Notice Prohibiting Judicial Officials from being Involved in Non-Trial Affairs* in 1998.

resistance often comes from gangsters and former "senior commune members". The gangsters do not have to use reason in their tax collection efforts. They may simply use force or threats. If anyone refuses to pay, they would play the "tough guy" card, which most villagers are still afraid of.

In their effort to collect tax, rural gangsters would not consider the rural household's financial situation or the reasonableness of people's requests; they do not have to worry about having their legality or legitimacy being undermined for the lack of consensus on whose requests is reasonable, because the villagers would not expect them to use legal or legitimate means. To use reason on the gangsters is useless, because they completely disregard reason. For them, all problems will be settled using their "fist". They do not bother themselves about the reasonableness or the legitimacy of anything, or bother to use reasonable or legitimate means. They would resort to brute force, for they are emboldened by their strong families and the enormous gangster network. Ordinary residents all fear them; people would say "if you can't beat them, avoid them". So people would immediately pay the due taxes when running into gangster collectors to avoid "using an egg to hit a rock". In contrast, ordinary officials are often treated with disdain by the shameless non-payers and may even face physical threats, and thus would abstain from doing anything. The result is the unreasonable outlaws could more efficiently accomplish tax collection tasks and appear much more competent compared with the honorable rural officials.

Many officials told me that gangsters were able to accomplish many things that failed to be done through normal procedures. Sometimes officials try very hard but the result is not as good as what gangsters may accomplish—that's an official's evaluation on gangster's performance in rural governance. Incorporating gangsters is indeed a low-cost alternative. A government official of *Xianning* City told me that, since it might be impossible to remove all gangsters and destroy the rural underworld, the most important thing would be to avoid serious events and in the mean time "use them for our good". Tax collection and land expropriation are examples for what the government fails to do but may be easily accomplished by gangsters. Therefore, it may be safely concluded that the grassroots government needs gangsters for rural governance and may easily ally with them out of common interest.

Naturally tax collection is not always easy for gangsters. They may also run into trouble. Occasional conflicts may happen and cause serious events. But after all, such events are nothing like farmer-government conflict, but are farmer-farmer conflict in nature. In other words, once gangsters are incorporated into the political system, rural officials may to some extent stay out of the serious tax-related conflicts and transfer conflicts to within the village or conflicts between the corrupt gangster-turned officials and ordinary residents. The most serious conflict may arise between the gangster-turned officials and non-cooperating tough guys or gangsters. Incorporating gangsters into the political system for tax-collection has also proved to be more effective than the government doing the job directly. Despite the potential risk of undermining the government's legitimacy, it takes time for such political risk to build up, whereas the government collecting taxes directly may cause immediate political risks. Once the government gets to stay out of the conflicts, it may act as the mediator. When tax collection causes conflicts or serious events, villagers may appeal to the county and town level government or take those corrupt and unethical village officials to court. But the county and town government is well aware that in an effort to collect taxes they have formed a symbiotic relationship of common interest with village officials. Therefore, instead of being the impartial mediator, the county and town government often become the village official's patron, because as far as the grassroots government is concerned, though corrupt and unethical, those gangster-turned officials have turned out to be very competent in tax collection.

It is worth noting that the above-mentioned town-village symbiosis has formed because they both have something to gain from accomplishing tax collection tasks. Other cases for the town and the village to form a symbiosis may include land expropriation and implementation of such policies as family planning and funeral reform. In affluent villages offering great potential collective benefits or illegal income for officials, it is also very likely for the town and the village to form a symbiotic relationship. Further research is needed for understanding such symbiosis. But the focus of this book is on pure agrarian villages. By far, over 80% of the Chinese villages still belong to this category, or are mainly supported by farming income. Before the lift of agricultural tax, the town-village symbiosis has been formed at most of such villages and have had a fundamental impact on the status quo of the village and people's life.

After the lift of agricultural tax, the town-village relationship has undergone great changes. The town does not need the village to collect taxes and fees, and rural officials' payrolls are covered by government transfer payment (instead of by tax revenues). Meanwhile, in many areas of the country, the pressure to carry out policies such as family planning and funeral reform has been greatly reduced for the town/township government. Rural officials are not as needed as they were to the town government—it may act alone or simply relying on the judicial system. In this way, the former town-village symbiosis is likely to break up. The town government would not have to depend on rural gangsters or tough guys in rural governance, hence no need for "strategic tolerance" with them. The town government may use the villager autonomy principle to their advantage and screen out those gangster-turned or gangster-related cadres in elections. Unfortunately, the academia and the government have failed to notice such failing town-village relationship, and thus the reforms they attempt to implement for the rural political system are very problematic. The political system is gradually retreating from or weakening at the grassroots level with the implementation of such measures as reducing the number of townships by merging townships with towns (*che xiang bing zhen*),[8] streamlining bureaucracy, merging villages and their corresponding production teams (*he cun bing zu*), and abolishing the position of village-group chief, etc. Measures have been taken to alleviate the village's financial distress after the lift of agricultural tax, and in the mean time, to "allow the entering of the people while government officials retreat" (*guan tui min jin*). But after the retreat of government officials, those that enter the scene

[8] Translator's note: When referring to political divisions of China, town is the standard English translation of the Chinese "*zhen*". The Constitution of the People's Republic of China classifies towns as third-level administrative units, along with townships (*xiang*). A township is typically smaller in population and more remote than a town.

are not the people but rather rural gangsters and illegal cult (*xiejiao*) organizations. Gang influence is growing in many rural areas. Besides, after the lift of agricultural tax, using tax resistance to involve the government in a negotiation is no longer an option for farmers, and the government is more inclined to disregard farmer's needs and preferences. And when gangsters attempt to take advantage of villagers, villagers may no longer pressurize rural officials into acting as the impartial mediator in the same way they have done before the lift of agricultural tax. What the grassroots government must do is simply to follow certain procedures and laws. As a result, the government will not be held accountable for its reluctance to solve real problems, and that has greatly compromised people's sense of security and well-being.

8.3 "Good Gangsters" in the Eyes of the Villager

On my field trips, an appellation that I've heard frequently is "good gangsters", which sounds quite contradictory on the literal level—people often hold negative opinion of gangsters whereas "good" is used to show positive opinion. Such a contradictory expression is indicative of an important phenomenon in rural governance, that is, most gangsters are bad people, but some are different from others. But the key question is what kind of gangsters may be regarded as "good gangsters"?

Wang San from *Baiyang* Village of *Jingmen* City is a case in point. He was a good student and intended to apply for a military school but got rejected by one of the "powerful figures" in the political investigation (*zhengshen*) process.[9] Being deeply frustrated, he lost faith in normal ways to climb the social ladder and started to hang out with local gangsters. When he became an official member of a local gang, he planned to get back at the "powerful person" that had got in his way, but got arrested for doing so and later was allowed to serve the sentence outside of prison. During non-custodial sentence, he served as the power supply manager (*guan dian yuan*) of the village. Power theft and electricity disputes were both very common. Electricity rates increased continuously, with the highest reaching 2 Yuan per kWh. *Wang San* was "entrusted" with the task of managing power supply "at this critical moment", but he was a firm hand: whoever used power must pay for it, or otherwise the power supply would be cut off; he insisted on severely punishing those stealing electricity. *Wang* was able to enforce the law by taking advantage of his infamy and influence as a former gangster. The rural officials and "senior commune members" who used to use electricity without having to pay for it detested him but did not dare to oppose him. While he served, he had almost eliminated all power thefts and ensured steady electricity rates. So people liked his "reign". *Wang San* was also very nice to his villagers. He did not take advantage of the poor households and allowed the impoverished households to delay payment or pay the bill by handing

[9] Translator's note: People applying for military schools would need to pass a "political investigation", or *zhengshen*, a process in addition to the normal testing procedure. The investigation covers the person's family backgrounds and connections to determine their political suitability.

over grains instead. Later he was elected as the director of the village's security committee (*zhibao*), and as a hardliner, regained control of a fish pond from the town government (Chuang 2006).

Another example would be *Liao* from the Fourth Village Group of *Shangwu* Village, *Jingmen* City. He may not exactly be a "good gangster" in the eyes of his fellow villagers, but he is not purely evil, for he has also done quite a few good things for the village and provided convenience to others whenever he can. Villagers told me about several "good deeds" he had done. The first one was using his connections for releasing water from an upper-stream reservoir when people had no water in a year of drought; the second was, when a local villager was injured in a dispute with an outsider, urging the outsider to compensate in full for his fellow villager's medical bill; the third was "straightening things out" with grain collectors of the grain station by asking them to treat his fellow villagers nicely—no declaring with short weight or under-evaluation, and "sorting out" the grain collectors if they refuse to give face. People of his village group all think of him as a good person (Leiming 2007; Caisong 2006; Lin 2006; Lina 2006).

Chen Jingqiao from *Yanjing* Village of *Jingmen* City is regarded as one of the "good gangsters", or may be described as the "guardian" of the village. As a gangster, he went to jail in the 80s for repeated thefts and was divorced for being a womanizer. He has a big family with altogether 18 brothers and cousins; he still maintains contact with "the mob"; several of his relatives are town-level officials; a few years ago he made some money by doing wood business and became one of the wealthiest among his fellow villagers. With these "strengths", *Chen* may be counted as one of the "wealthy and influential" people in the village and no one—regular residents or rural officials alike—dared to provoke him. However, being an upholder of justice and always ready to stand up for others, *Chen* becomes one of the stabilizing factors in the village. He does not bully his own folks; instead, he often offers a helping hand. Whenever seeing anything unjust, he would take up the cudgels for others. Whenever someone needs him, a simple phone call would bring him back, and he would do his best to make things work. The local villagers would love to seek help from him whenever in trouble. He does not mind being involved in family affairs, e.g. lack of filial piety. Those with a guilty conscience are often easily intimidated. For that reason, *Chen* is a very good problem solver. He does not intentionally cause trouble for rural officials; nor would he be easily intimidated by rural officials. Being an ardent advocate for the public, he once even beat up the village secretary. In the 1999 village committee election, when *Chen* found out about a villager assisting the incumbent with vote rigging, he gave the person a kick right away and called the person "a scumbag". In the 2005 village committee election, when the incumbent was found to have bribed voters, *Chen* asked people to sign a petition against election bribery and did not stop until it was delivered to the Organization Department of the Provincial Party Committee, significantly affecting the election result. For another time, when rural officials went to the village to discuss anti-drought affairs, *Chen* insisted that irrigation water for the village be made a priority, even though that meant bringing himself into conflict with rural officials. *Chen* is away doing business a lot, but he has never taken his mind off village happenings. So he would often come back to live in the village though he owns a nice town place. Considering that the village is largely inaccessible, he paid for the sand used for road-building (Yi 2006; Yuanyuan 2006).

The former Village Secretary *Liao* of *Huangnie* Village, *Jingmen* City is a gangster-turned official (Rong 2006). *Liao* is also regarded as one of the good ones, though not in the same way with *Chen Jingqiao. Liao* has been a gangster at *Jingmen* City since he was a teenager. In 1994, when *Liao* was fined 1,320 Yuan for having an unplanned birth with his under-aged wife, he planned to simply not do anything about it. So the town-level family planning office had to use his furniture and appliances for paying the penalty. For that reason, he took it out on the village officials by paying them many visits and even beating them up. In 1996, after one of *Liao*'s uncles was elected as the village secretary, he was appointed as the company commander of the village's militant group, admitted to the Party, and two years later, elected as the village secretary himself. In the villagers' eyes, as the secretary *Liao* was bold, resolute, corrupt and greedy at the same time. He was not the best official, but he's also done many good deeds. And for that reason at the end of his term, people were nostalgic for his old days as the secretary. Therefore, he may be counted as one of the "good gangsters". Here are a few good deeds he had done:

First, in 2001, under his leadership, villagers of the Fourth Village Group volunteered to build farmland path for farming vehicles, making the Fourth Village Group the one with the best farming paths—their paths used to be the worst. Second, before the lift of agricultural tax, with heavy tax burden and negligible income from farming, many villagers stopped farming and became migrant workers instead. This has led to a shortage of taxpayers for ensuring tax revenue. This thorny problem unsettled by his predecessors was easily solved by Secretary *Liao,* by contracting the deserted land to non-locals to grow yellow ginger. This generated tax revenue and increased total income for the whole village community. Third, when villages around the country were looking for ways to attract investment, Secretary *Liao* expropriated the land of the Sixth Village Group, giving the villagers a minimum level of compensation, and rented the land to a self-employed businessman from *Suizhou*[10] for growing black ear fungus. Several years later, when the *Suizhou* businessman had left, *Liao* converted the farmland to fish ponds and rented them to the villagers, which has greatly increased the village's revenue. Forth, in 1998, when the town government needed to purchase a large amount of yellow sand for road building, *Liao* used his private contacts and successfully took over the project. He also increased revenue for the village by coercing 5 households into relinquishing their farmland for sand digging.

During his incumbency, *Liao* would hang out in the town and got hooked up with a hair salon sister (*falangmei*).[11] In 2003, having been divorced and remarried, he moved to the town. To sum up, as the secretary, *Liao* has displayed resourcefulness

[10] Translator's note: *Suizhou* is a prefecture-level city in northern *Hubei* province.

[11] Translator's note: Women who work in places that offer commercial sexual services under the guise of massage or health and beauty treatments are referred to as "hair salon (*falang*) sisters (*mei*)".

and resolute leadership, done a good job in bringing revenue by undertaking projects, but probably has been involved in corruption. Though being a gangster, he did not act like one in his own village and generally treated his fellow villagers well—unlike other gangsters treating people "as if they were fish on the cutting board". That has given him a good reputation. Despite some of the coercive measures that have harmed the interests of certain rural households, prevailing opinion is in favor of him and would remember him for his resolute leadership and good deeds. Despite some of his corrupt practices, villagers would show understanding since "in this world, of all those bigwigs who is not corrupt?" People's positive opinion of him is further reinforced when he is compared with his successor Secretary *Kong*.

Secretary *Kong* is a senior Party member and a veteran. When serving as the village secretary from 1990 to 1994, he has won universal approval. People say that "for so many years as the secretary, he had never taken a pack of cigarettes". He is universally acknowledged as "an old nice guy" but he is weak and old-fashioned and his hands are tied. In 2005, the government's plan to build a road connecting the village with the urban areas required the demolition of several houses in *Huangnie* Village. And when one of the households refused to have the house torn down, out of people's expectation, Secretary *Kong* decided to do it himself. *Kong* got injured from a fall and did not recover until resting at home for half a year. In the years since his return to office, the village community has basically failed to provide any public goods; many problems have arisen in terms of road building and irrigation. These problems, according to *Kong*, have been caused by a lack of capital and labour, and a lack of subsidy for farmers when land was seized to build roads and dig channels—in other words, his hands were tied. Surely he did not mention these problems to evade responsibility since these were real difficulties. Before the lift of agricultural tax, villagers often refuse to pay tax in a way to coerce rural officials into doing things for them; but when tax is lifted, this effective "weapon" is gone, and rural officials would rather do things in their own ways (Xuefeng 2007h).

That being said, villagers of *Huangnie* Village came up with another way to keep rural officials involved. When those households with land on the upper stream of the irrigation channel refused to let water pass through to those on the lower stream, they clearly requested "the appearance of rural officials for allowing the water to pass through". They forced the other households to fetch rural officials. This trick has been played for several consecutive years, mostly by "bargainers". Some of the households requested that the village community keep its word by giving them land to compensate for their land lost to sand-digging; some requested reassessing the amount of grain given in compensation for lost land. Always being a nice guy, *Kong* has not been a good problem solver, which has evoked greater nostalgia for the former Secretary *Liao*. It occurred to them that if a rural official is capable of doing something for the villagers, a little corruption wouldn't hurt; that a corrupt but competent rural is better than one that gets paid for doing nothing. When comparing *Kong* with *Liao*, villagers felt increasingly strongly about *Kong's* weaknesses. They would be like, "If the people from the above do not apply pressure on them, and us living at the bottom do not make requests, all that the rural officials have to do is to take money", and "Anyone would love to do a job that pays several thousand Yuan for doing nothing!" So it does not matter whether *Liao* is really a "good gangster" and a good village official or not. What really matters is that he seems to be such a person in the eye of the villager.

Such prevailing positive opinion on the "good gangsters" may be mainly attributed to their competency to provide public goods to meet with people's needs. He may be a competent official who ensures the provision of public goods to the village by obtaining resources from outside and using force on the inside. He may be a gangster who has the spirit of "self-sacrifice" and uses his personal contacts to obtain benefits for all and uses force to settle disputes in an effort to ensure the provision of public goods. To sum up, good gangsters are characterized by their special way of using force, not for harming people, but for solving people's problems and providing public goods; or harming a minority group but protecting the majority's interests. But when we take a step back, we would ask why the provision of public goods requires the use of gangster's force. If such force is necessary, what role does it play? We must answer these questions by first investigating into the problems in public goods provision.

8.4 The Public Goods Provision Predicament and "Law Enforcement by Private Party"

The main feature of public goods is the public nature, or non-excludable and non-rivalrous in consumption. Non-excludability means that one's usage of the good does not reduce consumption of that good by others. Non-excludability means that it is technically impossible to exclude individuals from consumption, or technically possible but financially infeasible to do so since the costs of doing so outweigh the benefits. Public goods in rural areas include road building, drought relief, flood controls, hydraulic engineering, public security maintenance, environmental protection and sanitation, dispute mediation, etc.

The key problem in public goods provision is the "free-rider" problem, because people tend to maximize their benefits by not paying for the provided goods but still using them. Therefore, limit free riding is essential to public goods provision. In urban areas, public goods are almost entirely provided by the government, and the government limits free riding with taxation laws and regulations. In rural areas, public goods are provided mainly by the village community itself. That is still true after the lift of agricultural tax, even though the government has stepped up its effort to provide public goods to rural areas. The issue is further complicated by the fact that public goods provision to rural areas requires collective efforts from all walks of life.

In a traditional acquaintance society, people have long-term expectations for living in place, and thus hope to obtain others' positive opinion or "look good" to others. They would not be preoccupied with temporary financial gain or loss; nor would they go too far in pursuing their interests. With long-term expectations, people have repeated social interactions. One's one-time free-riding behavior may cause one to be punished in the next social interaction. So people refrain from being too calculating in a one-time encounter and would avoid reckless free-riding. Instead of focusing on certain one-time encounters, people would calculate their gain or loss based on repeated social interactions. Therefore, free-riding is a stupid behavior, similar to "killing the chicken to get the eggs". Gaining benefits in a one-time free-riding behavior does not bring real benefits in the long run. People would firstly consider the moral consequences of an act before taking it. This includes whether this act goes against others' expectations, the established code of conduct and accepted conventions. When people unconsciously or habitually honor traditions and social norms, free-riding does not occur frequently. Naturally free-riding still happens, but it is effectively handled by the acquaintance society.

One way of handling free-riders is to sever any relationship with them, ignore and isolate them. These are also common ways of punishing a person in the acquaintance society. If a villager get rejected and isolated by many people, his area of activity would undoubtedly be greatly limited, and any social interaction, cooperation and life assistance concerning the person would be limited. The result of being abandoned by the mainstream society and being rejected by the community is to eventually become "social waste". Being isolated means being abandoned by the village community. It is nothing short of "community-level death" (Hua 2008b), because compared with the village community claiming the moral high ground, individuals are fragile and insignificant with only negligible power, and are in no position to resist the village community. As far as the village community is concerned, isolation and rejection are necessary means to sustain itself. The village's control over individuals is essentially a kind of long-lasting power relation. Individuals have obligations and responsibilities for the village community and other people, and must obey the common will of the community. People lead an obligation-oriented life—they must fulfill their obligations in order to obtain their life space in the community. And when something goes against the collective will—free-riding undoubtedly so, the consequence must be punishment. An extreme example would be Regreg's tragedy on the island of Bali. "To leave the community of agreement is to lie down and die." (Geertz 1983) Free-riders in a Chinese village may not suffer a destiny as tragic as Regreg, but they would become a peripheralized person whose speech and act are not taken seriously by anyone. They would be treated as freaks and be ignored by most people. Whatever benefit that has been gained through free-riding is outweighed by the severity of such punishment. And that's why such punishment may effectively suppress people's impulse for free-riding and the chain reaction that results from free-riding. People of the acquaintance society generally hold the occasional free-riders in contempt and do not "stoop to their level" (*da pin zhang*) with them. Therefore, free-riding does not pass around; it does not affect the sustenance of the community; it does not affect the traditional mechanism for providing public goods.

Besides, the acquaintance society also has tangible penalties for free riding. At the end of the *Song* and *Ming* Dynasties, in areas where clans have been successfully rebuilt and there is no lack of the gentry class, clan and natural villages have the same boundary; within the village, public goods are mainly provided by the clans; outside the village, public goods are mainly provided by the gentry class. State power fails to penetrate to the rural areas and the government rules over the village mainly through clan organizations. When it comes to public goods provision, the clan's decision-making has a strong binding power on all people, because the clan is capable of coercion and has some judicial power. If someone violates norms established for cooperative efforts to provide public goods and makes a free riding attempt, the clan organization would punish the person, or coerce the person into fulfilling obligations (Yanhui 1979:153). A clan is legally approved to be binding and to use coercion. A potential free-rider may disregard punishments such as social isolation and rejection, but may not take the clan's coercive measures lightly. To sum up, in a traditional society, the nature of the village and the functioning political system have determined that free riding could not have become a serious problem.

However, having experienced several revolutionary movements in the twentieth century and received the blow of market economy since the 1980s, only a small number of villages remain the acquaintance society in nature and are self-sufficient in public goods provision. The rural areas of China have undergone fundamental changes: the clan is not as legally approved to use coercion as it was, and thus it could not use coercive measures on the free riders; the village has increasingly deviated from its ideal condition; people expect to live in the village for a shorter time; public opinion has a diminishing restraining power on the villagers; with punishment mechanisms increasingly ineffective, the village has lost its ability to punish free riders; people do not seek to "look good" in front of others—they may still care about face, but this face has changed drastically (Baifeng and Junxia 2007); people would no longer sacrifice immediate interests just because they expect to stay in the village. These changes are particularly true with people of the Double-lake Plains. People all hope to get out of the village. And because the village is nothing but a temporary shelter, people do not care about their names there or other villagers' opinions about them. Public opinion no longer serve as a punishment. People do not isolate the free riders, and the free riders do not care about being isolated. So people are increasingly encouraged to free ride and that's also what they do, but the village and other villagers may do nothing about free riding.

The contemporary Chinese rural society is characterized by two things: first, state power has irreversibly penetrated to the rural areas; second, market economy has caused farmer behavior to be reasonable. Because of the former, people are still aware of the clan but the clan is no longer legally approved to use coercion as it was. Because of the latter, farmer's intense feelings for a locality and the resulting potential use of personal attack—what used to be effective punishment—are no longer legally protected. (Xuefeng 2007f: 155). The contemporary rural society has gradually become an atomized flat organization, making financial benefits from free riding more attractive.

The modern state power is based on the fundamental principle of protecting individual rights. On one hand, it is very strong, since it affects all aspects of people's life and in daily life no one could break away from its monitoring. On the other hand, it seems quite weak, being subject to various limitations. State power must be cautious wherever and whenever it is used. State power ends just where individual rights begin. This will easily lead to what *Fei* has described as an evil outlaw being protected by the law, or the paradox of state power—it becomes the patron of evil organizations (Xiaotong 1992). On the contrary, villager's shared feelings are not protected by law, and as a result violence used by the collective group is regarded as dangerous in that such violence may infringe on other's individual rights. In other words, farmer's shared or group feelings are often described as outdated and ignorant.

In the mean time, the ubiquitous power of the market economy has reformed people's value system. Through redefining interpersonal relationships, it has established a value system that agrees with market economy principles. It seems like anything, including the value of life, may be measured by financial interest or monetary value. Under the market economy, anything may be bought with money, which makes money the only standard and medium of measuring everything. This has also been reinforced by the ubiquitous consumerism tools such as advertisement and fashion. Anything that does not bring direct financial interest—fame, face or prestige—will be regarded as valueless. Anyone that still look forward to living in the countryside will be looked at as "fools". Anything about rural life has become a one-off behavior. It is meaningless to think in the interest of others or one's offspring since it does not bring real interest.

In the contemporary society, when "group feelings" are already in a precarious situation, it will only become weaker since it is not protected by state power. As a result, it would be very hard for the public to form unanimous opinion or for the face-saving mechanism to function properly, making the village community unable to cope with the impulse and act of free riding. Policy-makers and scholars hope to ensure public goods provision by introducing the democratic system of "having ad hoc meetings on specific matters" (*yi shi yi yi*). This is apparently inappropriate. "*Yi shi yi yi*" means that public affairs are discussed at villager representative meeting or villager assembly. Public goods provision often requires each household to contribute either labor or capital. In other words, all people are involved in this. For some complex issues, the same issue may concern different people's interests. What often happens at such meeting or assembly is that people would engage in endless argument because nothing can make all people happy. We have been present at several such meetings at Double-lake Plains; in one of the most heated discussions, some of the villagers whose voice is not well listened to or those with unreasonable requests would simply walk away. That is no way for keeping the discussion going (Xuefeng 2007f:155). In theory, the staying villagers may continue with the discussion and come up with a decision. But the key question is, when a decision has been made, those dissenters would simply disregard this decision.

Since the decisions reached by "*yi shi yi yi*" meetings often requires villagers to contribute labor or capital, those not present at decision-making or those dissenters would come up with various excuses for not contributing anything. And that may lead to more people doing the same. Others often "stoop to the same level" (*da pin zhang*). They would be like "if he does not help out or offer money, why would I have to do it?" Villager self-governance based on the "*yi shi yi yi*" system may ensure democracy but lacks centralized power to enforce a decision. That means that the ideal principle of "majority decision" would not be supported when being enforced, and would eventually be reduced to "a minority decision". The opposition of a minority group would cause the decision to fail. At Double-lake Plains, none of the decisions reached by the *"yi shi yi yi"* system have been successfully enforced. Such success may still be seen in clan villages of south China, where with their established sense of identity with the clan, people still care about other's opinion. In clan villages, the internal face-saving mechanism and public opinion are still properly functioning, posing a kind of collective power to administer punishment.

The paradox of "good gangsters" may be well understood in the context of public goods provision predicament. They are referred to as "good gangsters" for using personal contacts to promote public welfare or for providing public goods while making self-sacrifice. Most importantly, they may forcibly settle disputes in public goods provision and ensure steady provision of public goods. Generally people would be harmed by the use of force, but that's not true with "good gangsters". They use force to settle problems and reconcile disagreements. Agreements made in public goods provision will benefit the whole community, whereas free riding only benefits individuals. Without the intervention of coercive power, the impulse and act of free riding may not be effectively inhibited. This is particularly true with hydraulic engineering, where free riding brings "trivial benefits" to certain individuals, but "significant benefits" may have been lost for the whole community. In rural areas of *Chujiang* City, because free riding is not effectively inhibited, villagers fail to agree on building anti-drought pumping stations and installing equipment (large facilities) for common usage and have to make do with well pumps (smaller facilities) in individual households. The cost of small facilities for one year equals to that of large ones for ten years (Xingzuo 2006). Once such difference in real benefits becomes clear to people, they would come to see that the problem-solving "good gangsters" are really a blessing for most villagers. Good gangsters would bring real benefits to the majority of the people and eventually also to those making failed attempts of free riding.

It may be safely concluded that "good gangsters" are "private law enforcers" (Xi 2005), except that the "law" does not refer to laws promulgated by the state, but rather collective feelings and regional conventions shared by the villagers and the village community (Baifeng 2008b). In the traditional acquaintance society, people are supposed to abide by collective decisions and contribute to public causes of the clan or the village community, but free riding goes against these basic principles and thus must be punished. Free riding is punished by clan organizations using tangible penalties, and by the villager community using soft punishment (in the form of public opinion). In the Commune era, state power starts to penetrate to the rural areas, undermining the clan's authority, and rural officials have taken the clan's place. Having inherited authority from traditional clan organizations, they take over the job of punishing lawbreakers. After reform and opening-up, state power starts to withdraw from rural areas, which gradually undermines the authority of rural officials. When it comes to the 1990s, rural officials do not have the needed authority for effectively punishing free riders. As a result, when enforcing regional conventions becomes increasingly difficult, and people are infuriated by rampant free riding and transgressions but may not do anything, a upright and chivalrous "gangster" who is willing to use his personal power for punishing free riders would be well received by villagers, and naturally becomes an enforcer of regional conventions. By resorting to violence or threats of violence, "good gangsters" have filled the void left by the withdrawal of state power, or have served the government's role as a law enforcer.

Though gangsters have become "private law enforcers", we still stand against some contentions about developing non-governmental forces to solve issues related to public goods provision. Such contentions look forward to "the withdrawal of the state and the entrance of villager rights", hoping that the complete withdrawal of state power from rural areas would give non-governmental force a good chance to be developed and allow the villages to solve problems on their own. Before the lift of agricultural tax, some contend that the grassroots government's excessive levies may be solved by establishing farmer association (Jianrong 2005). However, the status quo of rural China has proved that, what comes after the withdrawal of state power is not "villager rights" but the revival of the gray forces. The fact that laws are enforced by gangsters only and that only law enforcement by gangsters may achieve effect are exactly why the penetration of state power is indispensable to rural areas. We should not pin our hope on gangsters for ensuring public goods provision, because "good gangsters" are just a minority. If the government relaxes its control on the rural society and let "non-governmental forces" develop, the result would inevitably be rampant gangs and organized crime groups and their tightened control over the rural society. Apparently, under current circumstances, to convince others on legal grounds, to use reason or conduct democratic consultation are much more costly than simply using force by "private law enforcers".

After the lift of agricultural tax, the central government steps up its efforts to provide public goods to rural China, but that still fails to alleviate the severity of the problem. Through transfer payment, the central government compensates for the reduced tax revenue to the county, town and village levels, but that is not enough; under financial pressure, the grassroots government is forced to reduce government agencies and staff, reduce the number of townships by merging townships with towns (*che xiang bing zhen*), and merge villages and their corresponding production teams (*he cun bing zu*). These measures have further undermined the town and village level's ability to provide public goods. Being more inclined to do things in their own ways, the grassroots government is not paying enough attention to villager's preferences for public goods; and when levying taxes on villagers or conscripting statute labour is not a choice, its ability to provide public goods is seriously limited. The severe shortage of public goods has caused great losses to rural residents, which may have far exceeded the benefit of having the agricultural tax lifted. The village has to provide many public goods on its own, but without the intervention of the grassroots government, the village has difficulty in overall planning and mass mobilization. To convene a villager assembly or conscript statute labour are equally as difficult. The villagers commented by saying, "It's the time of the market economy now. We do not have to pay tax. Why would anyone be willing to do unpaid work? Who would work for nothing?" Under such circumstances, "good gangsters" are more welcome; many people become nostalgic about gangster's "reign".

8.5 Village Governance: Change and Predicament

Both "competent but unethical rural officials" and "good gangsters" have appeared in a way to cope with the rural governance predicament. Rural cadre's "competence" is shown by their ability to collect taxes and fees using coercive measures; and the "good" side of gangsters is also shown by coercing villagers into cooperation when solving problems related to public goods provision. Their commonness lies in the use of coercion to prevent free riding and handle holdouts (*ding zi hu*). However, the use of brutal force would by no means be considered legal.

Free riding has been a historical problem since ancient periods, and the key to preventing free riding is to handle the holdouts. The traditional acquaintance society uses public opinion, social banishing and clan violence as punishment for free riders. The three methods have provided a basic solution to the problem. After the founding of P.R.C., the government extended its agencies to the grassroots level in an attempt to provide public goods directly to rural China. In a climate favorable for revolutionary and moral ideals, there are basically no holdouts or free riders. After reform and opening-up, as the government continues its effort to tap resources from rural areas while abandoning the Commune system, more holdouts have appeared and free riding becomes a more prominent problem. When rural officials are faced with the pressure to collect taxes and fees, rural gangsters are incorporated into the political system; in other words, the lack of proper mechanism for public goods provision constitutes a loophole that has enabled the active involvement of rural gangs.

When rural gangs have moved from the periphery to the center of rural governance, they start to exert powerful control on rural order and villagers. Both "competent and unethical" rural officials and "good gangsters", through their use of violence or threats of violence in rural governance, have violated the face-saving and not-going-too-far principles. That indicates that the acquaintance society is dismantling and the behavior pattern of rural land has changed, but these are inevitable results as the village community attempts to cope with governance predicament. To some extent, changes to the behavior pattern are irreversible and may not be repaired in a short time. Therefore, the intervention of state power is much needed.

Having reviewed the changes in rural governance, we may arrive at the conclusion that state power has yet to penetrate to the grassroots level of the rural society. Ten years ago in their case study on "taking court sessions to the *Kang* (heated) bed" (*kang shang kai ting*), *Qiang Shigong, Zhao Xiaoli* et al. have revealed that the grassroots government's power is exercised by the village secretary; common tactics and technologies include using rural cadre's personal contacts, take advantage of their prestige (*mianzi*), "carrot and stick" (*yi da yi la*), and "good cop/bad cop" tactic, to name a few (Shigong 1997, 1998; Xiaoli 1997; Ge 1997; Liu 1999). My case study on the film "*The Story of Qiu Ju*" also reveals that due to limitations of state power, grassroots cadres must strike a balance between applying political doctrines and local knowledge; that the state must deliberately protect rural officials' face (*mianzi*) as a kind of social asset that they may use for fulfilling government functions and providing public services (Baifeng 2010). From a more macroscopic viewpoint, *Li*

(2000: 27) argues that the government exercises power through "delivering knowledge of law to rural areas" (*song fa xia xiang*) to gain some kind of control over the grassroots society. The fact that state power is constituted through knowledge or tactics shows that, balance of power between the state and the rural society has been disturbed, and that as it tries to effectively manage the rural society, the state's ability often falls short of its wishes. The emergence of both "competent but unethical" rural officials and "good gangsters" may be explained by the genealogical method. Research using the genealogical method reveals the status quo and the further decline of inadequate state power. Compared with case study by *Su Li* and other scholars, this chapter's discussion on the relationship between state power and the rural society has been based on my field investigations and has touched upon a more fundamental level of the issue.

Chapter 9
The Changing Behavior Pattern and Moral Ambiguity

Having discussed gang influence on different aspects of rural life from Chaps. 3 to 8, in this chapter I will continue the discussion in Chap. 2 on the behavior pattern of the acquaintance society in an attempt to revisit the issue proposed in the Chap. 1, i.e. understanding the nature of rural society during a transitional period by observing the interactions between rural gangs and rural life as opposed to rural life of the traditional acquaintance society. From Chaps. 3 to 8, I have investigated the development process of rural gangs in the *Double-lake Plains* during the past 30 years, changes and the predicament in public security, gang control on rural order, gang influence on interpersonal relationships, and the eccentric role of rural gangs in village governance. In Chap. 2, I use the acquaintance society as the ideal model for the traditional rural society and have discussed its mechanisms of the maintenance of public order and the basic behavior pattern (*xiang tu luo ji*). Against this model, this chapter will start with an investigation into the gang-related aspects of rural life, in an effort to analyze changes in public order maintenance mechanisms of the Double-lake Plains. The analysis is aimed at grasping changes in the behavior pattern and moral order and eventually at determining the nature of social changes in contemporary rural China.

9.1 Rural Gangs and the Acquaintance Society

The relationship between rural gangs and the acquaintance society has gone through three phases of change. In the traditional time periods, being embedded in the acquaintance society, rural gangs are subject to all kinds of restrictions; in the 1980s, rural gang's attempt to become disembedded from the acquaintance society has failed; in the 1990s, as they expand into organized networks, rural gangs have increasingly grown disembedded from the acquaintance society.

The origin of rural gangs and suchlike dates back to ancient time periods, but they used to be collectively repelled by the acquaintance society. Weber (1978: 324)

B. Chen, *The Underworld of Rural China*,
https://doi.org/10.1007/978-981-19-8710-6_9

wrote "the life of the peasant in a Chinese village should certainly not be pictured, to judge by all indications, as a harmonious patriarchal idyll", and "what one might call the 'effective' peasants (the *lao shih*, as they were called) were then typically simply handed over to the arbitrary will of the *kuang kun*". The "*kuang kun*" in Weber's writing are a group similar to rural gangs today, but the *kuang kun*'s relationship with the acquaintance society greatly differs from that of the rural gangs. Social history studies also reveal that, the "bare stick" or "rootless rascal" (*guang gun*[1]) in historic periods were invariably repelled by the acquaintance society, and that the government's judicial process was affected by such repulsion (Sommer 2000). Having reviewed Durkheim's (1982) views, *Suli* (1998) argues that crime consists of an action which offends accepted conventions, ethics and certain collective feelings; such counteraction happens on a daily basis and often gets punished; crime should be classified as among the normal phenomena of social life. I believe "*guang gun*" in historic periods should also be classified as among the normal phenomena of social life, but because their acts have offended rules and conventions accepted by villagers, they are abandoned by the mainstream of rural society and rejected by the community. Apparently, such repulsion is needed for maintaining social rules and regional norms, and as a result, *guang gun* became a kind of "social waste". In contrast, in today's rural areas (especially in north China), some gangsters have become social waste but they account for only a small share of the whole gangster group.

Normally the "rootless rascals" do not constitute a strong influence, but during the reign of a corrupt dynasty and when social control becomes increasingly relaxed, i.e. during times of war, they would form a gang, commonly referred to as "bandit groups of rascals". Weber (1978: 324) wrote about them like this: "Against this organization, any individual large property-owner, or even groups of such property-owners, were often completely defenseless and powerless." Smith (2005) has also noted similar organizations of ruffians and bandits in late *Qing* Dynasty. However, in the acquaintance society rural gangs do not always get their ways, not even during wars. In his research on *Taitou* Village of *Shandong* Province during the 1940s, *Maochun* (2001:162) has pointed out that "as a matter of fact, no matter how weak the family is, if the members behave honestly towards other villagers, the family will not be bothered by the powerful people or outlaws. The powerful ones or the outlaws may choose not to help out such families, but would never try anything bad about them. If the family is under unreasonable attack, the attacker will sooner or later find itself being punished by the whole village. We'd naturally think that rich families are the targets of burglary, but they are not bothered even during times of disarray, because members of these families have always treated their villager friends nicely. Such families, poor or rich, would not be attacked by burglars, because such attack will arouse the greatest popular indignation." It may be inferred that during traditional periods, rural gangs are well controlled by the acquaintance society even when the

[1] Translator's note: Scholars have used different terms for "光棍". For instance, Weber's translator describes "光棍" as "*kuang kun*" (transliterated), whereas Sommer uses "bare stick" or "rootless rascal" along with the transliterated term "*guang gun*".

society is in disarray, let alone during times of social prosperity. This may also be proven by widespread tales of ancient people's punishment for village transgressors. Therefore, we may conclude that in traditional time periods rural gangs are well embedded in the acquaintance society.

During the 1980s, when farmers have been freed from the people's commune system, idle young people in their adolescence form regional groups and relationship-based groups. They have formed a unique gangland though they do not have the intention to do so. They share a distorted view of heroism, mainly characterized by bellicosity and vanity. As they bring more serious harm to the society, the government has been looking for ways to deal with them. This includes the "legal education class", "help and education program" and "crackdown campaigns", all of which rely on the "local resources" within the acquaintance society (Li 1996). Therefore, it may be concluded that the rural underworld of the 1980s, from growth to decline, has been deeply embedded in the acquaintance society and governed by its mainstream values.

Since the beginning of the 1990s, the rural gangland has gradually become disembedded from the acquaintance society and unconstrained by it. Rural gangs expand through networking into stable organizational structures and form an alliance spanning both urban and rural areas. This organizational structure offers rural gangsters chances for social mobility or climbing the social ladder. Gangsters of all strata have their own methods and channels for making illegal gain. The upper stratum may effectively evade government punishment; the lower stratum may not be eliminated because juvenile delinquents, left-behind children and unemployed youth constitute a fertile source for it. When rural gangs develop into organized networks, they become the "super power", and have fundamentally changed the state of affairs in the village. Compared with the "super power" of the gang alliance, villagers seem to have meager power; village officials and community are generally impotent; sometimes even the grassroots government is rather weak. Social order in the village is controlled by rural gangs and interpersonal relationships affected by gang violence. Many rural gangsters have moved from being the marginalized groups to playing the dominant role in village governance. When rural gangs become disembedded from the acquaintance society, the government seems to have fallen into a rural governance predicament.

9.2 The Changing Behavior Pattern

An investigation into the interactions between rural gangs and rural life will reveal the changes to the behavior pattern of rural China. Chapter 2 argues that the traditional behavior pattern of the acquaintance society include four principles: the reciprocity, "not-going-too-far", discrimination and attachment principle. With gang intervention in rural life, violence is gradually taken to the next level, and all the four principles are in decline.

1. The decline of the reciprocity and discrimination principles. The underlying concept of the reciprocity principle is the "give and take" (reciprocity) of favors over a long term basis. The "give" and "take" of favors is similar to the modern concepts of "rights" and "obligations". In their daily encounter, people are restricted by their relationships of reciprocity, so they do not haggle over immediate benefits. Instead, they establish a balance of interests in the long term. The discrimination principle means that one may or must disregard the interests of strangers or outsiders while favoring acquaintances and locals. The establishment of these two principles is inseparable from the give and take of favors between acquaintances. Within the acquaintance society, based on communal living experience, people have established all kinds of kinship ties—close or distant, consistent ethical norms and reciprocity-based circles of acquaintance.

   However, when rural gangs have formed an alliance through network expansion, they become the "super power" that prevails over the acquaintance circles. They build up a social network, with gangsters as the key roles and the dominant forces. In other words, dictatorial power falls into the hands of well-organized gangsters, and therefore, becomes disembedded from the acquaintance society. Social atomization in the Double-lake Plains has made it easier for rural gangs to get into rural areas. As rural gangsters often resort to violence or threats of violence, interpersonal relationships in the acquaintance society increasingly deviate from the reciprocity-based behavior pattern of the rural land (*xiangtu luoji*). Acquaintances that used to respect reciprocity also become more accustomed to using violence. People treat their own folks—acquaintances that were once treated as their own parents and brothers—in the same way they treat strangers.

   Despite being "locally born and bred", local gangsters treat their fellow villagers as strangers. They have no respect for the reciprocity principle. They are supposedly acquaintances to the villagers but are becoming estranged. The "outsiders" have always been strangers to the acquaintance society, but they are not being discriminated against as they should have been. Instead, they get to blatantly do as they wish. So both local gangsters and "outsiders" have become strangers to the law-abiding villagers, or strangers who have "shown up at their doorsteps". They are strangers not because people do not know them, but because they use the discrimination principle on people who should have been treated as their acquaintances. They do not act in accordance with the traditional reciprocity principle. Therefore, their behaviors are entirely unpredictable. Because of these strangers at the villager's doorsteps, villagers that used to be familiar with each other are also growing apart. In this way, people no longer deal with frictions and conflicts according to the reciprocity and not-going-too-far principles, but instead turn to the "strangers on their doorsteps" and resort to violence. Clearly the rural gangs have not only changed the reciprocity-based behavior pattern, but have also undermined people's basic understanding of the rural land and their fundamental values.
2. The decline of the not-going-too-far principle. In the traditional acquaintance society, people expect to live for long in the village, so they do not abuse their

"rights" and advantages, but rather save them for the long term. A loss at this time can be earned back at another time; a loss in this matter can be earned back in another matter; a loss in this life can even be earned back in the next; a loss in one's own life can be earned back by one's son. Nowadays people do not expect to live for long in the village. When people are only focusing on today, who would think about tomorrow? The villagers do not respect the feelings they have built up by living together for generations, much less to intentionally develop feelings for the sake of their future generations. Instead, they stick to the rational man's principles in a market economy. As a result, people cannot stand the smallest loss and have no patience for the long term; going to extremes becomes people's daily routine; people no longer believe in reciprocal concessions; violence becomes more prevalent in the village; and everyone may go to "extremes". In a word, the not-going-too-far principle is in decline.

With the decline of the not-going-too-far principle, people act in their own self-interest—following the principle that prevails in a market economy. In today's Double-lake Plains, villagers' lives are not connected as they were. The demand for cooperation in hydraulic engineering and public security is not inelastic as it was. Anything may be bought in a market. People may still lead a good life away from others. When people do not need each other and do not expect to live for long in the village, they would focus only on immediate interests and would not hesitate to use their dominance to the fullest extent. Rural gangster's blatancy knows no limits. They boldly bring in outsiders to their village and treat the weak cruelly. All that may be explained with the decline of the not-going-too-far principle. Forbearance is no longer recognized as a virtue or a desirable way of life, and has gradually become the weak's method of self-protection against the strong's violence. People do not worry about having diminishing dominance. Instead of making mutual concessions in times of disputes, they would rack their brains to prevail over the weak. They are reluctant to take any risks that may jeopardize their present life, but does not consider whether or not their actions will jeopardize the life of their future generations. They do not worry about putting their future generations in the weak's position, because they believe no one may predict the future. What people used to believe in—"good deeds are done secretly to benefit the future generations" (*jiyingong*)—has become a superstition that people laugh at.

The decline of the not-going-too-far principle may also be explained by the falling level of acceptance of Confucian political theory by the government. The "doctrine of the mean" (*zhongyong*) is no longer regarded as affecting the order of all things; nor is it upheld as a moral standard for a benevolent man. On one hand, the political ideology favors "the building of a harmonious society"; on the other, it encourages those who are capable of "fighting for their rights" to "sue your neighbor". However, such a modern variant of the Confucian political ideology would be a highly risky game of gamble for both the winners and losers in the "fight". The initiation of the fight would invariably cause bad consequences for all people involved. After fighting for their own interests, villagers often get into endless conflict and hostility. It is not hard to imagine the level of harm inflicted on acquaintances since they are "bound to see each other very frequently and regularly". When sustaining harmony become increasingly difficult, the acquaintance society would gradually fall into disarray. Though the Confucian political ideology is still accepted by most people whereas its modern variant by only a minority, we may not tell for sure what consequences would be brought on people's ethics. But one thing is for sure, that is, as ethical standards change, the modern ideology that encourages "fighting for rights" has overturned villager's conventional wisdom.

3. The decline of the attachment principle. The traditional rural China is tied to the rural land. Their livelihood depends on land; their mind is also tied to the land—wherever they are, they could not forget the rural home land. People find the same sense of belonging in the land as what others find in religious beliefs. That is true for regular villagers and even more so for social elites who have left the rural land. However, in today's Double-lake Plains where great changes have occurred, people's attachment to the rural land has diminished: the serious outflows of human and capital resources, goods and materials are straightforward facts; whenever they can, villagers would leave the rural land and pick up the urban dweller's lifestyles. People are no long attached to the rural land. Instead, they would like to sever their relationship with the village. Rural life is not desirable to them, whereas the urban life is a destination that they look forward to getting to. People stay only because they are entirely or temporarily not capable of leaving. In other words, the rural society has shrunk to "rural land in erosion" (Xiaotong 2006: 161).

   With the decline of the attachment principle, the village is nothing more than a temporary habitation and has lost its place as a spiritual sanctuary. Those trying to become gangsters do not feel the pressure of public opinion or do feel it but not strong enough to be taken seriously. For them, the village is not a permanent space where they live from birth to death, but a temporary shelter that they have their mind set on leaving behind and may easily do so. Whoever has brought outsiders in does not get criticized for doing so; nor does this provoke public indignation against the outsiders. When elites leave the village, they are no longer concerned about the latest developments at the village. It's all the more so for the gangster-turned elites. Elites do not mind their reputation in the village, because they are outward looking—they look forward to discovering the value and meaning of life from the market economy in the outside world. All that they care about is their personal interests. In a word, with the decline of the attachment principle, village elites give up their attempts to go for reputation and prestige in the rural society and thus stop being the keeper of rural order.

   To sum up, in traditional time periods and especially the Commune era, reciprocity-based behavior pattern (*xiangtu luoji*) has ensured a virtuous cycle for maintaining the rural and moral order that underpin rural life. Nowadays, the rural society is increasingly governed by the principle of self-interest in a market economy and the jungle law where brute force enjoys superiority. With the prevalence of brute force and wide application of the self-interest principle, the behavior pattern of the rural land are undergoing gradual changes, making it increasingly difficult to maintain order and causing the collapse of moral order.

9.3 Disintegration of Intimate Community

The changing behavior pattern of the rural land is indicative of the acquaintance society in disintegration. Chapter 2 argues that the acquaintance society is characterized by "familiarity" and "intimacy". The disintegration of the acquaintance society is in turn characterized by the decreased level of familiarity and intimacy.

Since the launch of reform and opening-up, the level of familiarity among villagers of the Double-lake Plains has continuously decreased. This may be attributed to the change from group to individual work. In group work, people take pleasure in working together and may get to know each other while doing so. The change from group to individual work gives people more free time, but with it comes the gradual reduction of all kinds of public events. Farmer's free time has not been used for interpersonal communication, but has indirectly lowered the level of interpersonal familiarity. What used to be widely accepted forms of events—political assemblies and entertainment activities carrying ideological features—have continuously decreased or entirely disappeared. Though "indigenous public space within the village" has been created by the expansion of rural markets, the increased popularity of funerals and weddings, the penetration of the Internet and the revival of mutual aid by non-governmental agency (Hailin 2005), public events have largely been reduced compared with previously. At the Double-lake Plains, due to reduced group entertainments and celebrations on traditional holidays (such as dragon boat races, dragon lantern dances and local operas), people strongly feel that traditional holidays are "no fun" any more.

Since the 1990s, farmers in large groups migrate to urban areas to seek employment or do business. Increased mobility and changes in both personal and shared life experiences have lowered the level of familiarity between villagers and in turn that of the whole village. Many people work away from home for years and only come back during spring festivals or once in a few years; some have owned apartments in urban areas and thus only live in the village occasionally. That has caused people to be less familiar with each other. Estrangement is also exacerbated by the fact that people marry different groups of people. Women from other towns and counties or even provinces are married to the local people, whereas local women are married to people in the city or more remote places—more strangers have come to the village. In addition, diverse employment has caused disparity in villager's income: some have become better-off while others have not. This will lead to psychological estrangement and different perspectives on the same things. The rural society is undergoing a predicament caused by a lack of shared experience (Xuefeng 2008). In a word, the villager's employment, income, habits, hobbies, disposition and social networks have become diversified; they have developed personal spaces.

Decreased familiarity and increased strangeness will cause people to further grow apart. This in turn exacerbates the level of estrangement. People feel like a rift has been created between them. Their relationships are not as casual as before. Now that they may be unfamiliar with some of family members of the households they visit—for instance, they know nothing about a migrant wife's temper, personality, likes and dislikes, people often find it inconvenient and embarrassing to drop in and have a chat. People also have different arrangements for their time: the home owner may be watching TV or have other plans when being visited. Besides, due to improved living conditions, such as the wide availability of television, electric fans and air conditioners, people would rather spend their free time staying at home by themselves than visiting others and having chats.

Decreased familiarity will inevitably reduce intimacy between people. Chapter 2 argues that people's intimacy is closely linked with the reciprocity of favors (*renqing*), for it has created the power relations between any two individuals, and sustained the acquaintance society as an intimate community. The reciprocity of favors occurs in "expressive occasions", such as ceremonies and non-ceremonial occasions for expression of emotions, and also in "non-expressive occasions", such as routine cooperation and mutual assistance. Reduced intimacy is seen in both types of occasions.

In "non-expressive occasions", the level of mutual assistance and cooperation has been gradually decreasing. This is often described by the villagers as being "short-sighted" and "insensitive". This is particularly true to the period after the end of the 1990s, when mutual assistance between villagers of the Double-lake Plains in house building, weddings and funerals and agricultural production is next to nothing. Hiring acquaintances is also not a choice, for that makes one indebted to others. People find it inconvenient to owe others; they are also unwilling to be owed. So they would rather hire help in a job market. Having calculated the financial cost, many villagers believe that hiring acquaintances costs more for they must treat acquaintances well with good cigarettes and wine. In the past, people have nothing to do once farming is done and are more willing to help others out since their free time is not used for making money. Helping out others is also a kind of investment, one that does not produce immediate returns, but the time spent will be saved in the form of favors. In future others will offer help to return this favor. So providing help to others is like making the best use of one's free time. That's how people pull off busy farming seasons and massive tasks like building a house. However, nowadays helping out others is a waste of time for the time could have been used for earning more money. They would rather trade their time for money to be deposited in a bank than save the time in the form of favors, for with money they will afford help when they need it. The widespread introduction of the market mechanism to the rural areas has caused the method of "saving one's time and resources up in the form of favors" to fail. This very much resembles the decreasing cooperation between villagers in Southeast Asia as described by Scott (1985).

Decreased cooperation is also evident among villagers in the Double-lake Plains. Sharing of farm equipment becomes much less common—not even between brothers. That is indicative of the changing relationship between brothers and between villagers. Villagers told me two reasons for not sharing: firstly to avoid conflicts and disputes; secondly, people have the financial means for owning their own tools. We may possibly say that improved finances have allowed for the possibility of breaking up the traditional cooperation mechanism. With improved finances, the villagers have reduced demand for mutual assistance and cooperation, and that's also what's actually happened. Reduced assistance and cooperation will in turn reduce disputes caused by such assistance and cooperation. However, that does not mean villagers become closer to each other; instead, they will further grow apart. Without mutual assistance and cooperation, they each "till their own lands and have their own meals"; they have lowered expectations on each other; they no longer count on others to offer a helping hand in times of major setbacks. Without such expectation, people no longer maintain an intimate relationship. Others and the whole village become insignificant.

Reciprocity of favors in "expressive occasions" has become seriously distorted. Expressive occasions in interpersonal relationships have increased with time. People have various unconventional reasons for hold a banquet and gift-exchange (*gan renqing*) becomes a heavy burden. At *Chujiang* City, a regular rural household must prepare gifts for 20-odd or 30-odd occasions, with the total spending at around 4000 Yuan; at *Linhu* City, that number reaches 6 or 7 thousand. The talk of "*renqing*" (reciprocity of favors) causes the change of color of the face, but people cannot avoid being trapped in the network of interpersonal relationship. Evidently, "expressive occasions" serve more purposes. At the beginning of the 1980s, people hold banquets only for several conventional purposes, such as engagement, marriage, house building, death of the elderly, or welcoming a one-month-old baby to the family. Nowadays formal banquets are held also for entering a higher educational establishment, joining the army, birthdays of the elderly or the children, etc.

Exchange of favors is needed for more occasions and money is often needed for such exchange. Some villagers hold banquets as a way to raise money. Those who do not hold as many banquets would feel like being taken advantage of, and would do the same under false pretenses. At *Xinwang* Village of *Chujiang* City, when building a two-story house, some villagers even hold a celebration for each story built. At *Qiaotou* Village of *Linhu* City, some hold a celebration for building a small boat; the value of received gifts is several times higher than the cost of the boat. Even more ridiculously a banquet was held for "murder" at *Qiaotou*. The parents of several young men embroiled in a murder case filed frivolous proceedings in an attempt to overturn the death sentence. As they needed money for "going through the back door" (*zou hou men*), they decided to raise money through treating people to a nice meal and taking gifts. Another more absurd example is sham banquet. At *Qiaotou* Village of *Linhu* City, a couple with no offspring hired a child for holding a "birthday banquet" to take gifts to balance the gifts they had given to others before. Very soon, a young man followed their example and spent 200 Yuan on hiring a girl for his "engagement banquet". The girl left when the man had collected gifts from his guests. Several Village Secretaries of *Qiaotou* Village held sham banquets for house purchase. They sold their houses after having received gifts, which are often worth tens of thousands.

The reciprocity of favors could have enhanced the level of intimacy among villagers, reinforced the public nature of rural life and maintained local conventions. However, when more scenarios have been incorporated into the type of "expressive occasions" that requires the reciprocity (*renqing*) of favors and when people seek all means to use "*renqing*" for making money, "*renqing*" does not serve the purpose of bringing people closer but has become a force of alienation. Instead of maintaining the intimate community nature of the acquaintance society, such "expressive occasions" only serve to break down intimate relationships among villagers.

Decreased familiarity and intimacy are clear indications for the disintegration of the intimate community and the changing nature of the acquaintance society. In disintegration the intimate community becomes more susceptible to gang violence and the people in it are more inclined to pursue financial gain. It has also emboldened the rural gangsters and lowered the social costs of gangster's unlawful activities, and thus enabled rural gangs to become disembedded from the acquaintance society.

The change in social nature seems less conspicuous compared with the change in people's behavior pattern. Naturally we may not find empirical evidence to support an assertion about whether the disintegration of the intimate society has caused people's behavior pattern to change or vice versa, because they are different aspects of the same social change. The same may be said about the correlation between rural gangs and the integration of intimate community. The truth is probably that the two phenomena have been mutually influencing and reinforcing. Therefore, it may be meaningless to examine the link between rural gangs and the intimate community using genetic method. In some villages, the reason why people turn into or ally with gangsters to pursue financial gain is that the intimate community has disintegrated and thus does not exert enough control over rural gangs; in some other villages, the intimate community in disintegration may have become more susceptible to external gang influence and have caused the locals to bring in outsiders. Since the same phenomenon may cause entirely different consequences, we may not make assertions based on the genetic method. What we may do is simply to reveal a multi-dimensional picture of the rural society and analyze its differences compared with the ideal model of the acquaintance society.

9.4 Changing View of Work

Gang intervention on rural life and the decline of work and wealth ethic are also correlated phenomena. Since the beginning of 20th Century, the way agricultural production is organized and economic attitudes of farmers may be summed up in 3 stages of development, from "leisure economy", to "labor economy" and "consumption economy". Each transition is backed up by a specific underlying work and wealth ethic. In the era of "leisure economy", leisure and "profit by other people's toil" symbolizes social class and rank; in the era of "labor economy", people firmly believe the idea widely spread by government propaganda that "working is the most honorable thing"; in the era of "consumption economy", consumption and fashion symbolizes social class and rank and becomes the motivation for engaging in productive labor. As people's attitude towards the economy changes, their work and wealth ethic also changes with it.

The term "leisure economy" was used by *Fei Xiaotong* during his research on the *Lucun* Village of *Yunnan* Province to encapsulate a traditional farmer's economic attitude. It describes a basic principle that "they'd rather makes less than work hard". He identified two types of people in the village: those who do not need to work and those who keep working during slack farming seasons. "Those who squat down by the street smoking, laughing and chatting would not be busy during the busy season. When the time has come for winnowing, they simply change the place for squatting, say by the fields if not by the street. Those who use slack season for carrying salt would be even busier during the busy season. The difference between busy and lazy farmers seems particularly distinct at *Lucun* Village (Xiaotong and Zhiyi 2006: 42)." "The generation that has stayed out of field work has not used the effort they've saved for other business. Most of their time has been wasted in bed for smoking, or in a gamble house for gambling, in the street for chatting, and in the tea house for tea drinking. ((Fei Xiaotong and Zhang Zhiyi, 2006: 107)." In the traditional rural China, not engaging in farm work and being able to "reap without sowing" symbolizes status and prestige; one with idle time to kill is often regarded as a lucky person.

That being said, people are not averse to labor and still uphold the basic ethic of "diligence brings wealth". To "reap without sowing" requires ownership of vast tracts of farmland, which is limited to only a minority group. The majority is still struggling to survive and may only support oneself and one's family through hard work. Landowners may "reap without sowing" and stay idle, while farmers with little or no land have no other choice but to work diligently. In a traditional society, to become the "idle class" with wealth brought by diligence is an aspiration shared by many. Because of such shared aspiration and occasional successful cases for achieving it, people believe in the apparently contradictory concepts of "leisure economy" and "work ethic". For the same reason, in the traditional agrarian society, people have an ambivalent attitude towards gangsters (Ning 2005: 250). People have bad feelings about gangsters because gangster's reluctance to form households, pursue career or engage in agricultural production goes against the work ethic; people's feelings towards gangsters are not entirely bad because their reluctance to work seems to coincide with the concept of leisure economy and that makes them a new kind of "leisure class". Regular farmers all look forward to having some leisure time, but they are compelled by reality to suppress such desire.

Leisure economy becomes possible only when it's legally allowed to own land privately and retain earnings by virtue of ownership without working. Leisure economy is ruled out by the land reform[2] of the 1950s. To live off land and other

---

[2] Translator's note: From 1953 onwards, the Chinese Communist Party began to implement collective ownership of expropriated land through the creation of "Agricultural Production Cooperatives", transferring property rights from the former landlord class to the Chinese state.

means of production without working is regarded by the new regime as "exploitation", which is a shameful thing. Having leisure time does not symbolize status and prestige, but rather is looked at as the life habit of the exploiter class. The idea that "labor is glorious" is upheld by the new regime as an unshakable ideological and moral principle, and has also been widely accepted by people of the new society. Because the diligent work ethic is already well accepted by farmers in traditional China, when their past aspiration for having leisure time is defined by the government as something the exploiter class would typically do, they seem to have very quickly converted to a not-so-new attitude towards work—the kind of commonly accepted attitude in a "labor economy"—and come to believe in "the glory of hard work" and "the shame in reaping without sowing". For the new regime, "the glory of hard work" is not just the kind of economic attitude and ethical principle that the government promotes among farmers, but more importantly a form of propaganda to organize the economy. The new regime is faced with the arduous task of modernizing the country and attempts to accomplish the task by adopting the planned economy. What the planned economy does is basically exposing the country for "exploitation by the government through the economic system", which allows the government to quickly accomplish the basic accumulation needed for modernizing the country. The belief that labor is glorious serves as an "auxiliary device" to the planned economy. It encourages people to sympathize with the modernization task under the leadership of the new regime and to contribute to its accomplishment with their own diligence.

Since the launch of reform and opening-up, China has entered the global system with increased willingness to work with other countries and has also become more susceptible to the tide of globalization. Economic growth is highly unbalanced for different regions of China. Some parts of the country has been modernized, others more so, and still some others not so much. While western countries have marched into the stage of consumerism, being susceptible to globalization and other trends, the Chinese economy may not stay clean of the consumerism influence. The Chinese People have gradually accepted the values of the "consumption economy". *Fei Xiaotong* once compared the economic attitudes of a traditional Chinese farmer and a westerner by saying that "the result of reduced labor and consumption is leisure. In a western city, people are busy the whole day, busy with work or enjoying themselves. They do not rest even when they take the day off. Entertainment is like real business to them, which makes people equally tired. They would not have enjoyment unless they spend some money. But in rural China things are quite different. They know how to find enjoyment without enduring the pain. And that is leisure. The difference between leisure and consumption lies in the fact that consumption gives people enjoyment while resources are being depleted, whereas leisure consumes free time but nothing else" (Xiaotong and Zhiyi 2006: 112). The kind of westerner's economic attitude as described by *Fei* is gradually accepted by the Chinese farmers. While abandoning traditional values upheld by the "leisure economy" and "labor economy", they have been convinced by the pervasive television advertisement and start to view consumption and fashion as sources of happiness and success in the same way urban people do.

In contemporary Double-lake Plains, people are involved in a fierce competition for consumer goods and face (*mianzi*). The rural economy is increasingly affected by the monetary economy. People rack their brains for ways to make money. People do not respect the traditional work ethic and are gradually abandoning it. Also people do not respect heroic and virtuous people as they used to do. Instead, they respect those with the ability to make money and buy things, but do not bother over how the money is earned. For that reason, rural gangsters are often envied and commended for their ability to make money out of their bellicose and aggressive nature. At *Xinwang* Village of *Chujiang* City, when a middle-aged villager told me about his nephew's experience as a gang member and could not help but felt some pride while doing so, I did not get why. However, when a former member of the security alliance spoke to me, I had an epiphany. He said, "In the 80s, a gangster family member would bring shame on the whole family. The whole family would be embarrassed to see their relatives and friends. But things are different now. Having a gangster member in the family or even having a convict becomes some kind of honor or usable resource." Nowadays, instead of feeling ashamed to talk about a convict in the family, they talk about them in the same way they talk about heroes. Instead of showing contempt over one's bellicosity, they express admiration and compliment.

That makes the rural gangs very attractive to the youngsters. Those "resourceful" gangsters are able to quickly gather up more followers, including many youngsters. Many of them join the gang out of admiration for a famous member; some parents even sent their children to the gangs. At a village of *Linjiang* County, a farmer sent his son to a big boss and said, "Here! You can have my son! He must become somebody some day, or I'd rather see him dead!" When they join the gangs, they may find a special way of climbing the social ladder. It has become the shared life goal for many beginners to become a big boss some day. In my research, many beginners told me that being invited in a fight gave them "a purpose in their gang life", and they looked forward to "rising above average". When that happens, money rolls in and they just sit comfortably at home, while others run the business, seize a dock and conquer a territory on their behalf. The same message is contained in a popular ballad of the Double-lake Plains. The ballad means something like this: "It's better to join the mob than to taste the bitterness of schooling, for the mob offers one nice food, drinks and notes, and ladies to sleep with". When changes have occurred to people's work ethic, being a gangster carries the nature of a socially accepted profession, and rural gangsters have grown from being the "social waste" to "social ideal". For that reason, since the end of the 1990s more and more rural gangsters do not "retire when coming of age". When being a gangster becomes a profession, for one thing, they would receive a steady flow of professional benefits, which allows them to stay in the gang; for another, they would obtain benefits by virtue of the professional networks. That means that they do not have to do everything by themselves, hence no need to retire out of old age or inconvenient life stages.

The commonness of "not retiring when coming of age" may also be explained by the changing restraints family have on rural gangsters. This may still be attributed to the dissolution of people's work ethic. Since the end of the 1990s, people have developed fetishism for money: whoever earns money is a man of ability but how the money is earned is less important. For that reason, moral condemnation on gangsters diminishes. Because people's work ethic has changed, a gangster's wife does not exercise the same discipline on the gangster. Instead, she may even encourage the gangster to "rise above the average" among all the gangsters. *Wang* from *Chujiang* City used to be the boss of a gang clique active in the town in the mid-1980s. When he finally decided to quit, he said that he'd quit because he did not want to cause anxiety to his parents and wife. However, when I asked a few gangster wives whether they would be bothered by having gangster husbands and whether they would talk their husbands into quitting—many of them have been married for over ten years, most of them replied, "we don't bother about what they do outside of the house". Their reluctance to discipline their husbands may be explained by their rising status in the family, but more convincingly by their husbands' ability to bring home enormous benefits for the family. They may not agree with the manner of money-making, but they would be happy if their husbands are able to rise above the average. All that is indicative of the changing work ethic and the nonexistent moral order.

9.5 Moral Ambiguity in Rural China

Now that the Chinese rural society is undergoing a period of rapid change, our understanding of it should not stop with the classic model of the "acquaintance society" proposed by *Fei Xiaotong*. In our research on rural governance in central China, our efforts have been devoted to discovering regional differences by investigating different aspects of the rural society, including villager autonomy, mediation and hydraulic engineering, etc. We have also investigated the implementation of policies, laws and regulations to have a better grasp of the changing social nature of rural China. This dissertation also aims at a proper understanding of the social nature of rural China through an investigation into rural gangs and gang-related rural governance and through a regional-comparison perspective. Judging by the status quo of rural gangs and gang-related rural governance, we may not use the classic model of the acquaintance society to describe the social nature of the Double-lake Plains, because people's behavior pattern, intimate community and ethical principles have all undergone tremendous changes. Moral order in rural China is on the brink of collapsing. In other words, the rural society of the Double-lake Plains may not be referred to as the acquaintance society; and since a new public order based on law and reason (*fali*) has not been established, the rural society is also far from becoming a gesellschaft.

Classical sociological theorists often emphasize the traditional-modern and urban–rural dichotomies. Dichotomy exists between "status and contract" in Maine's terms, between the "militant type of society" and the "industrial type of society" in Spencer, "feudalism and capitalism" in Marx, "Gemeinschaft and Gesellschaft" in Tönnies, "mechanical solidarity and organic solidarity" in Durkheim, "traditional Protestant economy and rational capitalism" in Weber, "folk and urban societies" in Redfield, "sacred and secular societies" in Becker, and "rule of ritual and rule of law" in *Fei Xiaotong*. Dichotomies are still used in today's sociological analytic frames as "an Occam's razor" for obtaining an understanding of traditional versus modern and urban versus rural. It is generally believed that urbanites are sophisticated, calculating, indifferent, insensitive, whereas rural folks are simple, sincere, honest and genuine. After having reviewed Simmel (1910)'s views, *Jihong* (2003) pointed out that, the difference between the sophisticated urban people and the simple and honest rural people may be summarized as different mindsets brought about by the different urban–rural states of affairs. The urban mindset has formed on the basis of the stimuli that the urban areas are flooded with. The urban people are often standing at the crossroads with people passing by; they must deal with the fast pace and volatility in the economy, career and urban living. And that has nurtured their mindset in a way that is different from that of the rural people. In addition, the money economy well developed in urban areas also teaches people to be practical in the way they treat others. As money measures universal exchange value, people would measure everything in relation to monetary value. This is entirely different from personal feelings of intimacy. As a result, individuals in reasonable interpersonal relationships are treated as numbers, and all people seem to have become identical.

In Bourdieu and Wacquant (1992)'s view, "the social cosmos is made up of a number of such relatively autonomous social microcosms, i.e., spaces of objective relations that are the site of a logic and a necessity that are specific and irreducible to those that regulate other fields. We can indeed compare a field to a game and it follows certain regularities". These regularities define the space constructed by the society. In their social life, while trying to keep their life going, people attempt to establish a sense of security over their existence. People of the country live in a space made up of acquaintances and the social roles are limited to just two kinds: family and community members. Because their family and community largely overlap, the roles they play do not differ significantly. However, people of the city must enter and exit plenty of social structures and play different social roles. As they enter and exit the social structures, they may experience reversal of roles or conflicting roles, and therefore, they need practical logic that go beyond the level of social structure and field for obtaining a sense of security over their existence. When monetary economy turns people into the digital generation and their daily life into procedures, the mental structure of the urban population has also changed. With the assistance of advanced technology, urban people have a much larger area for their activity than rural people. They also have clear boundaries for the public and private spheres and are constantly trying to switch logic. In contrast, rural people have a much smaller area and stable relationships that require simple maintenance. They may employ the same set of logic for their personal and social life since the two areas almost entirely overlap. Differences between urban–rural mental structures have been caused by these structural differences of their worlds (Jihong 2003).

As the result of urbanization, the dichotomy-based sociological analytic frames have become explanatorily incompetent. On one hand, frequent and close interactions have taken place between today's rural society and the external world: farmers become detached from the land; people's life and livelihood are no longer restricted to the village area and have been integrated into market economy of a larger society; people's relationships and behaviors are not restricted to the local community (acquaintance society) or the grassroots market towns, but have been integrated into a larger social system. In the mean time, the rural society has increased mobility and heterogeneity, which will in turn lead to greater uncertainty. On the other hand, such constant and close interaction has caused great changes in people's values and perceptions of the world. Urban prosperity and television broadcasting have eroded the farmer's attachment to land. They depend on the larger market for livelihood, and look to the urban people as good examples for their acts and values. Today's farmers who have become detached from the land are fundamentally different from their ancestors. Both their life and world have changed greatly: rural people look forward to living a life like the urban people; people have grown more rational over interpersonal relationships; social connections are not restricted to the local community; public authority has been undermined. Everything about the rural society is being reconstructed, which has served as the basis for changes in the mental structure of rural residents. Rural and urban people's social psychologies are growing more alike. Rural people have become more reasonable, calculating and indifferent. However, despite the changes in rural resident's real and mental world, the kind of urban order based on the rule of law has not been established in the rural society. Instead, the rural society is thrown into disarray. The village community is breaking apart; the rural society is collapsing. The status quo is referred to by scholars as "structural confusion" (Leiming et al. 2008).

Such "structural confusion" is more specifically demonstrated by the following facts: though villagers still know each other, the behavior pattern between acquaintances (logic of the rural land) have been badly shaken; changes have happened to their once intimate villager relationship; social order in the acquaintance society is kept by aberrant devices; ethics that honor diligence and hard work have been gradually abandoned and about to be replaced by faith in brute force and the principle of self-interest; moral order in the rural society fails to be maintained. All of these changes are fundamental changes, which may be summarized as "the fading of the acquaintance society". And because the rule of law has not been established yet, and that the nature of the rural society may not be summarized by the concept of "half acquaintance society", I refer to it as "moral ambiguity in the rural society" (*hui se hua* literally means "turning gray and ambiguous"). The use of this term is associated with the metaphors for "gray" in Chinese. As an intermediate color between black and white, "gray" has two metaphorical meanings in Chinese: gloomy and ambivalent (Contemporary Chinese Dictionary, Sixth Edition 1983: 499). For one thing, the rural gangs secure their livelihood in the gray area: not in the way normal people do, nor does their livelihood blatantly disrupt social order in the way organized crime groups do. For another, they lead a gloomy and disappointing life; when they use unethical methods for making illegal gain, they have caused negative impacts on villager life and rural order. Therefore, I describe village order controlled by rural gangs as "morally ambiguous (gray)" and the changing social nature of the village as "a state of moral ambiguity".

The result of urbanization on farmer life and world is two-fold: it allows farmer life to break free from the social constraints imposed by village life; it also allows farmers to pick up urban ways of thinking, which in turn causes them to break free from the psychological constraints imposed by the acquaintance society. However, the old social and psychological constraints of the traditional lifestyle fail to be replaced by those of the urban lifestyle. That's what makes "moral ambiguity" a thorny problem. When farmer life and mindset are not restricted by proper constraints, the village becomes a gray area—neither the traditional acquaintance society nor the modern society based on the rule of law. The village deviates from the traditional-modern and urban–rural dichotomies of classic sociological theory.

"Moral ambiguity" indicates radical changes, which are unprecedented in the several thousand years of the Chinese history. To better grasp those changes, one must compare them with the decline of rural areas in the Chinese history. The state of moral ambiguity is very different from the decline of the rural areas during dynastic transitions. The latter is mainly caused by bankruptcy of small-scale peasant economy, while the former by the processes of modernization and urbanization. The latter is nothing more than a phase in an effort to quell social unrest. Social order in the rural society will be rebuilt in no time once social stability has been reestablished, as described by a Chinese saying that "the emperor may bring stability to the country by simply issuing an edict" (Yiling 1988). In other words, during dynastic transitions, the micro-structure of the society remains intact. The rural society may appear to be in disarray, but its structural connections and autonomy may not be affected by dynastic transitions. However, that is not the case with moral ambiguity. It seems like a one-way journey with no return trips. Moral ambiguity in the rural society may be irreversible.

If moral ambiguity is nothing but a phase as the rural society moves from the acquaintance society to the modern society ruled by law, and if the process is not too enduring, then perhaps we might as well bear with it. But at the current rate, it is conservatively estimated that it may take more than one or two generations to urbanize the rural society. For a very long time in the future, the village will continue to be the place for the Chinese farmers to build a life. Therefore, rural governance and grassroots efforts to establish rule of law must build a home for farmers, prevent rural gangs from going rampant and keep the rural society from further decaying. A proper understanding of the social foundation is needed for attaining these important goals. As China moves progressively towards modernization, to reverse the trend of moral ambiguity is to defend the village, and to defend the village is to defend the whole society!

Chapter 10
A Supplementary Chapter: Rural Gangs and Grassroots Governance

This book approaches the topic from 3 perspectives. First, the "rural gang" perspective describes the development of rural gangs; second, the "public order" perspective discusses how the grassroots government handles rural gangs and its consequent predicament; third, the "village" perspective describes the interaction between rural gangs and village life. The main line of research for this dissertation may be summed up as follows: through an investigation into rural gangs, the dissertation attempts to analyze changes in the public order maintenance mechanisms and social nature of the Double-lake Plains as opposed to the acquaintance society. Unfortunately the first edition of this dissertation did not enter into in-depth discussion on the relationship between rural gangs and grassroots governance, e.g. rural gang's role in grassroots governance, their influence on grassroots governance, and the relationship between rural gangs and local governments. This chapter is written to remedy this problem. This chapter will investigate new developments in rural gangs in the recent decade, discuss the roles and influences of rural gangs on current rural governance, and attempt to deepen our understanding of the status quo for rural governance.

10.1 Rural Gangs: Variability in Regularity

It has become a common trend for rural gangsters to enter the urban areas. Previous chapters have classified rural gangs into 4 generations. After the first generation has been wiped out by the crackdown of the 1980s, the second generation now occupies important positions of the rural society, holding onto important resources. When the third generation joins the gangs, they shift their attention to the urban areas, mainly for two reasons: first, the village does not have more resources and opportunities to offer and they find it very hard to fight against the second-generation over resources; second, their appetite may not be satisfied by the remaining meager resources in the village. From the third-generation on, rural gangsters have all entered the urban areas. That also includes a few from the second generation. The fourth and fifth

B. Chen, *The Underworld of Rural China*,
https://doi.org/10.1007/978-981-19-8710-6_10

generations are born to farmers who have entered the city to work or do business, making them the city generations. The rural gang's "going to the city" has happened coincidentally with urbanization.

Since the beginning of the new century, farmers around the whole country move to towns or cities if they may establish a foothold there; some live in towns or cities while still earning a livelihood in the village. The urban area has also become the centre of life for gangsters. The reason for them to enter the urban area is associated with their human nature. They are a self-indulgent group: they are no diligent and hard-working people and thus would not be content with staying in the village. They are avaricious: the urban area offers them more chances to meet all kinds of people and to make financial gain. In addition to various chances and benefits, urban life is exceptionally attractive for the youngsters because it also offers "red lanterns and green wine" (*deng hong jiu lv*). In contrast, the village life is dull and lifeless for it always remains the same. The city also offers them a chance to get away from the ubiquitous supervision by their parents and relatives and to indulge themselves in the benefits of being young and work off their surplus energy. The city is also a gathering place for gangsters from different places. They gather because they love the hustle and bustle of the city, and also because they get to make mischief together, such as using illegal drugs and gambling.

In the recent decade, the urban and rural areas are moving towards integrated development, but the social network of rural gangsters has basically remained the same. The rural underworld is a gigantic social network consisting of a large number of interactive cliques or sects. Each gangster belongs to a clique of their own, which serves as their life space and the basis for pursuing financial gain. Gangsters establish shared feelings through "exchanging favors on certain special occasions" in the same way acquaintances do. Gangsters who have come from the same village may know each other, but may be members of different cliques, because they have worked on their own since the beginning of their gang life and have been pursuing interests in different lines of business. People from the same locality may not necessarily stick together for they do not have the chance to do so. It takes an omnipotent "big boss" for people to stay in the same clique, but that person does not come by easily. They would meet different kinds of people by sheer coincidence and constantly reestablish the circles of connection. That keeps the rural gangs' memberships constantly renewed.

Due to stratification of the rural gangs, every gangster's network has someone of a higher or lower stratum, and also of the same stratum. Differences in chance, character and family background have caused them to take different paths in life, resulting in stratification. Some get to lead a dignified life with a steady income; some have no place to live and wander in the street; still some others have become political or business elites. One thing is common among the upper class: certain occasions allow them to use their reputation for making financial gain. Their positions in the gangs are not permanent. Some join the gangs, others retire; some get promoted, others get demoted. A steady flow of new members come in, mainly juvenile delinquents. Some of the elder members may choose to retire, but "refusal to retire when come of age" has become a more common phenomenon. They are tempted to stay as they look forward to exchanging their names for real benefits some day. Social mobility has never stopped within the gangs. A steady flow of gangsters would seize the opportunity and climb the social ladder. Attractive opportunities are offered by the growing market economy, industrial development and changes in agricultural policies.

Avarice has been and is still the essential feature for rural gangsters since the second generation. They avail themselves of every opportunity that brings them financial gain. In the recent decade, the village still offers some of such opportunities, and some new opportunities have been created by reforms or changes in policy. In the past, resources are controlled by the village community, but because the village community does not have the right knowledge for managing the resources, it fails to make the best use of them. The central government rolls out policies for introducing market forces in resource allocation and encourages non-governmental forces to step in and harness idle resources. When the resources are "up for grabs", considerable profits that are once earned by vested interest groups have been released, which will naturally lead to fierce competition. The rural gangsters, with an acute sense of smell for profits, would soon swarm to those places offering lucrative opportunities.

Some typical policies by the central government includes: contracted operations of hydraulic engineering projects and forest tenure systems reform. Hydraulic engineering projects that are once managed by the village community are now contracted out to individual farmers. However, collectively-owned reservoirs at over 80% of my fieldwork destinations are now controlled by rural gangs for fishery (or farmers related to gangsters), and ordinary rural residents looking to take over a reservoir would be harassed by gangsters. Forest tenure systems reform begins with holding households responsible for managing forests (with leasehold of 70 years). The reform allows farmers to transfer forest tenure, which in turn allows certain farmers to scale the operations. Scaling requires the joining of forest land and one-on-one negotiation with a large number of small farmers who have tenure. The result of such one-on-many negotiation is that, if all farmers are treated equally, the person would have to close a deal at the highest asking price. Anyone's fraudulent intent would quickly push up the transfer price. However, that is not the real case, because gangsters would often intervene in the transfer of forest land. Farmers are often coerced into transferring their forest to gangsters at a very low price. In a town of *Jiangxi* Province, after the forest tenure reform, 70% of the forest tenure has fallen into the hands of the gangsters. Many people commented that "the money that should have been earned by the collective community has been pocketed by gangsters."

With urban–rural integrated development, rural gangsters have shifted their attention to the urban areas for earning their livelihood and seeking profits. That being said, they still intervene whenever profits are offered by the rural areas. There are mainly 3 types of villages that offer long-standing profits and thus become easy targets for gangsters. The first type is villages with rich mineral resources. Easy opportunities are offered when mineral resources are available for contracted operation. The second type is the suburban areas where village officials fail to settle disputes related to land expropriation; rural gangsters would easily find unethical methods to intervene. The third type is those traditional agrarian villages as destinations for top-down projects.

The implementation of such projects also offers opportunities of profit sharing, which rural gangsters would not let go of.

10.2 Enticing Potential Economic Interests at Different Villages

10.2.1 Monopoly Profits at Resource-Abundant Villages

Resource-abundant villages harbor minerals such as coal, rare metal and other resources. Technically, mineral resources are owned by the government or the collective, which means that anyone is entitled to the benefits. However, some minority groups attempt to monopolize resources under legal pretexts or using illegal and unethical methods. And to safeguard profits generated by the monopoly over resources requires the use of violence, which leaves room for gang intervention.

Villages rich in mineral resources offer the most attractive monopoly profits. Though mineral resources are owned by the state, the exploitation of resources must be done on the village's land. Besides, the government may not exploit resources by itself, and to start a state-owned enterprise for resource exploitation is not always the solution. Most of the mineral reserves have not reached the required level for doing so. Mining rights are often sold to private companies or even to sole traders. In this process, both outsourcing and contract signing leave room for profit sharing. Gangsters generally would gain the mining right at a relatively low price by using violence for deterring dissenting views or getting a slice of the cake in the whole industrial chain, including mineral extraction, transport and processing. Gang involvement is a common phenomenon. For instance, at *Baijiamao* Village, *Lyuliang* City, *Shanxi* Province, the village community regained ownership of coalmines from a company by taking legal action. Though successfully safeguarding their rights, they have violated the "accepted practices of the local rural gangland" and got battered by gangsters hired by the company. Over a hundred gangsters charged into their valley and hacked and battered more than 40 unarmed and vulnerable villagers for over an hour, causing 14 of them to be severely injured and 4 dead.[1]

County and town-level officials also attempt to secure profits while monopoly is being established over resources. The likelihood of involvement by rural officials at different levels depends on the potential profits and the mining areas of mineral reserves. Mineral reserves that may generate attractive profits would lure higher level officials to get involved, whereas those with small profits are attractive only for village officials. For instance, "at *Loudi* City, *Hunan* Province, when looking for evidence of *Kang* (a coalmine owner at *Xinhua* County)'s economic crimes, the public security authority of *Loudi* City unexpectedly found a notebook of *Kang's*,

[1] Extracted from *Villagers at a Coalmine at Lyuliang City Shanxi Province Cruelly Murdered by a Hundred Gangsters*, by *Ge Weiying and Wu Liwei*, https://news.qq.com/a/20091111/000901.htm.

which clearly recorded the names of 32 *Xinhua* County's public servants who have bought into *Kang's* firm and their specific amount".[2] Though the case dated back to more than ten years ago, coal mine collusion still exists today, as revealed by many local officials and residents when our team went to *Shanxi*'s coal fields in recent years. Attractive monopoly profits are offered by mineral resource zones around the country, *Shanxi* Province being the most typical example. "Rural gangsters have gone from some fighting and blackmailing to colluding with the government. They are making an official entry in the economic area of coalmine."[3]

Besides mineral resources, other resources owned by the collective or government are also subject to various means of private occupation. Take the above-mentioned rural areas of *Honghu* City as an example. Gangsters that live in villages close to the lake area would—without being permitted to do so—enclose the lake for fish farming. As turning the lake into private enclosure keeps other villagers from fishing there, disputes would arise. Generally gangsters would triumph over villagers by using violence. The result is that villagers have lost the right to fish; profits from fishing in the lake are pocketed by the minority of gangsters. At a town in *Shandong* Province where limestone on the riverbanks is also valuable resources, a powerful gangster has monopolized the exploitation of limestone. Similarly, after forest tenure reform, abundant forest resources end up with being controlled by gangsters or their connections.

To sum up, in some areas of central and western China, though the general public is not well-off and do not have the kind of opportunity offered by a big city, underground mineral reserves or other resources above the ground may offer certain individuals a chance to become better-off through private possession of such resources or getting a slice of the profits in the process of resource exploitation. To do so, the minority groups must rely on the use of force or coercion to deter the majority's dissenting views.

10.2.2 Real Estate Profits at Suburban Villages

With the process of urbanization, more and more land is being used for non-agricultural purposes. The value created by non-agricultural land obviously far exceeds that created by agricultural land. At villages on the outskirts and the down-town segments of cities, the high demand for rental properties creates attractive profits. The first way of converting land and property into cash earnings is illegal construction for residential or commercial use. At villages on the outskirts of cities with convenient transportation, illegal construction has completely occupied vacant areas and is about to encroach upon arable land. Illegal construction quickly extends

[2] Extracted from *Unveiling Coal Mine Collusion*, by *Liu Feixiao*, http://www.southcn.com/news/community/shzt/miner/benefit/200511290676.htm.

[3] Extracted from *Shanxi Coalmines Seized by the Rampant Underworld*, https://wx.abbao.cn/a/16784-e09c17851cc7a744.html.

to where urban roads and industrial parks are built. Illegal construction requires simple design and low costs but may bring huge rental income.

Land and property may also be converted into cash by the process of land expropriation. Compared with farming income and migrant worker's wage, land expropriation offers generous compensation, which may be an equivalent of a villager's revenues generated by up to ten years of agricultural production. After compensation, villagers have a large amount of cash, which may be used for building houses or purchasing vehicles to significantly improve their life, and also for preliminary capital for business or investment. Inevitably land expropriation will bring about demolition of existing buildings. Though *zhaijidi* (land on which farmer's houses are built) is officially recognized as "land used for rural development", when farmer's houses are about to be demolished, they are often compensated at housing prices of urban areas, otherwise farmers may resist having their houses demolished. Because a large number of illegal buildings have been constructed at villages on the outskirts of cities, many farmers may obtain sizeable profits from demolition and "become rich over night". In some rare cases, land expropriation and house demolition may impoverish farmers by depriving them of their land, but that only happens to farmers from underdeveloped small counties in central and western China. Due to limited finances, the county government of these areas may not afford adequate compensation for land expropriation. And to attract investors, the county government often imposes a low land price to gain a competitive edge in investment solicitation, which eventually causes farmers to receive limited compensation. However, land expropriation and house demolition generally still means great profits for villagers.

Due to the complexity in processes of turning land resources into cash and granting villagers profits from land, rural officials may gain benefits from corruption. Collective ownership of land in rural China determines that farmers only have contract rights to land and that compensation for land expropriation must be shared by the village community and farmers, with the larger share going to the village community. While implementing village-level democracy is still faced with various difficulties, rural officials often have the say in the expropriation of collectively-owned land and profit-sharing. Under the collective ownership of land, any procedure in land expropriation cannot move forward without rural officials' permission and cooperation. In addition to initiating and implementing land expropriation procedures, rural officials also manage the compensation for the village community. In other words, rural officials act as the go-between, which makes it possible for them to make illegal gain. To do so, they may agree to unreasonable terms proposed by the grassroots government, or collude with developers; they may also line their own pockets with the compensation once it has been deposited in the collective account. Embezzlement has been a common problem with land expropriation in the more developed coastal areas and has been kept under control by corresponding improvements in procedures and management. Thus the likelihood of corruption has been greatly reduced for these areas. However, in central and western regions, the advancement of urbanization has offered very attractive land profits whereas the needed management measures are still not in place. Therefore, in those regions corruption remains a problem and is perhaps a common phenomenon.

To sum up, in the process of land expropriation and illegal construction, farmers may gain profits and rural officials may make illegal gain. In other words, fierce competition for land or housing profits takes place among the grassroots government, developers, rural officials, farmers and other players. Such fierce competition leaves room for gang involvement as gangsters also look to have their slice of the cake. Another case in point would be *Xianning* City, where illegal construction on the outskirts of the city failed to be effectively managed by the urban management department, the government simply "outsourced" (*chengbao*) urban management to a firm established by *Wang*, a local gangster. Later *Wang* basically kept illegal construction under control, which allowed *Wang* and his followers to gain substantial profits.

10.2.3 Program Profits at Agrarian Villages

By far, agrarian villages are still the overwhelming majority and are widely spread in central and western regions. Farmers there are generally equally not well-off. Most households' income consists of two parts: farming income and migrant worker wage. While working as migrant workers, farmers would continue their agricultural production to retain ownership of means of production and to obtain consumer goods. Farmer's manner of livelihood may be summed up as "division of labor between generations", i.e., the elder generation continues farming while the younger generation works as migrant workers (the so-called "half working and half farming" model). In a word, rural residents have become "active agents" in both urban and rural areas and this manner of livelihood has reinforced the Chinese-style urban–rural dichotomy. As a Chinese expression describes, for a regular rural household, "land per capita is 1 *mu* 3 *fen* and land per household no more than 10 *mu*".[4] With limited land, farming income is also very limited and may only keep the regular households on the level of subsistence. Though migrant worker's wage is relatively higher, it only allows the households to save up a limited amount of money. The village communities of central and western regions do not harbor resources for cultivation, and the land owned by farmers may only be used for agricultural production and may not be used for non-agricultural purposes for making profits. Compared with resource-abundant villages and villages at the outskirts of cities, villages there offer meager profits. If it were not for policy intervention, these villages offering meager profits would never become targets of rural gangs.

Government projects have become important policies for the government to provide benefits to rural areas of central and western regions. These projects redirected gangster's attention to these regions. As a powerful tool of national governance, "project-based development" has been used by government at all levels for industrial

[4] Translator's note: This is the literal meaning of the Chinese expression. "1 *mu* 3 *fen*" and "10 *mu*" are not exact numbers but a rough estimation. It is used for describing limited land for farmers of rural China. We may convert 1 *fen* into 0.1 *mu*, and 1 *mu* into 666.7 m^2.

development. It has also become a common means of development in rural areas of central and western regions. These projects inject national resources into rural areas. Examples of project-based development include "extending cement roads to each village" (*cuncuntong*) and the launching of hydraulic engineering projects. Government projects tend to be concentrated in certain villages: a minority villages or "star villages" of a county or town. There are many reasons for such concentration: having a good cadre-based foundation: competent rural officials are able to implement the projects; having a good people-based foundation: project implementation does not encounter resistance or cause conflict of interest with the general public; at villages where previous projects has already been implemented, more projects would be introduced because it is easier to produce project-based results there.

A large amount of resources are brought into villages by establishing "points of contact" by government officials. Points of contact are established for implementing precise poverty alleviation, pilot projects for rural rejuvenation and land reform pilot study. Obviously projects are introduced because resources are readily to be invested, but more importantly, government officials introduce projects to certain villages to highlight good performance, because that is the best way to "refresh the old looks"[5]: a must-do when managing their "points of contact". A point of contact in central and western regions that is established by a provincial official would successfully attract investment worth 10 million Yuan, and the amount for those established by prefecture-level officials may be at several million, and that for county-level at almost one million.

Enormous potential benefits are brought by government projects and would naturally lure rural gangs to get involved. Benefits are created by both bidding and implementation of projects. To succeed in the bid for the limited number of projects requires careful management and often involves frequent exchanges of favors, thus leaving room for gang involvement. Gangsters may become involved in government projects when problems have arisen in the implementation of projects, such as handling the holdouts, ensuring public goods provision, etc. This is particularly true with villages where many projects are being implemented at the same time. Gangsters there may have formed a steady network established on project-based interests.

[5] The practice of "establishing points of contact" (*bandian*) is entirely different from "hands-on practice" (*dundian*) adopted by the first-generation Party members. "Hands-on practice" means that through living and working side by side with farmers, Party members would discover and solve problems and draw lessons for wider application. In contrast, "establishing points of contact" is to invest top-down resources in a village and change their looks in short periods of time. Those points of contact established by important leaders would change in appearance very quickly because large amounts of resources have been poured in. Those points of contact may look very different, and may also offer valuable lessons, but that's often the result of pooling resources (Xuefeng 2013).

10.3 The Mechanism of Gang Interference in Village Governance

10.3.1 Resource-Abundant Villages: Disturbance and Oppression

At resource-abundant villages, nominally resources are owned by the collective or the government, but benefits are really pocketed by a minority group for seemingly legitimate reasons or simply illegally possessed. If village affairs are handled democratically, villagers would definitely oppose such private possession. To deter opposing views, those who possess resources would have to rely on an act of violence or threat. By battering people with opposing or dissenting views, they would stop people from voicing their opinions, which would in turn put an end to opposing or dissenting views. In other words, the possession of profits at resource-abundant villages depends on rural gangster's acts of violence. That makes rural gangs the pillar for crushing opposition. When social atomization has become the prevailing trend, and that the rural society has a diminishing level of social connectedness, the villagers are not likely to unite into an effective deterrent against rural gangs. That in turn reinforces rural gangster's overwhelming power. The gangster's use of violence creates significant psychological intimidation for villagers, keeping them from publicly expressing dissenting views.

Rural gangsters may oppress villagers in two ways. First, ally with resource holders and help them retain control over resources. That's why gangsters always gather around business owners. The exploitation of mineral resources often entails fierce competition between miners or between miners and villagers. Business owners often need gangster protection. At our fieldwork destination in *Shanxi* Province, miners used to share profits with local gangsters. These gangsters later developed into organized crime groups and were wiped out by the "crackdown" campaign. Later, gangsters who have come from northern China quickly take their place because miners still need protection. Second, directly hold onto valuable resources. At a town in *Shandong* Province where limestone on the riverbanks is also a valuable resource, a powerful gangster has monopolized the exploitation of limestone. A group of gangsters are gathered to ensure that limestone exploitation may go on uninterrupted. However, limestone exploitation does not take place in the village. In addition, because the gangster treats his own folks very nicely and does not bring direct harm to interests of the local villagers, he is well received by his own folks and is later elected Secretary of the village. In rural areas of *Honghu* City (Double-lake Plains), profits from the lakes are also shared by gangsters who have first violently seized the public water areas and safeguarded their interests with force or threats of force.

At resource-abundant villages, it is very unlikely for gangsters to stay out of the profit competition. Even when business owners have acquired the right to earn income from the resources using legal methods or based on a democratic decision, which supposedly leaves no room for profit-sharing by gangsters, gangsters would still interfere in lawful operations of the business, e.g., stopping a vehicle transporting minerals and demanding compensation on the grounds that it has crushed the road. They may have disrupted the efficient transport of the company's resources, but this is just a civil or public security dispute and the police may not do anything about it. However, with frequent gang interference, the business's efficient operation may not be guaranteed and clients would be lost to other businesses. To minimize losses, business owners would have to share profits with gangsters. To ensure safety in business operations, owners often provide for a group of gangsters. As gangsters may interfere in any stages of production, such as mineral extraction, transport and processing, potential profits for them are widely spread in all the stages of production. Because the grassroots government may not effectively stop the gangsters and thus fails to provide a good business climate, it does not stand against the business owner's practice of supporting gangsters. It would tolerate the gangs as long as they do not cause "major disturbances", since they would ensure the basic order in resource management and a steady source of tax revenue for the government.

In some rural areas, if mineral resources are coveted simultaneously by several business people or gangsters, they may belong to different smaller cliques and cause mutual disturbances. When one of the cliques gains the dominant power to run the mineral reserve under the protection of certain cliques, other cliques would still cause trouble for them. Fierce competition may escalate into the kind of rough and tumble between mafia groups. The threat of force is needed for handling regular villagers, but to deal with business people protected by other gangsters, the threat of force will not do and that may cause violence to escalate. In coal-bearing regions like *Shanxi* Province, bloody battles between organized crime groups are commonplace and may lead to serious events. Rural officials had to intervene, and as a result of such involvement, a factual alliance has been formed by government officials, business people and rural gangs. Having formed such an alliance, some gangsters are emboldened to act even more recklessly. In other words, once the gangsters collude with the government, the economic sector is further infiltrated by organized crime, and their use of violence for protecting their financial interests becomes more blatantly reckless.

10.3.2 Suburban Villages: Competing for Economic Interests

As land is highly profitable, substantial profits are concentrated on the outskirts of cities. Illegal construction and land expropriation are both important ways of exchanging land for profits. Sharing profits from land and housing is a multilateral game participated by the grassroots government, developers, rural officials and other stakeholders. This is particularly true with the sharing of illegal profits. The fact that the game involves multiple players leaves room for gangster participation.

To curb illegal construction often requires the assistance of gangsters. Take *Mazhen* Town of *Xianning* City as an example. Curbing illegal construction has always been difficult: more illegal buildings continue to be constructed and control efforts have been ineffective. In 2014, the town government established a 5-man law enforcement squad for controlling illegal buildings with meager effect. Later the government hired 10 villagers at 2500 Yuan per month for scheduled inspections. Once they find illegal construction, they would try to stop it. If they have failed to stop it, they would report it to the town-level, which will forcibly tear down the illegal building. These measures were effective at the beginning, but later became less and less so, because the hired inspectors—being related to or having favor-based relationships with their fellow villagers—are easily bribed. This kept the control efforts from moving forward. Later the government had to dismiss the team, and outsourced the control efforts to a demolition contractor *Wang* at 600,000 Yuan. *Wang*'s demolition team is made up of mainly local gangsters.

The gangster demolition team turned out to be much more successful than their predecessor. Despite their regular attempts of site investigations, patrol and propaganda, or outright termination, the squad members were at their wit's end whenever real trouble had arisen. To construct a three-storey house built with steel-reinforced concrete in the rural areas takes about 300,000 Yuan. Illegal construction often requires a regular rural household to invest with money saved up for years. So the squad's forcible attempts of tearing down the house would naturally encounter fierce resistance from the villagers. Imaginably efforts to curb illegal construction would be very difficult. In contrast, the gangster's "reasoning" by threat of force would be much more effective. At the front line, earnest reasoning by government officials would often prove futile, whereas the slightest inclination to "toughen up" against villagers would cause their petitioning, violent public outbursts and even intentional ingestion of pesticides. Government officials are caught in a dilemma: leniency would cause them to be "vetoed", whereas non-tolerance would cause social disturbances and thus would also cause them to be "vetoed".[6] In contrast, gangsters do not have to be reasonable with the villagers; nor are they afraid of others making a scene. When they are in charge of curbing illegal buildings, they do not have the same psychological pressure for government officials. Because it seems like they would stop at nothing, the villagers are often intimidated by their force or threats of force.

Next to *Macun* Village, at *Lipu* Village, when the former town-level (*Mazhen* Town) law enforcement squad detected illegal construction and sent their men on persuasive efforts, the house owner threatened to "chop them up" with an axe. Later when the town government hired *Wang Changqing*'s demolition team for the job, a gangster arrived at the site and upon arrival used a brick to fight against the axe-holding house owner and immediately quelled him. As gangsters have proven to be more successful while being unreasonable at the same time, they are called by the villagers "a pack of dogs raised by the government".

Similar problems are encountered in land expropriation. When the government and developers decide on a certain land parcel, the government is in charge of land

---

[6] Translator's note: When an evaluation is done on government official's performance, certain targets are binding targets with veto-power (*yipiao foujue*). If these targets fail to be attained, the official would be disqualified.

expropriation and demolition before allowing the developer to have it. Before land development generates any profits, massive government expenditure goes to infrastructure and land compensation, etc. Holdouts would generate tremendous costs for the government. Thus the land expropriation process is highly risky. If the government fails to acquire land or demolish housing, it would have to pay compensatory damages to the developer. In some serious cases the developer would have to give up the project altogether. If that happens, the local government would have wasted all its efforts in the early stage of the project and would not be able to recover the huge loss or make tax revenues as expected. Therefore, in the process of land development, only by delivering the land to developers as soon as possible can the government gain an upper hand in the game against developers. The government would do its best to handle the holdouts and in that gangsters have an important role to play.

When time is pressing and does not allow for repeated negotiation with villagers, the government may resort to compulsory enforcement, but it must apply for such measures by adopting judicial procedures. However, the judicial system has stringent requirements for compulsory enforcement, and to avoid disrupting social stability, takes a cautious attitude towards such application. This makes the application a tiring and time-consuming process and leaves room for gang involvement. It has become the usual practice for the government to directly hire gangsters to ensure successful demolition. Under some circumstances, the government even hires gangsters for battering holdouts. Gangsters are employed by the government and the developer in a way that allows for "commercial transaction of violence". At a city of *Hubei* Province, a non-local developer failed to reach agreement with some holdouts and had to put the land development on hold. Eventually the developer decided to hire a local gang for "eradicating obstacles of land development". In exchange, the developer transferred shares of the development project to the gangsters. The local gang used to run a night club and is very experienced in handling villains. Therefore, they know their way around holdouts and can manage with ease.

To ensure compulsory enforcement of land expropriation, gangsters may use all sorts of violence-based or intimidation-based methods—some unethical others outright illegal. Violence may be used directly or indirectly to threaten or force holdouts into leaving and accepting the compensation plan. For instance, at a county of *Jiangxi* Province, when an elderly refused to move, gangsters played horrifying music right outside the man's house at late hours for several nights in a row, and the man had to move because he could not bear with such harassment any more. At a county of *Hubei* Province, gangsters bought several non-venomous snakes and once in a while snuck one of them into the house of a holdout, which caused the holdout to move out of fear. At a city of *Hubei* Province, a demolition firm run by gangsters used an excavator to slowly dig at the corner of a house. The dwellers were "terrified" and had to agree with the demolition plan. Of course, for most people there is no need for such unusual methods because simply verbal threats will do.

At villages on the outskirts of cities, the main reason for gang involvement in the sharing of land and housing profits is government tolerance. And the reason for that it is gangsters are able to "work miracles" with problems the government fails to solve. This is because the government must adhere to political principles, e.g.

rule of law, and is supervised by public opinion and the general public. Compulsory enforcement must be both legal and reasonable. It should not go against the law and the code of conduct for the Party, or it will surely encounter resistance. Sometimes legal and reasonable compulsory enforcement may still encounter resistance, and the government may do nothing about it. However, though gang violence is both illegal and unreasonable, people do not dare to voice out their discontent, or even if they dare to do so, may not be able to collect evidence needed for taking gangsters to court. Besides, "speaking up against gangsters" is a highly costly act, because gangsters would probably retaliate. Villagers may not use the same requirements on the government for gangsters, so they often suffer in silence.

Currently, when implementing land expropriation on the urban outskirts, the local government often enters into "administrative contracts" with the village committee, outsourcing certain tasks and delegating certain powers to it. This includes land measurement, compensation calculation, boundary surveying and dispute handling. This will allow the government to avoid direct encounter with rural households as much as possible and transfers some of the government-people conflict to inside the village. If the rural households want to increase compensation, they would have to turn to the village committee instead. This "administrative contract" system also causes conflicts and confrontations to become the normal state of affairs between villagers and the village committee. To protect their own interests or fight for more, villagers must engage in a long-standing rivalry against other competitors. The settling of such regular conflicts and confrontations often requires the village committee to use special means in their power to effectively handle the villagers' unreasonably enormous demands. In some more developed rural areas where the market offers handsome profits, rural officials may use all sorts of benefits and opportunities to keep the villagers in check. But rural officials in central and western China do not have the same means as their counterparts in developed areas. For them, to hire gangsters for help is in itself an optimal choice. Therefore, on the urban outskirts rural officials may easily form an alliance with gangsters or the position of rural officials is directly held by those with gangster connections.

10.3.3 Agrarian Villages: Sharing Program Profits

The collective resources of agrarian villages are often limited, and this is why state resources need to be invested to such villages. Special funds for rural development are allocated into the village community in the form of government projects, which bring huge potential benefits, but at the same time, have been a great lure to gangsters. There is room for gangsters to act in the government's project portfolio management and project implementation, and therefore there is room for them to make profits. Moreover, many problems may arise in project implementation, and gangsters have an advantage in solving these problems. In the process of supplying state resources for the village, the project system firstly is faced with the problem of how to match the supply side with the demand side. Different levels of government departments have all kinds of projects that may be "brought down" to the village, but both project portfolio management and project implementation requires bridging the information gap and the assistance of intermediaries. Compared with rural officials, gangsters have a wider social network and may act as an intermediary between the outsourcing party and the demand side. When a village gets a project, it must ensure its implementation and effectively handle holdouts and overcome other obstacles that may arise. The gangsters have their own strengths and may guarantee project implementation.

When state resources are being allocated into the countryside, because gangsters may take a share in project-based profits, they have been highly motivated to intervene in village affairs and have had a greater impact on village governance. *Zupei* (2016) summarized three mechanisms for gang involvement: running villages, running projects and acting as "gray brokers". Respectively the three mechanisms have caused gangster's unilateral dominance in village-level power structure, obstructed the smooth input of state resources, and undermined the effectiveness of rural governance. This suggests that gang involvement in government projects is in fact an intervention in village governance. Gang involvement has been caused by problems in state-countryside relationship and communication, which is in itself a governance problem. In my opinion, the mechanisms proposed by *Li Zupei* roughly correspond with the mechanisms of gang involvement in rural governance at different stages of project implementation.

First, before project implementation, gangsters would assist the government or village committee with "handling holdouts". Most of the top-down projects aim at improving rural infrastructure and conditions of production to facilitate agricultural production and improve villager life. But project implementation will invariably require demolishing some housing and damaging some young crops, or simply occupying some land for passing through an area and piling up construction materials. As these processes in project implementation may jeopardize farmer interests, it is reasonable for them to be compensated. But the problem is some may make exorbitant demand, the kind that is much higher than the reasonable level. If such demands were satisfied, those successful holdouts would become the hero and others would attempt to keep up with their level of compensation. Projects are often delayed or even simply abandoned because the contractor fails to meet the holdout's demands. Government projects are originally designed to benefit the people and should have been welcomed by the people, but when holdouts stand in the way, it looks as if the people are standing in the way of project implementation. The original intention of winning over the people fails to be fulfilled. Therefore, when rural gangsters use unethical methods to assist the grassroots government or village committee with compelling holdouts to submit, they would actually move the projects forward and boost project efficiency, in a way similar to how rural gangsters assist government or village committee with handling holdouts in land expropriation.

Second, in the stage of project undertaking, gangsters would directly undertake projects from grassroots government and ensure project implementation at the village. If rural officials fail to handle holdouts or settle disputes, outsourcing projects to rural gangs seems a good choice. In the early stage of implementing government projects, rural gangsters would forcibly undertake projects from government or village committee, who would often agree to do so to avoid conflict. Some gangsters would also use their connections in the government for undertaking projects. Later, as it becomes gradually clearer that rural officials are actually "incompetent", the grassroots government would directly outsource projects to the more "competent" gangsters. These gangsters do not necessarily perform the work. They may subcontract the project and make profits while doing nothing. In some places, projects fails to be outsourced following normal tendering procedures. As it is easier for rural gangs to implement the projects, they have an obvious advantage compared with other potential contractors. For instance, the same farmland irrigation projects implemented in different ways at two towns of the same city have produced entirely different results. The one implemented directly by the government failed to produce positive results, because of rural cadre's inaction and failure to confront farmers making exorbitant demands. The other project outsourced to gangsters has been implemented successfully, because gangsters easily solved whatever problems that have arised and farmers had to show them respect. Anyone with disrespectful behaviors would end up being threatened or battered.

Third, in the stage of project implementation, gangsters would act as rural officials and ensure project implementation. When resources are brought down to the countryside, the grassroots government is inclined to invest resources collectively into one or two villages and build them into models, for that kind of investment produces remarkable results. One of the many prerequisites to becoming one of the models is successful implementation, which requires competent rural officials to effectively resist holdouts. However, holdouts are a common problem for many projects; being fully aware of this problem, many rural officials are often inactive in undertaking projects. That gives rural gangs room for involvement by acting as competent rural officials. When rural gangs become active undertakers of projects, successfully implement projects and profit from doing so, they have actually formed a "new town-village symbiosis" with the higher authority, different from its predecessor that has formed right after the lift of agricultural tax. When the grassroots government intends to build up model villages, gangster-turned rural officials have seized top-down national resources through managing government projects. In an attempt to build up models, these rural officials would "seize village power, settle in-village disputes, obtain funding for project, and make profits". The village is an "enterprise" that the gangsters are managing and they would use all stages of project implementation for making gain.

Fourth, before project launch, gangsters would act as the intermediary. Lucrative profits are offered by the process of allocating agricultural funding. Only a limited number of projects will be funded, and to obtain funding requires exchanges of benefits, which leaves room for gang involvement. What distinguishes gangsters from ordinary farmers is that they are skillful players in many social groups: they have acquaints in the political circle; they also know someone in the business circle. In other words, they are the receiver of information unavailable to farmers or rural officials. With such information, they may act as the intermediary between county/municipal government (the project launcher) and the village, or profit from passing on information and undertaking projects. The higher the level of government is, the larger the gap between the decision-makers and the village. The gap makes it harder for the government to receive the information about facts or needs since the number of villages is numerous. But the rural gangsters have often become acquaintances with the county/municipal government; in the mean time, they know the village well; so they may act the intermediary. They often use all sorts of connections they have developed to make sure that projects are launched in their village, which will naturally bring with it all sorts of resources. Gangsters may also undertake some projects themselves, or subcontract the projects to others. *Chen* from *Xianning* is a case in point. Because he has a brother who works for the government, he becomes acquainted with many government officials and would easily know about government projects. He would recommend villages to the government for project launch and help some villages to apply for the launch. He would be designated as the contractor for most of the projects—an agreed term during the application process. For those projects he fails to undertake directly, he would subcontract them to other firms. The fact that he will benefit from acting as the intermediary is "understandable" to the government, village officials and subcontractors.

10.4 Power-Responsibility-Interest Structure for Gang Involvement in Village Governance

At all kinds of villages, the grassroots government would tolerate gang intervention in rural governance and even use gangs to their own advantage. At resource-abundant villages, gangsters would, in collaboration with rural officials and entrepreneurs, rely on the use and threat of force to seize and exploit mineral resources and gain monopoly profit while suppressing opposition and dissenting views. Those businesses that are not protected by gangs would fail to run smoothly because they are constantly bothered by gangs. The grassroots government chooses to tolerate gangs because it is highly costly to deter gangs and it also fails to ensure a good business climate. The bottom line is gangsters do not cause serious events that undermine social stability. At villages on the outskirts of a city, the grassroots government would outsource difficult tasks—e.g. handling holdouts in land expropriation—to firms protected by gangs. Village officials who have connections or ally with gangs would more effectively handle unreasonable requests by villagers, keep disputes under control and ensure smooth progress with land expropriation. At agrarian villages, as government projects that inject funds also requires expropriating land, holdouts are also one of main issues to be dealt with, and the village committee would use gangs to their advantage or even appoint gangsters as officials. Gangsters turn out to be quite effective at coercing holdouts into cooperating. As gang intervention will ensure smooth progress and boost efficiency with projects, the grassroots government would love to see good results produced by gangs.

In fact, the grassroots government would avoid closeness with gangs by generally keeping some distance with them. The government would not directly hire gangsters for a task, which is quite different from what happens during the 1990s, when the government used to directly hire gangsters for collecting sanitation fee in the urban areas, for public security alliance tasks or for any thorny problem that may have arisen in land acquisition. For the current time period, gangsters are not directly hired. They are still involved in rural governance, but not directly appointed as rural officials, but instead they act as market players or social agents. Since government agencies, market players and social agents are all involved in grassroots governance, gangsters get involved in rural governance through the market or the society (instead of through the political system). Hiring gangsters as rural officials belongs to the sphere of villager autonomy; but if gangsters are hired by rural officials or entrepreneurs, they become involved as social forces in rural governance; and if the government outsources tasks to gangsters or entrepreneurs under gang protection, they become involved through the market. In a word, now government tolerance and its use of gangs occur in the market and social fields.

This approach to handling its relationship with gangs has allowed the grassroots government to establish a firewall against risks, and has actually become a non-official arrangement in the political system. The government needs gangsters for handling some difficult tasks, but many problems and risks may arise, so it must be extra cautious. Allowing gangs to be involved through the market or society will "get the job done" and improve performance, and in cases of serious destabilizing events, the government may get to stay out of such events and evade responsibility. We may regard this approach as a non-official innovation to the political system that allows the government to use non-governmental forces to its advantage. This innovative approach may be attributed to a mismatch of rights and responsibilities for administrative authorities. The grassroots government has no choice but to come up with new ideas, thus tolerating gangs or using them to its advantage. The result is that gangs participate in rural governance through the market or society.

In China's administrative system, "meager rights and heavy responsibilities" are very common. In the recent decade, I've met many grassroots officials on trips to my investigation destinations, and they all complained to me about this problem. It seems to have occurred in all historic periods and at all departments of grassroots government. Before the lift of agricultural tax, collecting taxes and fees is the rural government's top priority, and as it becomes increasingly difficult to do so, to collect taxes in full becomes an impossible mission. Back then because the country has a relatively low level of industrialization, county-level government finance in most regions mainly relies on agricultural tax, which covers the pay of government officials and school teachers. To ensure collecting taxes and fees in full amount, county and town governments would delegate the task to town and village officials and set deadlines for them. Those who fail to accomplish the task would be punished. Town and village officials must complete the task while ensuring that no serious events are caused in the process. When persuasion turns out to be ineffective, they have no other choice but to think out of the box, e.g. using coercive measures such as confiscating villager property, restricting personal freedom, etc. However, such coercive measures may cause resistance, suicide and mass incidents. Officials who are held accountable for such consequences would be punished. In other words, when no effective legal means are available, the grassroots government must "ensure task fulfillment without using coercive measures". No wonder officials would complain about having meager rights while bearing great responsibilities"—they have spoken the truth.

At agrarian villages where government projects have become an important tool for grassroots governance, the grassroots government must secure top-down funding before it may save farmers from destitution and improve village appearance. Whether or not it may secure government funding has become an important tool for delivering good performance and an important yardstick for performance measurement. For many villages, if the village official does not intend to deliver good performance and does not undertake government projects, but instead handles only the routine affairs, the grassroots government would probably become "disembedded[7]" from the rural society. However, most town/township governments would not be content with simply maintaining the status quo and would try their best at securing government funding, which will not only improve rural infrastructure, benefit the rural residents and society, and will also help the rural officials achieve a better result at their performance. When projects funded by different departments of the government have been launched at a few villages, they become "the center of attention" for performance reporting and showcase. They are often called "demonstration centers for poverty elimination", for "building a new country", or "improving farmland irrigation and drainage".

The key issue in "governance by top-down projects" is how to ensure project implementation. All projects may benefit some and distress others. Even when all benefit, people do not benefit all equally. Those who suffer a loss or benefit not as much as others would try to obstruct project progress or demand for excessive compensation. Unable to handle these problems, some rural officials would choose to withdraw from the competition for government funding and simply keep things the way they are. Still some others would try to settle these problems and handle those people with rural gangsters' assistance. Those rural officials who are good at such problem-solving would naturally develop a closer relationship with the higher authorities. Once they do, they would be able to profit from the project implementation process, among other things. Because of this, rural gangsters develop a keen interest in becoming rural officials and would participate in elections of village officials either through legal means or by bribing their way into the election. Since they may bring tangible benefits to the village or villagers, they would easily gain villager's support. This will drive those "inactive" officials out and gradually rebuild the whole village political system. The new-generation is gangster-turned officials and their rule relies heavily on gangsters.

Things are similar to villages on the outskirts of cities, where the key issue is implementation of all sorts of development projects. Development projects there entail land acquisition and house demolition. Fierce competition for land or housing

[7] Translator's note: Disembedded is the translation of the Chinese expression "*xuanfu*", which means "hanging and floating". It is also sometimes translated into suspension. Disembedded means the opposite to "embedded". Embedded or embeddedness is often used to describe state-society relations.

profits takes place among farmers, the village community, developers, and the grassroots government. Such fierce competition may best be shown by the fact that some villagers resist having their houses demolished. Thus handling holdouts and ensure completion of demolition as scheduled are essential before development projects may be implemented. The grassroots government would then rack their brains for ways to get the job done. The job may be outsourced to demolition firms that are often run by gangsters. They often lure or coerce holdouts into cooperating. The grassroots government would also give some other difficult tasks to the village committee to do and hold it accountable for tasks such as land measurement and structure pricing. And for the village committee to effectively handle excessive demands by holdouts, it also requires the use or threat of force. Either the rural officials used to be a gangster or have connections in the gangs, or that the officials have a group of gangsters for them to use on those difficult holdouts. In a word, rural gangs' use or threat of force has become a common means adopted by the grassroots government and village officials to handle difficult tasks. It has been used in all kinds of manners for all kinds of working stages.

As the holdout problem exists in implementing both government projects at agrarian villages and development projects at villages on the outskirts of cities, the grassroots government with "meager rights and heavy responsibilities" has no other choice but to tolerate indirect gang involvement in rural governance. Therefore, it may be concluded that the mismatch between rights and responsibilities has provided room for gang involvement. And such mismatch may be traced to the universal top-down pressure in the political system. Superiors demand that their subordinates complete work tasks as scheduled and, based on the level of task success, punish and award subordinates. Evaluation of project performance is based on standards preset by the project sponsor. At villages on the outskirts of cities, the land acquisition task is divided into several stages and evaluated against quantitative indicators. Task assignment is inflexible—whatever task has been assigned must be completed, but superiors do not bother about whether task completion is possible or how and what it takes to get the job done. Superiors also demand that the process of task completion is satisfactory to the people and may not cause any serious events that undermine social stability.

In the administrative system, division of rights and responsibilities between superiors and subordinates has not been clearly regulated by legal stipulations. Downward task assignment mainly relies on the hierarchical superiority, particularly superior's control over subordinate promotion, salary and bonus payment. Task assignment is not accompanied by delegation of related powers or financial authority. Basically powers are seized by superiors while responsibilities are assigned to subordinates, hence the mismatch between powers and responsibilities. Technically the administrative system's efficient and effective functioning relies on the matching of power and responsibility for each level of administrative personnel. Rights and responsibilities of public servants must be clearly defined to clarify expectations and keep them highly motivated. Because responsibilities come with powers, rights are constrained by accompanying responsibilities, and that constitutes the basic check on powers. If rights have been clearly defined, responsibilities should be clearly defined too; the same is true the other way around. Apparently, based on the presuppositions of the Chinese bureaucracy, the current mismatch of rights and responsibilities that exist between superiors and subordinates in China's administrative system may not be a long-standing arrangement. However, this arrangement that should not have lasted has actually existed for a long time and also has quite effectively completed all sorts of top-down assignments.

Many scholars have described or explained the mismatch between rights and responsibilities in grassroots governance. Some of them explain it as this: the promotion mechanism together with other incentives has caused town and township heads to re-delegate their powers and responsibilities to their subordinates to alleviate pressure from their higher authority; as different levels of authority have the same interests, town and township heads may selectively carry out orders from the higher authority by "bending the rules" or "inaction". Some argues that during the transitional period for grassroots governance, as many questions remain unanswered, such mismatch offers a solution to problems such as information asymmetry and the lack of clearly-defined rules and regulations in the political system; and that with the mismatch, the higher authority "forces" the lower-level government to come up with creative solutions to problems—such is the political system that has supported China's growth miracle. These researches have proven the mismatch to be the status quo for China's administrative system and also discussed the involved player's coping tactics with the mismatch. When it comes to the topic in discussion for my dissertation, it may also be concluded that the right-responsibility mismatch has caused the political players to come up with coping tactics, which has in turn led to rural gang's general involvement in rural governance.

When the mismatch has become the status quo in the political system, the lower the political position is on the administrative hierarchy, the smaller powers and benefits but heavier responsibilities it has. Such status quo forces the grassroots government to think creatively or to make so-called systematic and institutional innovations. Allowing gangsters to be involved may be regarded as a kind of innovation. For instance, when government projects are about to be launched, the grassroots government would avoid direct involvement in the uneven sharing of project benefits; in certain stages of land acquisition, the grassroots government delegates some disputable tasks to village officials or outsource them to the private sector. The grassroots government retreats to the backstage and by using the responsibility system (*baogan zhi*), keep village officials at the forefront or by using the outsourcing arrangement, keep the private sector at the forefront. The grassroots government is well aware of the fact that both village officials and the private sector rely heavily on gang assistance but does not intend to interfere, because what the grassroots government needs is a solution to problems. What really wants is to get the job done, and gang involvement actually constitutes a kind of institutional innovation. Certainly there is more to such innovation. For instance, when certain tasks have been outsourced to the private sector, the contracting parties may in turn contract the task to the village committee. Such institutional innovation has, on one hand, motivated wider social forces, and on the other, allowed some social forces to obtain benefits in the gray areas. This includes the private sector, village officials and rural gangsters.

To some extent, the mismatch of rights and responsibilities has proven to be quite effective for handling internal relations in the administrative hierarchy during a period of rapid growth. Properly handling inaction from lower levels of government is very much needed to support economic expansion, but doing things the usual way does not fulfill that purpose. Because of the information asymmetry between the higher and lower levels of government, it is very hard for the higher authority to assess the degree of compliance at the lower level of government. The lower level may make a perfunctory effort and use all sorts of reasons to exaggerate the difficulties. The higher authority may not check the facts or fact-checking may be too costly. With the right-responsibility mismatch, the higher authority only focuses on whether a task has been done but not how it has been done. This is the simplest possible tactic for the higher authority. In addition to keeping the lower level of government highly motivated, this tactic also holds the lower level of government accountable if anything goes wrong.

During a period of social transformation, out of various reasons, conventional and legal means may not be effective and the government has to seek for new methods in the gray area. The completion of all top-down government decisions and growth targets eventually requires compliance by the general public at the grassroots level. However, the public does not form a united front, but instead, are broken into groups with their own interests. So, the grassroots government must bring these atomized groups together towards fulfilling the same target. But that is a major challenge. To rise to the challenge takes more than simply "playing by the rules" and requires creative thinking, i.e. resorting to some unethical methods. That explains why many problems are solved by measures and means that deviate from legal requirements and social conventions. Many so-called innovations to the political system have been used as a legitimate pretext for unethical methods. Institutional innovations such as outsourcing (*chengbao*) and the responsibility system (*baogan*) will actually allow the government to use rural gangs to its advantage while avoiding risks by indirect involvement. With the right-responsibility mismatch, the higher authority devolves its responsibility downwards, and the lower level of government and officials will more likely be held accountable to any serious event that may occur. Therefore, through "institutional innovations", the grassroots government will try its best to devolve responsibility to non-governmental forces.

10.5 The Involution of Grassroots Governance

Since the launch of reform and opening-up, the government has been in need of rural resources for sustaining its urbanization process. However, the conflict between farmers and the local government has been escalating and climaxed at the beginning of the new century. When the goal of using rural resources to support urban and industrial development has been basically attained, the government made a resolute decision to enact taxation reform, which eventually led to the abolition of agricultural tax. In the mean time, the government started its movement to "build a new socialist countryside", encouraging its industries to support agricultural development and urban areas to support rural development. By then the government has started to invest resources to rural areas, which should have increased public satisfaction with the government, because resource injection is expected to alleviate farmer-government conflict and benefit the farmers. However, public satisfaction only briefly rebounded before it experienced a period of constant falls. This means after the lift of agricultural tax, as more resources are injected into rural areas, farmer satisfaction does not really take a turn for the better. The legitimacy of the grassroots government, or public satisfaction with it and public compliance with it, has been constantly challenged. The fact that resource injection fails to bring expected benefits may be referred to as "involution in rural governance".

The term "involution" was coined by American anthropologist Alexander Goldenweiser to describe an artistic or cultural system that become more elaborate without innovation. When the artistic or cultural system reaches its definitive status, inspirations will be lacking, and they exhibit development in a baroque elaboration of the existing pattern. Clifford Geertz applied "agricultural involution" to his study of the self-conquering process of Javanese agricultural development. With limited capital and land resources, the labor force was continuously absorbed in agricultural production. Agricultural production thus grew more elaborate, complex and productive to keep the marginal productivity of labor from declining. Philip Huang used a pattern of "involutionary growth" to describe China's agricultural development. The increase of total agricultural output and output value relies mainly on the intensive input of labor per unit of land, which involves re-employment of inactive labor, while the rate of return for a marginal increase in labor diminished—in other words, this is "growth without development".

When it comes to the field of grassroots governance, Prasenjit Duara used the term "state involution" to illustrate China's expanding state structure and its modernizing process in the first half of the twentieth century. In northeast villages, "profiting brokerage" took the place of "protective brokerage", delegitimizing the traditional cultural nexus of power. Profiting brokers levied harsh taxes and imposed repressive rule on the people while lining up their own pockets. In other words, state power expansion through profiting brokerage failed to bring the state more revenues, causing "state involution". Tax revenues accumulated by legitimate tax-levying agencies did not increase; the increased revenue was pocketed by the informal agencies. The state had little control over these agencies and had to suffer the consequence of "collecting increased total revenue but with diminishing actual returns". In other words, though the state had increased power to exploit the countryside, it could not stop the unofficial agents from feathering their nest. Instead of increasing returns to existing agencies or establishing new agencies, the state had to rely on replicating its social relations to expand its administrative power.

The lift of agricultural tax marks a watershed in rural governance. Before the lift of agricultural tax, the government has been drawing on rural resources; afterwards, the government has been injecting resources into rural areas. Grassroots government power has shown signs of involution before the lift of agricultural tax. During the period of time after the collapse of people's Commune system and before tax reform, state power has been trying to gradually withdraw from the rural society, but still need rural resources for developing its economy. So for this period tapping rural resources becomes more difficult and expensive. When the town (township) government needs village officials for collecting taxes and fees and treats rural cadre's free-riding behavior with tacit agreement, the town and village levels have formed a steady "interest-driven symbiosis". As tax collection has become more difficult and requires more human power, the number of rural officials at the town and village levels continuously increased, leading to excessive bureaucracy. In other words, the grassroots government is experiencing "increased tax revenues and reduced benefits", and that is "an involution of state power".

"The involution of state power" is also reflected by the fact that resource injection has backfired by actually undermining the legitimacy of grassroots government. Resources allocated to rural areas have not been shared by all farmers, but farmers have to share the resources with many other people such as rural officials and other social groups. In fact, the lion's share has been taken by rural gangs. Farmers would naturally feel aggrieved and blame the grassroots government. Whenever the farmer obtains some benefits from resource allocation, more benefits are seized by rural gangs. Or worse, sometimes the benefits have been entirely pocketed by rural gangs. In some other cases, the farmers do benefit from injected resources, but they are still dissatisfied for rural gangs are gaining even more. The injection of increased resources would exacerbate farmer dissatisfaction with the grassroots government. Public satisfaction earned by the lift of agricultural tax soon starts to diminish.

Before the lift of agricultural tax, the grassroots government in some areas used to rely on gangsters for collecting taxes, thus incorporating rural gangsters into the town-village symbiosis. After the lift of agricultural tax, rural gangsters are not needed for tax collection but still for other governance purposes, i.e. handling holdouts. The holdout problem may arise in land expropriation, house demolition and project implementation, and thus rural governance may not get around the holdout problem. This means rural gangs are useful resources for grassroots governance. From the standpoint of the grassroots government, it might as well use rural gangs for handling holdouts to "get things done" while saving a lot of trouble. As rural gangs may be used for rural governance purpose, the grassroots government would treat them with "strategic tolerance". However, rural gangs will not be content with the meager gain obtained for being the "thug for hire" of the government. Once they have got a foothold in the rural society, they would seek substantial financial gain, using unethical or illegal methods.

Rural gangs gain from government projects mainly by directly undertaking a project, acting as rural officials to ensure project implementation, or acting as the intermediary for project launch. On the outskirts of cities, the expansion of city and town requires using rural land for non-agricultural purposes, which may bring substantial benefits and rural gangs may also benefit in the process. But as urbanization continues and more land is needed for non-agricultural purposes, to ensure grain security the Chinese government set up a minimum level of 1.8 billion *mu* (120 million hectares) of arable land (known as the "red line"), limiting the expansion of non-agricultural land. In other words, expropriating agricultural land at the town and village levels is subject to strict limitations. Non-agricultural land is a lever the central government may pull to redistribute land conversion quotas. Naturally the redistribution process has some profits to offer. Several complicated problems arise, such as how the profits are shared between the grassroots government and farmers, between the village community and individual households, and between contracting households and ordinary households. As the profit-sharing is a highly costly process, the grassroots government may hire rural gangsters or make gangster-turned rural officials fully responsible for handling affairs related to land expropriation and house demolition. That explains active gang involvement in such projects. Many gangster-turned rural officials become active players in governing rural areas on the outskirts of cities. Rural gangs are lured by huge profits into getting their cut in the cake, but instead of gaining from the grassroots government, they would turn to rural households to make gain.

Once rural gangs are incorporated into the rural governance system, they would seek other profits, e.g., by appropriating collective resources. Because the village community lacks the needed management skills for making the best use of collective resources, central policy has decided to adopt market-based solutions, encouraging non-governmental groups to cultivate idle resources. This provides a chance for gang involvement. Typical examples include gang involvement in above-mentioned contracted operations of hydraulic engineering projects and forest tenure systems reform.

To sum up, "bringing resources down to the countryside" has produced two consequences. First, farmers are not the sole receiver of benefits; instead of benefiting the farmers, sometimes resource injection may harm farmer interests, whereas rural gangsters get to make substantial gain. Second, as a result of the first consequence, resource injection also leads to serious psychological impact on farmers. When rural gangsters make gain by directly depriving farmers of what they already have, farmers may choose to bear with it, but some may also stand up against the gangsters. When rural gangs do not make gain through direct deprivation, but through seizing government resources, farmers would also feel aggrieved. Either way, farmer satisfaction with the government would be undermined. Farmers believe that it is the grassroots government's duty to ensure rural order and justice. But they've come to realize that rural gangs' profiting is generally based on the grassroots government's "strategic tolerance", or worse, government-gang collusion. That is, as more resources are brought down to the rural areas, farmer view on central policy takes a turn for the better while that on grassroots government officials for the worse. In other words, farmer satisfaction with the central government increases, while their satisfaction with the grassroots government diminishes.

When the government is viewed as a whole, the function of the grassroots government lies in implementing policies made by the central government. The central policy of "bringing resources down to the countryside" should have facilitated rural development through "urban areas supporting rural development" and "industries supporting agriculture", and should have increased farmer approval and legitimacy of the government. However, because of gang involvement, this central policy has backfired. If collective resources had not been "activated", they could have been harnessed by farmers in the future; but now that resources have been grabbed instantly by a minority group as a result of the central policy, farmers would only become more and more discontented. The fact that allocation of increased resources has actually undermined the legitimacy of the grassroots government may be referred to as "an involution of rural governance".

10.6 Managing Grassroots Governance

With the involution of rural governance, the government may not attain its governance goals. On my field trips, I was told by many villagers that "central policies are getting better, but local cadres are still as bad". This comment is indicative of the fact that the undermining effect of involution on public satisfaction and government legitimacy is limited only to the grassroots level. The legitimacy of the high-level (especially the central) government has not been undermined, and that will probably serve as the cornerstone of good rural governance. By bringing resources down to the countryside, mobilizing existing rural resources, the government aims to improve farmer life and maintain rural order. These policies contain the government's best intentions. But it is still faced with an important problem of how to establish the right rural governance mechanism and principles that best suits China's rural society. The urgent need for the current stage is to find a way out of involution, and that requires effectively managing rural gangs.

In January 2018, in a joint notice on *Launching a Special Campaign to Crack Down on Underworld Forces* (the Sweeping Black Campaign), the Central Committee of the CPC and the State Council pointed out that in response to the current new trends in issues involving crime and vice, we must effectively combine special governance with systematic governance, comprehensive governance, governance by law, and source governance, combine the fight against crime by underworld forces with high-level anti-corruption and "swatting flies" at the grassroots level, and combine the eradication of crime and vice with strengthening the construction of grassroots organizations. We must not only effectively crack down on the criminal activities of underworld forces and form an overwhelming situation, but also effectively eliminates the breeding grounds of underworld forces and form a long-term mechanism. We must adhere to strict punishment in accordance with the law, striking early before problems grow, and doing everything possible to eliminate vice. We must constantly maintain a high-pressure situation of a severe crackdown on crimes by all kinds of underworld forces. Political and legal institutions must further clarify the boundaries of policies and laws, unify law enforcement thinking, and strengthen coordination and cooperation. They must not only persist in cracking down on crimes by all kinds of underworld forces, but also insist on handling cases strictly in accordance with the law. They must ensure the uniformity of case handling quality and efficiency and ensure the uniformity of political, legal, and social effects.

We must actively adapt to the reform of the criminal procedure system centered on trials, strictly check facts, evidence, procedures and legal applicability, strictly forbid torture to extract confessions, prevent unjust, false, and erroneous cases, and ensure that every case is an iron case (Central Committee of the CCP and the State Council 2018).

The Central Committee of the CPC and the State Council are determined to "sweep away dark and evil forces (*hei'e shili*)". However, the effect may not be overestimated. In addition to "striking hard dark and evil forces", the central authority also mandates "handling cases strictly in accordance with the law". But most of gangsters may not be classified as "black and evil forces" because their activities fall into the gray area. The underworld is also divided into several echelons. At the upper echelon, the greatest winners of the rural underworld and the earner of the lion's share of the profits have turned themselves into entrepreneurs; instead of directly using violent means, they often rely on violence-based intimidation and thus may not be grouped into black and evil forces. Therefore, under the guidance of "strictly case handling that requires strict checking of facts, evidence, procedures and legal applicability", the campaign may not shake the underworld to its foundations.

The status quo of the underworld may be described as "prosperous", "vigorous" and "attractive", which are signs of an unhealthy society. Therefore, we must identify the causes and provide solutions. Solutions may be found by dealing with the following three problems.

First, "the black society" has provided a channel for climbing the social ladder for its members. With limited education and family finances, many gangsters do not have the required financial and social capitals for climbing the social ladder. But having "ventured into the black society" (*hunshi*), they may exchange their status in the underworld for illegal gain. They may become the middle or capitalist class and lead a happy life. To deal with this problem, the government must try its best to provide healthy ways for upward mobility, ensure wider access to social mobility, and make sure that more people are encouraged to move to a higher social class by doing the right thing.

Second, the education system has provided an endless supply of human resources for the expansion and development of the black society. To ensure order in teaching, instead of trying to keep, help or save juvenile delinquents in everyday teaching and management, schools go out of their way to release these students to "the black society" as soon as possible. That has caused many juvenile delinquents to join the gangs at an early age. But the education system has no other choice. It has been pressurized by public opinion and administrative responsibility into shouldering an excessive load of responsibilities but has been given very limited powers. School education is reduced to nothing more than lecturing and becomes ineffective at handling juvenile delinquents. When schools and teachers face risks that should not have been inflicted upon them, getting rid of the bad students becomes the best way to evade such risks. Obviously this problem may not be solved without the government. It must reestablish a healthy environment of public opinion on education and encourage an objective view on the rights, responsibilities and risks of the education system. The public should be taught to accept that responsibilities must be consistent with powers. It must empower schools and teachers and return schools to its former looks by ensuring school responsibilities and powers are balanced.

Third, the rule of law has not been established for profit sharing in certain markets and societies. The "black society" becomes prosperous because gangsters have gained substantial profits, and the sharing of such profit is not governed by the law. Economic expansion under the market economy has inflated the aggregate social profits, and has also led to legal gaps and vacuums. The legal system does not offer a complete safety net for regulating all sorts of economic and social interests. The government may find its hands to be tied when it comes to law enforcement. That suggests that the law has not been established in certain areas. And that's also where rural gangs may swoop in to make profit. To deal with this problem, we must ensure validity of laws and close legal gaps by perfecting the legal system and ensuring that all economic and social benefits are regulated by the law; we must improve the judiciary and law enforcement systems and enhance the government's judiciary and law enforcement capacities, so that unethical and illegal benefits may not go undetected.

To improve rural governance, we must build up a "firewall" against the rural underworld. We may take the following measures.

First, establish stronger village resistance against rural gangs, by implementing villager autonomy, enhancing village official's accountability, increasing group activities and strengthening village cohesion. Cohesiveness has a great impact on the village's resistance against rural gangs. Gangsters may seem reckless when facing individuals, but still have to be careful around government officials, upright elites and united villagers. However, government officials and upright elites may find it very hard to deal with the unreasonable gangsters inclined to use underhanded methods. Therefore, to put up resistance against rural gangs, village officials and elites must make concerted effort with villagers to pose psychological intimidation on gangsters. Implementing villager autonomy and enhancing village official's accountability aim at preventing official-gangster collusion and encouraging officials to set an example for standing up against gangsters. Increasing group activities and strengthening village cohesion will provide the required manpower for officials and elites in resistance against gangs.

Second, prohibit the grassroots government from turning to gangsters for help, and severely punish those who do. That is the only way to prevent interference and involvement by rural gangs and prevent them from gaining by doing so. Effective supervision is needed for the grassroots government and officials. Specific measures include improving government transparency, establishing administrative supervision, promoting inner-Party democracy and strengthening inner-Party supervision. These measures aim at preventing self-interest in grassroots government, which will in turn prevent the government from turning to gang assistance. Former Township Party Secretary *Li Changping* told me that during his incumbency he would be extremely careful when senior Party members and senior government officials were around, for they would not hesitate to voice direct criticism even to officials of a higher rank. With a good atmosphere for inner-Party democracy, township officials used to act very carefully. Obviously, democracy and supervision are effective constraints on the government and officials. When both government and official are subject to strict supervision and acts of turning to gangsters for help are severely punished, rural gangs would not be “strategically tolerated”; gang interference in rural governance is no longer tolerated by the political system.

Third, step up crackdown upon rural gangs. Specific measures include improving information flow, improving risk management and strengthen supervision over all parties involved in rural governance. The purpose is to establish an effective and efficient rural governance system that provides clear definitions of rights and responsibilities. As the author has touched upon this topic in the previous context, the author does not intend to go into detail herein.

We must take these measures to establish a rural governance system and political rules that suit rural China. Only by doing so can we eventually ensure good governance or set rural governance on the politically correct track. That is also the only way to ensuring the legitimacy of the Party and the government.

Glossary

People

beibatian	北霸天
Bing'er	兵儿
Cai Minghui	蔡明辉
Cai Shengqin	蔡圣勤
Cao Hailin	曹海林
Cao Jinqing	曹锦清
Caoshangfei	草上飞
Chang Zhongyuan	常仲元
Chen Baifeng	陈柏峰
Chen Feng	陈锋
Chen Gonghu	陈恭虎
Chen Hongxin	陈洪新
Chen Jinglong	陈京龙
Chen Jingqiao	陈井桥
Chen Jinliang	陈景良
Chen Xiaojin	陈孝金
Chen Xing	陈兴
Chen Xunqiu	陈训秋
Chen Yinqiu	陈银球
Chen Yufan	陈玉凡
Chen Zixiang	陈自翔
Cheng Xi	成希
Dai Yanhui	戴炎辉
Dai Yanhui	戴炎辉

(continued)

B. Chen, *The Underworld of Rural China*,
https://doi.org/10.1007/978-981-19-8710-6

(continued)

Deng Xiaoping	邓小平
Deng Yanhua	邓燕华
Deng Yunchu	邓云楚
Deng Zhenglai	邓正来
Di Jinhua	狄金华
Di Zhongdong	翟中东
Ding Mingsheng	丁明胜
Ding Wei	丁卫
Ding Zuoming	丁作明
Dong Biwu	董必武
Dong Leiming	董磊明
dongbatian	东霸天
Fan Mingqing	凡明清
Fan Xiaoyin	凡孝银
Fan Zhongxin	范忠信
Feng Dejun	冯德军
Feng Xiaoshuang	冯小双
Fok Ying Tung	霍英东
Fu Jianfeng	傅剑锋
Fu Lipeng	付立鹏
Fu Yiling	傅衣凌
Fukutake Tadashi	福武直
Ge Weiying	葛维樱
Gong Chunxia	龚春霞
Gong Zhihui	龚志辉
Guangzhou	广州
Guo Junsheng	郭军生
Guo Junxia	郭俊霞
Guo Liang	郭亮
Han Hong	韩鸿
Han Min	韩敏
Hatada Takashi	旗田巍
Hirano Yoshitaro	平野义太郎
Hiroaki Terada	寺田浩明
Hou Meng	侯猛
Hou Yijie	侯义阶
Hu Dingxing	胡定兴
Hu Jintao	胡锦涛

(continued)

(continued)

Huang Chongdao	黄崇道
Huang Chuang	黄闯
Huang Falong	黄发龙
Huang Guangguo	黄光国
Huang Lao Da	黄老大
Huang Lao Liu	黄老六
Huang Lao Wu	黄老五
Huang Wengao	黄文高
Huaung Lao Si	黄老四
Ji Weidong	季卫东
Jiang Chengshu	蒋成树
Jiang Lihua	江立华
Jiangnan qiguai	江南七怪
Jin Shouxin	金守信
Kaino Michitaka	戒能通孝
Kang	康某
Ke Daxia	柯大侠
Kong	孔
laoda	老大
laoer	老二
Laokouzi	老口子
Lei Hong	雷洪
Lei Zehong	雷泽红
Li Changping	李昌平
Li Changshuan	李长栓
Li Derui	李德瑞
Li Dongping	李东平
Li Guoqing	李国庆
Li Houping	李后平
Li Jian	李剑
Li Jihong	李继宏
Li Junfa	李俊发
Li Liangkun	李良坤
Li Liu	李六
Li Ming	李明
Li Pinnan	李品南
Li Shifan	李十凡
Li Xinmin	李新民

(continued)

(continued)

Li Yuanxing	李远行
Li Yuting	李玉廷
Li Zhi	李志
Li Zijiang	李子江
Li Zupei	李祖佩
Liang Shuming	梁漱溟
Liao	廖某
Lin Duan	林端
Lin Lin	林琳
Lin Yutang	林语堂
Liu Can	刘灿
Liu Daorong	刘道荣
Liu Dingsheng	刘丁生
Liu Feixiao	刘非小
Liu Hong	刘宏
Liu Honggao	刘宏高
Liu Lin	刘林
Liu Qin	刘勤
Liu Renshan	刘仁山
Liu Weimin	刘为民
Liu Yanwu	刘燕舞
Liu Ye	刘爷
Liu Yue	刘岳
Liu Zhuo	刘焯
Lu	鲁某
Lu Huilin	卢晖临
Lu Xun	鲁迅
Luo	罗某
Luo Gan	罗干
Luo Ruiqing	罗瑞卿
Lv Dewen	吕德文
Ma Wenhong	马文鸿
Ma Yuanyuan	马源源
Mao Gangqiang	毛刚强
Masao Uchiyama	内山雅生
Nie Liangbo	聂良波
Ou Yangjing	欧阳静
Pan Zhanlin	潘战林

(continued)

(continued)

Peng batian	彭霸天
Pi Jiasong	皮家松
Qi Haibin	齐海滨
Qiang Shigong	强世功
Qin Dongxiong	秦东兄
Qu Pei	瞿沛
Ren Enshun	任恩顺
Rong E	荣娥
Rong Jingben	荣敬本
Sang Benqian	桑本谦
Shan Yongkang	单永康
Shen Deying	沈德银
Shen Duanfeng	申端锋
Shen Xianbing	申献兵
Shimizu Morimitsu	清水盛光
shuishangpiao	水上飘
Shunzi	顺子
Song Lina	宋丽娜
Su Li	苏力
Su Renzhong	苏仁中
Sun Guangzu	孙光组
Sun Q iuyun	孙秋云
Tan	谭某
Tan Tianguo	谈天国
Tang Huangfeng	唐皇凤
Tang Jiping	汤吉平
Tang Jisong	汤吉松
Tang Luowen	汤骆文
Tang Xiao'en	唐孝恩
Tang Yunxiang	汤云翔
Tao Siju	陶驷驹
Tian Chunying	田春英
Tian Xianhong	田先红
tiezhangwan	铁掌玩
Tong Zhihui	仝志辉
Wan Debi n	万德斌
Wan Huaqing	万华清
Wan Hui	万晖

(continued)

(continued)

Wan Jin	万进
Wan Liang	万良
Wan Liu	万六
Wan Min	万民
Wan Ming	万明
Wan Qing	万清
Wan Rong	万荣
Wan Sanyuan	万三元
Wan Shan	万山
Wan Sheng	万胜
Wan Shuai	万帅
Wan Xin	万新
Wan Yi	万义
Wan zi	丸子
Wang Changqing	王长清
Wang Hui	王会
Wang Mingming	王铭铭
Wang Qiliang	王启梁
Wang Weiwu	王为武
Wang Wu	王五
Wang Ximing	王习明
Wang Zhongfang	王仲芳
Wen Guangxin	文光兴
Wu Changhua	吴昌华
Wu Liwei	吴丽玮
Wu Ye	武爷
Xia	夏某
Xia Tao	夏涛
Xiang Jiquan	项继权
Xiao Desheng	肖德生
Xiao Hongmei	肖红梅
Xiao Qing	肖青
Xiaocun	小村
xibatian	西霸天
Xie Jiamei	谢加美
Xie Shouyin	谢守银
Xu Caisong	徐财松
Xu Chenglong	许成龙

(continued)

(continued)

Xu Diyu	徐涤宇
Xu Shan	徐山
Xu Xi	徐昕
Xue	薛某
Yan Liangquan	严良全
Yan Qingfeng	鄢庆丰
Yang	杨
Yang Chunmei	杨春梅
Yang Daigen	杨代根
Yang Guang	杨广
Yang Hua	杨华
Yang Maochun	杨懋春
Yang Shanhua	杨善华
Yang Shuxiao	杨书校
Yang Suixing	杨遂行
Yao Shougui	姚守贵
Yi Jiming	易继明
You Chenjun	尤陈俊
Yu Jianrong	于建嵘
Yu Shengtian	余生田
Yuan Song	袁松
Zeng Fanhua	曾凡华
Zeng Jian	曾建
Zeng Weiping	曾维平
Zengji	曾集
Zengling	曾岭
Zhai Zhongdong	翟中东
Zhang Chang'e	张嫦娥
Zhang Chuanrong	张传荣
Zhang Demiao	张德淼
Zhang Fei	张飞
Zhang Guangbao	张光豹
Zhang Guangfu	张光富
Zhang Guanghu	张光虎
Zhang Guilin	张桂林
Zhang Jicheng	张继成
Zhang Letian	张乐天
Zhang Li	张莉

(continued)

(continued)

Zhang Mingcan	张明灿
Zhang Ning	张柠
Zhang Shiyong	张世勇
Zhang Si	张思
Zhang Tingguo	张廷国
Zhang Xiuli	张秀丽
Zhang Zhengping	张正平
Zhang Zhiyi	张之毅
Zhao	赵某
Zhao Xiaofeng	赵晓峰
Zhao Xiaoli	赵晓力
Zhao Yi	赵义
Zhao Yi	赵翼 (诗人)
Zheng Haolan	郑浩澜
Zheng Yi	郑谊
Zheng Yinnian	郑银年
Zhou Dajun	周大军
Zhou Jinzhi	周金芝
Zhou Yongkang	周永康
Zhu	祝家
Zhu Bingxiang	朱炳祥
Zhu Suli	朱苏力
Zhu Xianglin	朱祥林
Zou Dang	邹谠

Places

Anhui	安徽
Anjia	安加
Anxian	安县
Anyang	安阳
Baiyang	白洋
Baolian	宝莲
Binhu	滨湖
Changde	常德
Changde	常德
Changjiang	长江
Chen Jiafang	陈家坊

(continued)

(continued)

Chencun	陈村
Chenwan	陈湾
Chuanxi Plain	川西平原
Chujiang	楚江
Dabie Mountain	大别山
Dengwan	邓湾
Dingxiang	定向
Dongfeng qiao	东风桥
Dongting Lake Plain	洞庭湖平原
Dongxi	董西
Feixi	肥西
Fucun	付村
Fugou	扶沟
Fujian	福建
Fulou	付楼
Gangkou	港口
Guangdong	广东
Guantai	观台
Guizhou	贵州
Guohan	郭韩
Guozhuang	郭庄
Heli	河里
Henan	河南
Hongjiang	洪江
Huaibei	华北
Huang Zongdang	黄宗荡
Huangcun	黄村
Huangmei	黄梅
Huangnie	黄聂
Huaqiao	花桥
Huaxian	滑县
Hubei	湖北
Huchang	湖场
Hunan	湖南
Ji'an	吉安
Jianghan Plain	江汉平原
Jianli County	监利
Jiati	甲堤

(continued)

(continued)

Jinglan	井栏
Jingmen	荆门
Jingshan	京山
Jingzhou	荆州
Jinjiahu	金家湖
Lianghu	两湖
Lifu	李复
Liji	李集
Linhu	临湖
Linjiang	临江
Linsha	临沙
Linzhuang	蔺庄
Lipu	里铺
Liu Shiba	刘什坝
Liuyi	六一
Lixin	利辛
Liyu	李圩
Longyou	龙游
Longzhou	龙洲
Loudi	娄底
Ludong	路东
Luochang	罗场
Luoci	罗祠
Luoping	罗平
Lyuliang	吕梁
Macun	马村
Maocun	毛村
Maohan	毛韩
Mazhen	马镇
Meixian	梅县
Minzu	民主
Ningling	宁陵
Pingyu	平豫
Puxing	普兴
Qiaocun	桥村
Qiaotou	桥头
Renmin	人民
Runan	汝南

(continued)

(continued)

Sanhe	三河
Sanlidian	三里店
Sanquan	三泉
Shagang	沙岗
Shandong	山东
Shanxi	山西
Shanxiang	山湘
Shaqiao	沙桥
Shawan	沙湾
Shenzhen	深圳
Shicao	石槽
Shima	石马
Shiping	石坪
Shuanghe	双河
Shuicun	水村
Shuiyuean	水月庵
Songzhuang	宋庄
Suizhou	随州
Taiping	太平
Taishan	太山
Taitou	台头
Taiwan	台湾
Tangwan	塘湾
The Huai River	淮河
Tianmen	天门
Tianxin	田心
Tongxin	同心
Wan	皖
Wancheng	万城
Wangcun	王村
Wanghan	王韩
Wangzhen	旺镇
Wuchang	吴场
Wusan	五三
Wusi	五四
Xianan	下南
Xiangtan	湘潭
Xianning	咸宁

(continued)

(continued)

Xiaochi	小池
Xicun	溪村
Xinfa	兴发
Xinhe	新禾
Xinqiao	新桥
Xinwang	新王
Xuzhuang	许庄
Yangsi	杨泗
Yanjiang	沿江
Yanjing	盐井
Yincheng	鄞城
Yuejin	跃进
Yueshan	岳山
Yueyang	岳阳
Zengling	曾岭
Zhaoba	赵八
Zhejiang	浙江
zhen	镇
Zhengzhou	郑州
Zhonggulou	钟鼓楼
Zhongnan	中南
Zhuqiao	竹桥
Zhushan	株山

Chinese Terms

Anti-Hooligan Act	检肃流氓条例
bandian	办点
banggong	帮工
baogan zhi	包干制
baojia zhang	保甲长
bianzhi	编制
bingzhuang	禀状
buxiao	不孝
chaxu geiju	差序格局
che xiang bing zhen	撤乡并镇
chefei luba	车匪路霸
chengbao	承包

(continued)

(continued)

chengguan	城管
chi hei ma	吃黑码
chou shui	抽水
chuangshou	创收
chulaoqian	出老千
chunpu	淳朴
cuncuntong	村村通
cundui	村队
cunmin xiaozu	村民小组
cunmin zizhi	村民自治
Cunzhi	村治
da hunhun	大混混
da pin zhang	打拼帐
da si	大私
daigeng	代耕
daizui ligong	戴罪立功
Dang Guan De	当官的
dao de ren	道德人
datao	搭套
deng hong jiu lv	灯红酒绿
ding zi hu	钉子户
diuzi hankuan	丢字喊款
diyi shijian	第一时间
diyuan	地缘
Dongfeng	东风
dundian	蹲点
e ba	恶霸
er liu zi	二流子
falangmei	发廊妹
fali	法理
Falun Gong	法轮功
fang zhi	房支
fangong daosuan	反攻倒算
fanshen	翻身
Farmer's Life and Values	《农民生活及其价值世界》
fen	分
fenzi	份子
Fumu Guan	父母官

(continued)

(continued)

gan renqing	赶人情
gaomaozi	高帽子
gei mian zi	给面子
geiyu	给与
geng du chuan jia	耕读传家
gongfen	工分
guan dian yuan	管电员
guan tui min jin	官退民进
guang gun	光棍
guanxi	关系
guofa	国法
guoji	过继
he cun bing zu	合村并组
he li	合理
he qing	合情
Hei'e shili	黑恶势力
heju	合具
heqing heli	合情合理
hnhun	混混
hong jin long	红金龙
huaifenzi	坏分子
hui se hua	灰色化
hukou	户口
hun	混
hunhuner	混混儿
hunhunzi	混混子
hunshi	混世
huzu	户族
jiaguo tonggou	家国同构
jiang ren qing	讲人情
Jianghu	江湖
jiazu	家族
jidongtian	机动田
jiedao	街道
jiefang	街坊
jingzhang	警长
jiyingong	积阴功
ju	矩

(continued)

(continued)

kan qing hui	看青会
Kang	炕
kang shang kai ting	炕上开庭
keji fuli	克己复礼
kuang kun	光棍
kuiqian	亏欠
Kuomingtang	国民党
lai hai zi	赖孩子
lan gan zi	烂杆子
lan tong zi	烂桶子
lao shih	老实
laogai	劳改
Legal Management of the Watchdog Role of the Media	《传媒监督的法治》
Li	礼
li xing ren	理性人
Lianghu	两湖
Liangzhan	粮站
libeng yuehuai	礼崩乐坏
Lide	立德
Ligong	立功
Liji	《礼记》
Lisu	礼俗
liu da gui	流打鬼
liu zi	流子
Liumang	流氓
Liushi mowang	混世魔王
liyan	立言
Liyun	《礼运》
maiming	卖命
mazai	马仔
mazhuang	码庄
Mihunzhen	迷魂阵
Mimi huishe	秘密会社
minjing	民警
mu	亩
neiyou waihuan	内忧外患
pai	派

(continued)

(continued)

Production brigades	生产大队
production teams	生产小队
qing	情
qing li fa	情理法
qing miao fei	青苗费
Qing Tian Da Lao Ye	青天大老爷
qingfei fanba	清匪反霸
Qingfen	情分
qingfen	情分
qingsuan	清算
qingyi	情义
quntixing shijian	群体性事件
qunzhong luxian	群众路线
Renmin	人民
renqing	人情
Rural Grassroots Governing System in China	《中国农村基层政权》
Rural Justice	《乡村司法》
San Nong	三农
San Wu	三五
shangmen nvxu	上门女婿
Shanxi Coalmines Seized by the Rampant Underworld	《山西煤炭曾经的阴暗一面 黑道横行堪比无间道现实版》
shehuixing daofei	社会性盗匪
Shengli	胜利
shengyuan	生员
shishen shiliao biji	士绅史料笔记
shiye danwei	事业单位
shushe	锄社
song fa xia xiang	送法下乡
Suanzhang	算账
suozhang	所长
Tanqiao	滩桥
the Ming Dynasty	明朝
The Story of Qiu Ju	秋菊打官司
tianli	天理
tianxia	天下
tiezhangwan	铁掌玩
tizhinei	体制内

(continued)

(continued)

tongqing dali	通情达理
tongyangxi	童养媳
Tou ji mo gou	偷鸡摸狗
tuanti geju	团体格局
tundan	吞单
Unveiling Coal Mine Collusion	《查出来的大多是"小萝卜头"官煤勾结有黑幕》
Villager's Self-Government System	《中国农村村民自治》
Villagers at a Coalmine at Lyuliang City Shanxi Province Cruelly Murdered by a Hundred Gangsters	《山西吕梁煤矿百人殴砍村民血案》
Violence and Order	《暴力与秩序》
wenbao	温饱
Wuyan jian jiangdong fulao	无颜见江东父老
xianda	显达
xiang	乡
xiangcun hunhun	乡村混混
Xiangcun Jianghu	《乡村江湖》
xiangcun zhili	乡村治理
Xiangtu luoji	乡土逻辑
Xiangtu Zhongguo	《乡土中国》
xiao qin zu	小亲族
xiao si	小私
Xiashi	侠士
xiaxiang	下乡
xiejiao	邪教
xijie	洗劫
xuanfu	悬浮
xueyuan	血缘
yamen	衙门
yi	义
yi	谊
yi da yi la	一打一拉
yi shi yi yi	一事一议
yihua	异化
yijin huanxiang	衣锦还乡
ying jian de you xi	迎检的游戏
yipiao foujue	一票否决
yiqi	义气
you mianzi	有面子

(continued)

(continued)

youwei	有味
youzi	游子
yu zhai	圩寨
Zhaijidi	宅基地
zhengdun	整顿
zhengshen	政审
zhibao weiyuan	治保委员
zhibaohui	治保会
zhidaoyuan	指导员
zhiguo pingtianxia	治国平天下
zhiqing	知青
Zhitiao	治调
Zhong Yong	《中庸》
zhongda qingkuang zhaibao	重大情况摘报
Zhongguo Guanxing Diaocha Baogao	中国惯行调查报告
zhongyong	中庸
zongzu	宗族
zou hou men	走后门
zulongzi	做笼子
zuo shan guan hu dou	坐山观虎斗
zuofeng	作风

Bibliography

Arthur Henderson Smith. 2005. *Village Life in China: A Study in Sociology*. Routledge.
Beattie, Hilary. 1979. *Land and Lineage in China*. Cambridge: Harvard University Press.
Blau, Peter M. 1964. *Exchange and Power in Social Life*. New York: Wiley.
Bourdieu, Pierre., and Wacquant, Loïc. J. D. 1992. *An Invitation to Reflexive Sociology*. University of Chicago Press.
Central Committee of the Chinese Communist Party and the State Council. 2018. *The Central Committee of the CCP and the State Council Issue the Notice on Launching a Dedicated Struggle against Crime and Vice*, CSIS interpret China, https://interpret.csis.org/translations/the-central-committee-of-the-ccp-and-the-state-council-issue-the-notice-on-launching-a-dedicated-struggle-against-crime-and-vice/.
Chungli, Chang. 1955. *The Chinese Gentry: Studies on their Role in Nineteenth-Century Chinese Society*. Seattle: University of Washington Press.
Decker, Scott, and Barrik Van Winkle. 1996. *Life in the Gang: Family, Friend, and Violence*. New York: Cambridge University Press.
Duara,Prasenjit. 1988. *Culture, Power, and the State: Rural North China, 1900–1942*. Stanford University Press.
Elizabeth, J. Perry. 1980. *Rebels and Revolutionaries in North China, 1845–1945*. Stanford University Press
Émile, Durkheim. 1982. *The Rules of Sociological Method*. Translated by Lukes/Halls. New York: Free Press.
Emile, Durkheim. 1997. *The Division of Labour in Society*. Translated by W. D. Halls, intro. Lewis A. Coser. New York: Free Press.
Eric, J. Hobsbawm. 2010. *Bandits*. New Press
Fei Xiaotong [Hsiao-Tung Fei]. (1939). *Peasant Life in China*, London: Routledge.
Fei, Xiaotong. 1992. *From the Soil: The Foundations of Chinese Society* (A Translation of Fei Xiaotong's *Xiangtu Zhongguo*). Translated by Gary G. Hamilton & Wang Zheng. University of California Press, p. 176.
Foucault, Michel. 1995. *Discipline and Punish*. Translated from the French by Alan. Sheridan. New York: Random House.
Geertz, Clifford. 1983. *Local Knowledge: Fact and Law in Comparative Perspective*. London: Fontana Press.
Giddens, Anthony. 1985. *The Nation-State and Violence*. Cambridge: Polity Press.
Goffman, Erving. 1959. *The Presentation of Self in Everyday Life*. Garden City, New York: Doubleday.
Hinton, William. 1966. *Fanshen: A Documentary of Revolution in a Chinese Village*. Monthly Review Press.

B. Chen, *The Underworld of Rural China*,
https://doi.org/10.1007/978-981-19-8710-6

Hsiao, Kung-ch'üan. 1960. *Rural China: Imperial Control in the Nineteenth Century*, Seattle: University of Washington Press.
Huang, Philip. C. C. 1988. *The Peasant Economy and Social Change in North China*. Stanford University Press.
Huang, Philip. C. 1990. *The Peasant Family And Rural Development in the Yangzi Delta, 1350–1988*. Stanford University Press.
In: *Local Knowledge: Further Essays in Interpretive Anthropology*. New-York/N.Y./USA etc.: Basic Books, pp. 167–234. http://hypergeertz.jku.at/GeertzTexts/Local_Knowledge.htm.
John King Fairbank, (1983), *The United States and China*. Harvard University Press.
Kuhn, Philip A. 1970. *Rebellion and Its Enemies in Late Imperial China; Militarization and Social Structure, 1796–1864*. Cambridge, Mass.: Harvard University Press.
John, Hagedorn., and Perry, Macon. 1998. *People and Folks: Gangs, Crime and the Underclass in a Rustbelt City*. Chicago: Lake View Press.
Mann, Michael. 1986. *The Sources of Social Power*. Cambridge University Press.
Maurice, Freedman. 1965. *Lineage Organization in Southeastern China*. London: University of London, The Athlone Press.
Moore, Joan. 1991. *Going Down to the Barrio: Homeboys and Homegirls in Change*. Philadephia: Temple press.
Myers, Ramon. H. 1970. *The Chinese Peasant Economy: Agricultural Development in Hopei and Shantung 1890–1949*. Harvard University Press.
Nonet, Philippe., and Selznick, Philip. 2001. *Law and Society in Transition Toward Responsive Law*. Routledge.
Ping-ti, Ho. 1962.*The ladder of Success in Imperial China: Aspects of Social Mobility, 1368–1911*. New York: Columbia University Press.
Polanyi, Karl. 2001. *The Great Transformation: The Political and Economic Origins of Our Time*. Boston: Beacon Press.
Riis, Jacob. 1892. *The Children of the Poor*. New York: Charles Scribner's Sons.
Schoppa, Keith. 1982. *Chinese Elites and Political Change*. Cambridge: Harvard University Press.
Scott, James C. 1985. *Weapons of the Weak: Everyday Forms of Peasant Resistance*. New Haven: Yale University Press.
Simmel, Georg. 1910. How is Society Possible?. *American Journal of Sociology* 16(3): 372–391. published by: The University of Chicago Press. https://www.jstor.org/stable/2763090?seq=1.
Sommer, Matthew. 2000. *Sex, Law, and Society in Late Imperial China*. Stanford: Stanford University Press.
Thrasher, Frederic. 1927. *The Gang: A study of 1313 Gangs in Chicago*. Chicago: University of Chicago Press.
Weber, Max. 1978. *Selections in Translation*. Cambridge University Press.
Weber, Max., Baehr, Peter. R. and Wells, Gordon. C. 2002. *The Protestant Ethic and the "Spirit" of Capitalism and Other Writings*, Penguin.
Whyte, William F. 1993. *Street Corner Society*. Chicago and London: University of Chicago Press.
William Skinner, G. 1964. Marketing and Social Structure in Rural China. *The Journal of Asian Studies 24*(1): 6–32.
Yunxiang, Yan. 1996. *The Flow of Gifts: Reciprocity and Social Networks in a Chinese Village*. Stanford University Press.
Yutang, Lin. 2006. *My Country and My People*. Hesperides Press.
Zedong, Mao. 1965a. *Selected Works of Mao Tse-Tung*, vol. 1. Peking: Foreign Language Press.
Zedong, Mao. 1965b. *Selected Works of Mao Tse-Tung*, vol. 3. Peking: Foreign Language Press.
Zedong, Mao. 1977. *Selected Works of Mao Tse-tung*. Peking: The People's Publishing House.
中国社会科学院语言研究所词典编辑室. 1983.《现代汉语词典》(第2版) [*The Contemporary Chinese Dictionary* (2nd ed)]. 北京: 商务印书馆.
中国社会科学院语言研究所词典编辑室. 2005.《现代汉语词典》(第5版) [*The Contemporary Chinese Dictionary* (5th ed)]. 北京: 商务印书馆.

于建嵘. 2003. 农村黑恶势力和基层政权退化 [*The Rise of Black and Evil Force in Rural Areas and the Withdrawal of Grassroots State Power*].《战略与管理》第5期.
于建嵘. 2005. 当代中国农民维权组织的发育与成长 [*Birth and Growth of Farmers Rights Organizations in Contemporary China*].《中国农村观察》第2期.
仝志辉、贺雪峰. 2002. 村庄权力结构的三层分析 [*An Analysis of the Three-layered Power Structure at the Village*].《中国社会科学》第1期.
仝志辉. 2004.《选举事件与村庄政治》[*Election Affairs and Village Politics*]. 北京: 中国社会科学出版社.
任恩顺. 1996. 梅县农村社区警务改革新思路 [*New Thoughts on Reform of Community Policing in Rural Areas of Meixian County*].《人民公安》第23期.
何清涟. 1998.《现代化的陷阱》[*The Traps of Modernization*]. 北京: 今日中国出版社1998年版.
傅剑锋、成希. 2007a. 水面下的秘密力量 [*Secret Forces Under the Water Surface*].《南方周末》8月9日.
傅剑锋、成希. 2007b. 刀尖上的'无间道' [*Dangers for Being an Informant*].《南方周末》8月9日.
傅衣凌. 1988. 中国传统社会: 多元的结构 [*The Traditional Chinese Society: A Diverse Structure*].《中国社会经济史研究》第3期.
内山雅生. 2001.《二十世纪华北农村社会经济研究》[*A Study of the Social Economy of Rural North China in the Twentieth Century*]. 李恩民、邢丽荃译, 北京: 中国社会科学出版社.
冯小双. 2002. 阅读和理解转型期中国乡村社会 [*Understanding Rural China During a Transitional Period*].《社会学研究》第1期.
刘世定、邱泽奇. 2004. '内卷化'概念辨析 [*Formulating the Concept of "Involution"*].《社会学研究》第5期.
刘大可. 2004.《闽台地域社会与族群文化新探》[*A New Study into Regional Societies and Ethnic Cultures at Fujian and Taiwan*]. 北京: 方志出版社.
刘灿. 1994. 巡访制: 农村基层公安工作新探索 [*Inspections: New Mechanism in Public Security*].《人民公安》第10期.
卢晖临、李雪. 2007. 如何走出个案? [*How Should We Move From Individual Cases?*].《中国社会科学》第1期.
卢晖临. 2005. 社区研究: 源起、问题与新生 [*Community Research: Origin, Problem and Rebirth*].《开放时代》第4期.
叶贵仁. 2007. 乡镇长权责不对等研究: 以一个城关镇为个案 [*On the Mismatch of Rights and Responsibilities for Town/Township Leaders: Based on a Case Study of a Chengguan Town*].《江西师范大学学报 (哲学社会科学版) 》第6期.
吕德文. 2006a. 村治研究的传统与现状 [*Traditions and Status of Research on Village Governance*].《华中科技大学学报(社会科学版)》第4期.
吕德文. 2006b. 村庄政治中的庄外的庄里人 [*Local Villagers That Stay Out of the Village in Village Politics*]. 三农中国网.
吕德文. 2007. 乡土如何本色 [*The Inherent Qualities of the Rural Land*]. 三农中国网.
吕德文. 2008. 在钉子户'与'特困户'之间 [*Caught in the Middle Between "Holdouts" and "Impoverished Households"*].《中国农业大学学报 (社会科学版) 》第1期.
吴毅、李德瑞. 2007. 二十年农村政治研究的演进与转向 [*The Development and Changes in Rural Politics Researches for the Past Two Decades*].《开放时代》第2期.
吴毅、贺雪峰、罗兴佐等. 2005. 村治研究的路径与主体 [*The Approach and Main Content of Village Governance Researches*].《开放时代》, 第4期.
吴毅. 2002.《村治变迁中的权威与秩序》[*Authority and Order in Changing Village Governance*]. 北京: 中国社会科学出版社.
吴毅. 2005. 农村政治研究: 缘自何方, 前路何在 [*Researches on Rural Politics: Its Origin and Prospect*].《开放时代》第2期.
吴毅. 2007.《小镇喧嚣》[*The Rustle and Bustle of a Small Town*]. 北京: 三联书店.
吴重庆. 2005. 乡土儒学资源的再生 [*The Rebirth of Confucianism in Rural Areas*].《天涯》第4期.

周荣德. 2000.《中国社会的阶层与流动》[*Stratification and Mobility of the Chinese Society*]. 上海: 学林出版社.

周飞舟. 2006. 从汲取型政权到"悬浮型"政权——税费改革对国家与农民关系之影响 [*From a Profiting State to a Disembedded State—The Impact of Tax Reform on State-Farmer Relations*].《社会学研究》第3期.

唐皇凤. 2007. 常态社会与运动式治理 [*The Norms of the Society and Crime Control through Campaign*].《开放时代》第3期.

夏柱智、贺雪峰. 2017. 半工半耕与中国渐进城镇化模式 [*Half Working and Half Farming: the China Model for Gradual Modernization*].《中国社会科学》第12期.

夏柱智. 2014. 乡村合谋视角下的混混治村及后果 [*Village Governance by Gangs and its Consequences: a Collusion Perspective*].《青年研究》第1期.

孙远东. 1999. 论乡村地痞对农村基层行政的影响 [*On the Impact of Local Gangsters on Grassroots Governance*].《开放时代》第3期.

季卫东. 1999.《法治秩序的建构》[*The Establishment of Order Ruled by Law*]. 北京: 中国政法大学出版社.

宋丽娜. 2006. 三泉村的村庄治理 [*Village Governance at Sanquan Village*]. 董磊明主持:《荆门仙居村治调查》, 华中科技大学中国乡村治理研究中心, 2006年10月.

寺田浩明. 1998. 权利与冤抑.《明清时期的民事审判与民间契约》[*Civil Trials and Contract in Ming and Qing Dynasties*]. 北京: 法律出版社.

张厚安、徐勇、项继权等. 2000.《中国农村村级治理》[*Village Governance in Rural China*]. 武汉: 华中师范大学出版社.

张厚安. 1992.《中国农村基层政权》[*Rural Grassroots Governing System in China*]. 成都: 四川人民出版社.

张厚安. 2001. 三个面向, 理论务农: 社会科学研究的反思性转换 [*Engage in Agriculture with Theories: a Reflective Transition in Social Sciences Research*].《华中师范大学学报 (人文社会科学版)》第1期.

张思. 2005.《近代华北村落共同体的变迁》[*Changes to Village Community in the Modern North China*]. 北京: 商务印书馆.

张柠. 2005.《土地的黄昏》[*The Dusk of Land*]. 北京: 东方出版社.

张继成. 2002. 事实推定的逻辑基础 [*Logic Foundations for Presumption of Facts*].《北京科技大学学报 (社会科学版)》第2期.

张莉. 2001. 论社区警务与群众路线 [*On Community Policing and the Mass Line*].《辽宁公安司法管理干部学院学报》第1期.

强世功. 1997. '法律'是如何实践的 [*How are Laws Implemented?*]. 王铭铭、王斯福主编:《乡村社会的公正、秩序与权威》, 北京: 中国政法大学出版社.

强世功. 1998. '法律不入之地'的民事调解 [*Civil Mediation in "the Land of No Laws"*].《比较法研究》第3期.

强世功. 2001. 惩罚与法治: 中国刑事实践的法社会学分析 (上) [*Punishment and Laws: A Socio-legal Study of Criminal Law Practice (1)*].《刑事法评论》第9卷, 北京: 中国政法大学出版社.

强世功. 2003.《法制与治理》[*Rule of Law and Governance*]. 北京: 中国政法大学出版社.

徐勇. 1991. 重心下沉: 90年代学术新趋向 [*Sent Theorists Down to the Grassroots Level: The New Academic Trend in the Nineties*].《社会科学报》11月14日.

徐勇. 1992.《非均衡的中国政治: 城市与乡村比较》[*Chinese Politics in Disequilibrium: A Comparative Study of Urban and Rural Areas*]. 北京: 中央广播电视出版社.

徐勇. 1997.《中国农村村民自治》[*Villager's Self-Government System*]. 武汉: 华中师范大学出版社.

徐昕. 2005.《论私力救济》[*On Relief Efforts by Private Parties*]. 北京: 中国政法大学出版社.

徐财松. 2006. 三泉村治 [*Village Governance at Sanquan Village*]. 董磊明主持:《荆门仙居村治调查》, 华中科技大学中国乡村治理研究中心, 2006年10月.

戴炎辉. 1979.《清代台湾之乡治》[*Village Governance at Taiwan in the Qing Dynasty*]. 台北: 联经出版事业公司.

明恩浦. 2006.《中国乡村生活》[*Rural Life in China*]. 陈午晴、唐军译,北京: 中华书局.

曹锦清、张乐天、陈中亚. 1995.《当代浙北乡村社会文化变迁》[*Social and Cultural Changes to Contemporary Villages of North Zhejiang Province*]. 上海: 上海远东出版社.

曹海林. 2005. 乡村社会变迁中的村落公共空间 [*On Village Public Space During Social Changes in Rural Society*].《中国农村观察》第6期.

曹锦清. 2000.《黄河边的中国》[*China Along the Yellow River: Reflections on Rural Society*]. 上海: 上海文艺出版社.

朱晓阳. 2003.《罪过与惩罚》[*Crime and Punishment*]. 天津: 天津古籍出版社.

朱晓阳. 2004. '延伸个案'与一个农民社区的变迁 [*An Extended Case and Changes to a Farmer Community*].《中国社会科学评论》第2卷, 北京: 法律出版社.

朱晓阳. 2007. '语言混乱'与法律人类学的整体论进路 [*"Confusion of Tongues" and a Holistic Approach to Legal Anthropology*].《中国社会科学》第2期.

朱晓阳. 2008.《面向法律的语言混乱》[*On the Confusion of Legal Tongues*]. 北京: 中央民族大学出版社.

朱苏力. 1998.《自由秩序原理》读书笔记 [*Reading Notes for The Constitution of Liberty*]. 民俗学论坛, https://www.chinesefolklore.org.cn/forum/viewthread.php?tid=9410.

李国庆. 2005. 关于中国村落共同体的论战 [*The Debate About China's Village Community*].《社会学研究》第6期.

李文治、江太新. 2000.《中国宗法宗族制和族田义庄》[*China's Clan System and Clan Fields*]. 北京: 社会科学文献出版社.

李昌平. 2002.《我向总理说实话》[*Being Truthful with the Premier*]. 北京: 光明日报出版社.

李祖佩. 2016. '项目进村'过程中的混混进入 [*Gang Involvement in the Process of "Government Projects Brought Down to Villages"*].《青年研究》第3期.

李继宏. 2003. 城乡心理和生活世界 [*The Minds and Worlds of Urban and Rural Areas*].《人文杂志》第4期.

杨代根、谈天国、曾维平. 2003. 改革创新, 整合警力资源 [*Innovation and Reform in Reorganizing the Police Forces*].《四川警官高等专科学校学报》第5期.

杨华. 2007a. 地方性、国家性与村庄构成 [*Regionalism, Nationalism and the Composition of the Village*]. 未刊稿.

杨华. 2007b.《传统村落生活的逻辑》[*How Traditional Rural Life Works?*]. 华中师范大学硕士论文.

杨华. 2008a. 家族、公私观念与村庄主体性建构 [*Concepts of Family, Public and Private Spheres and the Establishment of Village Identity*].《开发研究》第2期.

杨华. 2008b. 初论'血缘共同体'与'关系共同体' [*A Preliminary Discussion on "Kinship-based Community" and "Relationship-based Community"*].《开发研究》第1期.

杨华. 2009. 乡村混混与村落、市场和国家的互动 [*Rural Gangs and Their Interaction with the Village, the Market and the State*].《青年研究》第3期.

杨善华. 1999.《当代西方社会学理论》[*Contemporary Western Sociology Theory*]. 北京: 北京大学出版社.

杨懋春. 2001.《一个中国村庄: 山东台头》[*A Chinese Village: Taitou Village of Shandong Province*]. 张雄、沈炜、秦美珠译, 南京: 江苏人民出版社.

杨柳. 1999. 模糊的法律产品 [*The Result of Vagueness in Law*].《北大法律评论》第2卷第1辑, 北京: 法律出版社.

林琳. 2006. 三泉印象 [*Impressions on Sanquan*]. 董磊明主持:《荆门仙居村治调查》, 华中科技大学中国乡村治理研究中心, 2006年10月.

桑本谦. 2005.《私人之间的监控与惩罚》[*Supervision and Punishment by Private Parties*]. 济南: 山东人民出版社.

梁漱溟. 1987.《中国文化要义》[*Fundamentals of Chinese Culture*]. 上海: 学林出版社.

梁漱溟. 1999.《东西文化及其哲学》[*Eastern and Western Cultures and Their Philosophy*]. 北京: 商务印书馆年版.

欧阳静. 2008. 农民公私观念与国家观念的南北差异 [*Farmer Conception of Public and Private Spheres and the North-South Disparity in Farmer Views on the Government*].《开发研究》第2期.

汪雄涛. 2007. 汪村宅基地的占有与使用 [*The Ownership and Usage of House-Building Land at Wangcun Village*]. 未刊稿.
沈延生. 1998. 村政的兴衰与重建 [*The Rise and Fall of Village Politics and the Rebuilding of the Village*].《战略与管理》第6期.
渠敬东. 2012. 项目制: 一种新的国家治理体制 [*Government Projects: A New National Governance System*].《中国社会科学》第5期.
王习明. 2007.《乡村治理中的老人福利》[*The Elderly's Benefits from Rural Governance*]. 武汉: 湖北人民出版社.
王仲芳. 1989.《中国社会治安综合治理的理论与实践》[*Theory and Practice in China's Public Security Administration*]. 北京: 群众出版社.
王小军. 2009.《转型之痛——赣中南路东村调查》[*The Pains of Change—Based on an Investigation on Ludong Village in South Jiangxi Province*]. 济南: 山东人民出版社.
王铭铭. 2004.《溪村家族》[*The Lineage of Xicun*]. 贵阳: 贵州人民出版社.
王铭铭. 2005.《社会人类学与中国研究》[*Social Anthropology and the China Study*]. 桂林: 广西师范大学出版社.
申端锋. 2007a. 中国农村出现伦理性危机 [*An Ethical Crisis in Rural China*].《中国评论》(香港) , 第3期.
申端锋. 2007b. 农民行动单位视阈中的小亲族 [*Small Kinship Group: A Unit of Action Accepted by Farmers*].《江海学刊》第4期.
申端锋. 2008. 具体的私与抽象的公 [*The Specific Public Sphere and the Abstract Public Sphere*].《开发研究》第2期.
瞿同祖. 2003.《清代地方政府》[*The Local Government in the Qing Dynasty*]. 范忠信、晏锋译, 何鹏校, 北京: 法律出版社.
罗义云、陈涛. 2007. 荆门罗祠村调查报告 [*Report on an Investigation of Luoci Village at Jingmen*].《"转型期农村社会弱势群体生存状态的区域比较与治理对策研究"调研报告》, 华中科技大学中国乡村治理研究中心, 2007年1月.
罗兴佐. 2002. 第三种力量 [*A Third Power*]《浙江学刊》第2期.
罗兴佐. 2006.《治水: 国家介入与农民合作》[*Managing Floods: State Intervention and Farmer Cooperation*]. 武汉: 湖北人民出版社.
罗兴佐. 2007. 乡村社会中的混混 [*Gangs in the Rural Society*]. 三农中国网.
罗瑞卿. 1957. 关于中华人民共和国治安管理处罚条例草案的说明 [*Some Explanations on the Regulations on Penalties for Public Security 1957*]. 10月22日.
翟中东. 2004.《犯罪控制》[*Crime Control*]. 北京: 中国政法大学出版社.
耿羽. 2016. 征地拆迁中的混混暴力市场化 [*Market Trading of Gang Violence in Housing Demolition for Land Acquisition*].《中国青年研究》第7期.
肖业炎、张艳. 2002. 农村地痞恶势力类型、危害及其治理 [*The Types, Dangers and Management of Rural Evil Forces*].《江西公安专科学校学报》第1期.
苏力. 1996.《法治及其本土资源》[*Rule of Law and the Required Local Resources*]. 北京: 中国政法大学出版社.
苏力. 2000.《送法下乡》[*Sending the Law Down to the Countryside*]. 北京: 中国政法大学出版社.
苏力. 2007. 费孝通、儒家文化和文化自觉 [*Fei Xiaotong, Confucian Culture and Cultural Self-consciousness*].《开放时代》2007年第4期.
荣娥. 2006. 走向原子化的村庄 [*The Atomized Village*]. 董磊明主持:《荆门仙居村治调查》, 华中科技大学中国乡村治理研究中心, 2006年10月.
荣敬本等. 1998.《从压力型体制向民主合作体制的转变》[*From Pressure to Democratic Collaboration*]. 北京: 中央编译出版社.
董磊明、陈柏峰、聂良波. 2008. 结构混乱与迎法下乡 [*Structural Disintegration of Rural Society and Welcoming the Rule of Law to the Country*].《中国社会科学》第5期.
董磊明. 2006a. 村庄纠纷调解的研究路径 [*Research Approaches to Village Dispute Resolution*].《学习与探索》第1期.

董磊明. 2006b. 农村调解机制的语境化理解与区域比较研究 [*Contextual Understanding of Mediation Mechanisms in Rural Areas and Regional Comparative Studies*].《社会科学辑刊》第1期.
董磊明. 2006c. 新庄村的纠纷调解 [*Dispute Resolution at Xinzhuang Village*]. 未刊稿.
董磊明. 2007. 村将不村——湖北尚武村调查 [*The Village is in Peril—An Investigation into Shangwu Village of Hubei Province*].《中国乡村研究》第5辑, 福州: 福建教育出版社.
董磊明. 2008.《宋村的调解: 巨变时代的权威与秩序》[*Mediation in Songcun Village: Authority and Order in a Era of Radical Change*]. 北京: 法律出版社.
蒋成树. 2000. 巡访村警制与农村派出所改革 [*Village Inspection Visits and Reforms on Rural Police Stations*].《湖南公安高等专科学校学报》第5期.
袁松. 2007. 豫东付楼村村治模式 [*Village Governance Model for Fulou Village of Eastern Henan Province*]. 未刊稿.
谭同学. 2007.《治理转型中的道德、权力与社会结构》[*Morality, Powers and Social Structure in Governance Reform*]. 华中科技大学博士论文.
费孝通、张之毅. 2006.《云南三村》[*Three Villages in Yunnan*]. 北京: 社会科学文献出版社.
费孝通. 1990. 人的研究在中国 [*Anthropology in China*].《读书》第10期.
费孝通. 2006. 乡土重建 [*From the Soil: The Foundations of Chinese Society*].《乡土中国》[Reconstructing Rural China], 北京: 上海人民出版社.
贺雪峰、仝志辉. 2002. 论村庄社会关联 [*On Social Relations in Villages*].《中国社会科学》第3期.
贺雪峰、董磊明、陈柏峰. 2007. 乡村治理研究的现状与前瞻 [*Research Status and Prospects on Rural Governance*]. 高翔主编:《中国社会科学学术前沿 (2006–2007)》, 社会科学文献出版社.
贺雪峰. 2003b.《乡村治理的社会基础》[*Social Foundation for Rural Governance*]. 北京: 中国社会科学出版社.
贺雪峰. 2004. 熟人社会的行动逻辑 [*The Behavior Pattern of the Acquaintance Society*].《华中师范大学学报 (人文社会科学版)》2004年第1期.
贺雪峰. 2005a. 为什么要深度农村社会调查 [*Why Conduct Careful Investigations into the Rural Society?*].《博览群书》第10期.
贺雪峰. 2005b. 论村级负债的区域差异 [*On Regional Differences in Village-Level Debts*].《管理世界》第11期.
贺雪峰. 2005c. 现代化进程中的村庄自主生产价值能力 [*The Village's Intrinsic Level of Productivity for Economic and Ethical Values in the Process of Modernization*].《探索与争鸣》第7期.
贺雪峰. 2006a. 私人生活与乡村治理研究的深化 [*Private Life and Deepening Research on Rural Governance*].《读书》第11期.
贺雪峰. 2006b. 乡村治理区域差异的研究视角与进路 [*Perspective and Approach of Researches on Regional Difference in Rural Governance*].《社会科学辑刊》第1期.
贺雪峰. 2006d. 公私观念与农民行动的逻辑 [*Conceptions of Public and Private Spheres and Farmer Behavior Pattern*].《广东社会科学》第1期.
贺雪峰. 2006e. 乡村关系研究的视角和进路 [*Perspective and Approach of Researches on Rural Relationships*].《社会科学研究》第1期.
贺雪峰. 2007a. 二十世纪中国乡村治理的逻辑 [*How Rural Governance Works in the 20th Century*].《中国乡村研究》第5辑, 福州: 福建教育出版社.
贺雪峰. 2007b. 个案研究与区域比较 [*Individual Cases and Regional Comparison*].《华中科技大学学报(社会科学版)》第1期.
贺雪峰. 2007c. 关于中国农村区域类型的想像 [*My Thoughts on the Types of Rural Regions in China*]. 三农中国网.
贺雪峰. 2007d. 川西平原的院子 [*Courtyards in Western Sichuan Plain*]. 三农中国网.
贺雪峰. 2007e. 川西平原乡村善治的关键 [*The Key to Good Governance in Western Sichuan Plain*]. 三农中国网.
贺雪峰. 2007f. 农民行动逻辑与乡村治理的区域差异 [*Farmer Behavior Pattern and Regional Differences in Rural Governance*].《开放时代》2007 年第1期.

贺雪峰. 2007g.《乡村的前途》[*The Prospects of the Village*]. 济南: 山东人民出版社.
贺雪峰. 2007h. 二十世纪中国乡村治理的逻辑 [*How Rural Governance Works in the 21st Century*].《中国乡村研究》第5辑, 福州: 福建教育出版社.
贺雪峰. 2007i. 取消农业税后的国家与农民关系 [*State-Farmer Relations After the Lift of Agricultural Tax*].《甘肃社会科学》第2期.
贺雪峰. 2008. 农村的半熟人社会化与公共空间的变化 [*Rural Society Turning into Half Acquaintance Society and the Changes in Public Space*].《中国乡村研究》第6辑, 福州: 福建教育出版社.
贺雪峰. 2009a.《村治模式: 若干案例研究》[*The Model of Village Governance: A Multi-Case Study*]. 济南: 山东人民出版社.
贺雪峰. 2009b.《村治的逻辑: 农民行动单位的视角》[*How Village Governance Works: From the Perspective of the Unit of Farmer Action*]. 北京: 中国社会科学出版社.
贺雪峰. 2010.《地权的逻辑》[How Land Rights Work]. 北京: 中国政法大学出版社.
贺雪峰. 2013. 办'点'的政治学 [*The Political Practice of "Establishing Points of Contact"*].《决策》第11期.
贺雪峰. 2015. 行政体制中的责权利层级不对称问题 [*The Mismatch of Rights and Responsibilities in the Administrative Hierarchy*].《云南行政学院学报》第4 期.
贺雪峰. 2017.《治村》[*Governing the Village*]. 北京: 北京大学出版社.
贺雪峰. 2003a.《新乡土中国》[*Rural China in the 21st Century*]. 桂林: 广西师范大学出版社.
贺雪峰. 2006c. 公私观念与中国农民的双层认同 [*Conceptions of Public and Private Spheres and Chinese Farmers' Double Units of Action and Identity*].《天津社会科学》第1期.
赵晓力. 1997. 关系—事件、行动策略和法律的叙事 [*Relationship-Incidents, Strategies and Narratives in Legal Texts*]. 王铭铭、王斯福主编:《乡村社会的公正、秩序与权威》, 北京: 中国政法大学出版社.
邓小平. 1993.《邓小平文选》[*Selected Works of Deng Xiaoping*]. (第三卷) , 北京: 人民出版社.
邓燕华. 2006. 地下'六合彩'在农村社会的运作逻辑 [*How Underground Mark Six Lottery Works in Rural China*].《社会》第1期.
邹谠. 1994.《二十世纪中国政治》[*Chinese Politics in the 20th Century*]. 香港: 牛津大学出版社.
郑戈. 1997. 规范、秩序与传统 [*Regulations, Order and Traditions*]. 王铭铭、王斯福主编:《乡村社会的公正、秩序与权威》, 北京: 中国政法大学出版社.
郑浩澜. 2006. '村落共同体'与乡村变革 [*"The Village Community" and "Rural Reforms"*].《乡村中国评论》第1辑, 桂林: 广西师范大学出版社.
郑谊. 2006. 盐井村调查报告 [*Investigation Report of Yanjing Village*]. 董磊明主持:《荆门仙居村治调查》, 华中科技大学中国乡村治理研究中心, 2006年10月.
郭亮. 2015. 土地征收中的"行政包干制"及其后果 [*Administrative Responsibility System in Land Acquisition and its Consequences*].《政治学研究》第1期.
郭俊霞、陈柏峰. 2016. 乡村混混进城后的分化和流动 [*Stratification and Mobility of Rural Gangs after Entering the City*].《中国青年研究》第7期.
郭俊霞. 2010. 农村安全供给的基本需要与制度保障 [*Basic Requirements and Systematic Guarantee for Rural Security*].《重庆社会科学》第2期.
金守信. 1995. 农村派出所勤务制度改革新尝试 [*New Attempts at Reforming Rural Police Station Services*].《公安研究》第1期.
钟瑞庆. 2001. 农民负担问题中所体现的冲突解决方式 [*Manners of Conflict Resolution in Farmer Burden Issues*].《中外法学》第5期.
陈宝良. 1996.《中国的社与会》[*China's Organizations and Associations*]. 杭州: 浙江人民出版社.
陈宝良. 2005.《明代儒学生员与地方社会》[*Confucian Shengyuan in Ming Dynasty and Local Society*]. 北京: 中国社会科学出版社.
陈柏峰、董磊明. 2009. 乡村治理的软肋: 灰色势力 [*The Weakness of Village Governance: The Gray Forces*],《经济社会体制比较》第4期.
陈柏峰、郭俊霞. 2007. 也论面子——村庄生活的视角 [*Also on Face—The Rural Life Perspective*].《华中科技大学学报 (社会科学版) 》第1期.

陈柏峰. 2005.《暴力与屈辱: 鄂南陈村的法律民族志》[*Violence and Disgrace: Ethnographic Studies of Legal Phenomena at Chencun Village of Southern Hubei Province*]. 武汉: 中南财经政法大学法学院硕士论文.
陈柏峰. 2006. 村落纠纷中的'外人' [*"Outsiders" in Village Disputes*].《社会》第4期.
陈柏峰. 2007. 法律民族志与当代中国法律社会学的使命 [*The Mission of Legal Ethnography and Sociology in Modern China*].《法哲学与法社会学论丛》第11卷, 北京: 北京大学出版社.
陈柏峰. 2008a. 规则之治时代的来临? [*The Arrival of the Rule-of-Law Era?*],《法律和社会科学》第3辑, 北京: 法律出版社.
陈柏峰. 2008b. 地方性共识与农地承包的法律实践 [*Local Consensus and Legal Affairs in the Contracting of Farmland*].《中外法学》第2期.
陈柏峰. 2008c. 北方村庄的派性政治与日常生活 [*Factional Politics and Daily Life at Villages of Northern China*].《开发研究》第1期.
陈柏峰. 2009a. 村庄生活中的面子及其三层结构 [*Face in Village Life and Its Three-level Structure*].《广东社会科学》第1期.
陈柏峰. 2009b. 纠纷解决与国家权力构成 [*Dispute Resolution and the Composition of State Authority*].《民间法》第8卷, 济南: 山东人民出版社.
陈柏峰. 2010. 秋菊的'气'与村长的'面子' [*Qiuju's "Anger" and Village Chief's "Face"*].《山东大学学报 (社会科学版) 》第3期.
陈柏峰. 2011.《乡村江湖——两湖平原"混混"研究》[*The Underworld of Rural China*]. 北京: 中国政法大学出版社.
陈柏峰. 2016. 征地拆迁上访的类型与机理 [*Classification and Mechanism of Appeals in Housing Demolition for Land Acquisition*].《华中科技大学学报 (社会科学版) 》第1期.
陈柏峰. 2017. 党政体制如何塑造基层执法 [*How the Party Ensures Grassroots Law Enforcement*],《法学研究》第4期.
陈玉凡. 1996. 新形势下的刑事特情工作 [*Criminal Intelligence Under New Circumstances*].《河南公安高等专科学校学报》第4期.
陈锋. 2015. 分利秩序与基层治理内卷化 [*The Hierarchy of Profit Division and Grassroots Governance Involution*].《社会》第3期.
陶驷驹. 1993. 重温毛主席的教诲, 做好新时期的公安工作 [*Reviewing Chairman Mao's Teachings, Ensuring Public Security in the New Era*].《人民日报》12月22日.
韩敏. 2007.《回应革命与改革》[*A Response to Revolution and Reform*]. 陆益龙、徐新玉译, 南京: 江苏人民出版社.
马源源. 2006. 盐井村调查报告 [*Report on an Investigation of Yanjing Village*]. 董磊明主持:《荆门仙居村治调查》, 华中科技大学中国乡村治理研究中心, 2006年10月.
马源源. 2007. 安徽肥西滨湖村调查报告 [*Report on an Investigation of Binhu Village, Feixi County, Anhui Province*].《"转型期农村社会弱势群体生存状态的区域比较与治理对策研究"调研报告》, 华中科技大学中国乡村治理研究中心, 2007年1月.
鲁迅. 1981.《鲁迅全集》(卷三) [*Complete Works of Lu Xun* (vol. 3)], 北京: 人民文学出版社.
黄光国. 2006. 中国人的人情关系 [*Reciprocity-based Relationships of the Chinese People*]. 文崇一、萧新煌主编:《中国人: 观念与行为》, 南京: 江苏教育出版社.
黄海. 2007. 湖南P县D乡黄村调查报告 [*Report on an Investigation of Huangcun Village of D Township, P County, Hunan Province*].《"转型期农村社会弱势群体生存状态的区域比较与治理对策研究"调研报告》, 华中科技大学中国乡村治理研究中心, 2007年1月.
黄闯. 2006. 荆门白洋村调查报告 [*Report on an Investigation of Baiyang Village of Jingmen City*]. 董磊明主持:《荆门仙居村治调查》, 华中科技大学中国乡村治理研究中心, 2006年10月.
龚春霞. 2007. 村庄中的'私物'、'他物'和'公物' [*Private Goods, Others' Goods and Public Good in Villages*]. 三农中国网.

GPSR Compliance
The European Union's (EU) General Product Safety Regulation (GPSR) is a set of rules that requires consumer products to be safe and our obligations to ensure this.

If you have any concerns about our products, you can contact us on

ProductSafety@springernature.com

In case Publisher is established outside the EU, the EU authorized representative is:

Springer Nature Customer Service Center GmbH
Europaplatz 3
69115 Heidelberg, Germany

www.ingramcontent.com/pod-product-compliance
Ingram Content Group UK Ltd.
Pitfield, Milton Keynes, MK11 3LW, UK
UKHW021822051026
15588UKWH00010B/649
* 9 7 8 9 8 1 1 9 8 7 1 1 3 *